Diapause in the Crustacea

Developments in Hydrobiology 114

Series editor

H. J. Dumont

Diapause in the Crustacea

A compilation of refereed papers from the International Symposium, held in St. Petersburg, Russia, September 12–17, 1994

Edited by

Victor R. Alekseev and Geoffrey Fryer

Reprinted from Hydrobiologia, vol. 320 (1996)

Kluwer Academic Publishers
Dordrecht / Boston / London

Library of Congress Cataloging-in-Publication Data

A C.I.P. Catalogue record for this book is available from the Library of Congress.

ISBN-13:978-94-010-7246-5 e-ISBN-13:978-94-009-1667-8
DOI:10.1007/978-94-009-1667-8

Published by Kluwer Academic Publishers,
P.O. Box 17, 3300 AA Dordrecht, The Netherlands.

Kluwer Academic Publishers incorporates
the publishing programmes of
D. Reidel, Martinus Nijhoff, Dr W. Junk and MTP Press.

Sold and distributed in the U.S.A. and Canada
by Kluwer Academic Publishers,
101 Philip Drive, Norwell, MA 02061, U.S.A.

In all other countries, sold and distributed
by Kluwer Academic Publishers,
P.O. Box 322, 3300 AH Dordrecht, The Netherlands.

Printed on acid-free paper

This volume is dedicated to the memory of
Professor Aleksandr Danilevski – founder of
the Russian school of photoperiodism
and a pioneer in the experimental
study of diapause in invertebrates

Contents

Preface

Diapause is an event of great ecological significance in the lives of a wide range of invertebrates. The underlying physiological and genetic mechanisms pose many intriguing questions and have important evolutionary implications. Some of the most dramatic examples of diapause are provided by crustaceans. Most branchiopods produce resting eggs that can withstand intense desiccation and extremes of temperature and pressure far beyond those to which they are ever subjected in nature. Furthermore, they can remain in a dry condition, displaying no measurable metabolism, for many years. Such abilities enable them to inhabit infrequently-filled pools in arid regions. When the pools fill with water these dormant eggs hatch and give rise to animals that grow quickly to maturity and produce a new generation of resting eggs. There is every reason to believe that primitive branchiopods have employed diapause in this way since Palaeozoic times and that their abilities in this direction have played a key role in their continued success as freshwater animals.

It is now apparent that diapause is of more frequent and more widespread occurrence in the Crustacea than was long supposed. For example, it is an integral feature of the life cycle of many freshwater copepods, whose diapause sometimes involves resting eggs, sometimes larval stages, sometimes adults. Furthermore, it is not restricted to species that frequent habitats that dry out, but is common in many planktonic species. Nor is it restricted to freshwater species: many marine copepods are now known to undergo diapause.

It is now also evident that there are degrees of diapause among the Crustacea. The intense diapause of some branchiopod, ostracod and copepod eggs represents one end of the spectrum. At the other end, larval stages and adults of some copepods and decapods undergo less intensive "active" diapause. Their metabolism is slowed down but they may remain motile, and even feed, but at lower intensities than do active individuals. In a symposium held in St. Petersburg, Russia in September 1994, of which this volume includes some of the papers presented, all grades of diapause were considered. Most of the papers naturally deal specifically with diapause in the Crustacea and cover a wide range of topics. These include the nature of resting stages, their physiology, the induction and termination of diapause and its ecological and evolutionary significance, as well as such unusual aspects as the way in which diapause in copepods impinges on the interactions between fishes and their helminth parasites. Because diapause occurs not only in other aquatic animals, whose employment of the strategy parallels that of crustaceans, but also elsewhere, we have felt it justifiable to include a paper that deals largely with the environmental control of seasonality in insects that is relevant to diapause in other invertebrates. Also included is a paper that deals largely with rotifers, another group of organisms some of whose freshwater representatives display striking ecological parallels to certain crustaceans. What is learned about one group can throw light on events in another and it is helpful for students to be aware of what is known about different animals.

We extend our thanks to all who contributed to the success of the symposium; our colleagues at the laboratories of freshwater hydrobiology and of experimental entomology of the Zoological Institute, the organisers in St. Petersburg, the Russian Academy of Sciences which made facilities available, the Russian Foundation of Fundamental Investigations which sponsored the symposium, and the following who, in addition to several of the participants, acted as referees of the various contributions:

Nick Aladin (Russia), Gary Carvalho (United Kingdom), Hugh Danks (Canada), Valentina Galtsova (Russia), Aleksei Ghilarov (Russia), Marina Ivanova (Russia), Petter Larssen (Norway), Andrei Monakov (Russia), Gerhard Maier (Germany), Mark Ohman (U.S.A.), Takashi Onbe (Japan), Sergei Podlipaev (Russia), Jury Rudjakov (Russia), Nickolai Smirnov (Russia), Mauritius & Catarina Tauber (U.S.A.).

VICTOR ALEKSEEV & GEOFFREY FRYER

Hydrobiologia **320**: 1–14, 1996.
V. R. Alekseev & G. Fryer (eds), Diapause in the Crustacea.

Diapause, a potent force in the evolution of freshwater crustaceans

Geoffrey Fryer
Institute of Environmental and Biological Sciences, Lancaster University, Bailrigg LA1 4YQ, UK

Key words: diapause, freshwater Crustacea, antiquity, diversity, evolution

Abstract

After a brief historical review of the discovery of diapause in freshwater crustaceans, its dramatic nature in certain cyclopoid copepods, in which diapausing individuals may occur at densities of $>10^6$ per m^2, is used to illustrate the enormous ecological significance of the phenomenon. Some of the problems presented by diapause in cyclopoid copepods are noted, including the different behaviour in different lakes of what appears to be a single species. Different physiological cues or different genetic endowments are clearly involved.

The wider incidence of diapause in freshwater copepods and ostracods is noted.

Among freshwater crustaceans it it the Branchiopoda that have universally adopted diapause, always at the egg stage. Even such an ancient order as the Anostraca, perhaps the most primitive of all crustaceans, produces elaborately constructed resting eggs that are capable of cryptobiosis, can remain viable in a dry state for long periods, and can tolerate extreme conditions. The nature of branchiopod resting eggs is briefly reviewed. Of these, only those of the Anomopoda are protected by containers derived from the parental carapace. These are mechanically complex in the most advanced species but, as shown by fossils, are extremely ancient structures.

Factors initiating the onset and termination of diapause in branchiopods are briefly noted, and the process of hatching of resting eggs is outlined.

Introduction

Although diapause is important in the lives of many freshwater crustaceans and has played a key role in the adaptive radiation and ecological success of the three major groups (Branchiopoda, Ostracoda and Copepoda), the very existence of the phenomenon in these animals long escaped detection. For example, although the production of drought-resistant resting eggs is a characteristic feature of anostracans and notostracans, and although observations on these animals date back to the early years of the 18th century, this attribute was not immediately apparent because the eggs of at least some species sometimes hatch without being dried or undergoing an appreciable period of rest. This was clearly the reason why diapause for long remained unknown in *Chirocephalus diaphanus*, the earliest mention of which was made as long ago as 1709 by Petiver whose brief diagnosis is accompanied by the terse statement 'I have hatched these from their eggs'. The same is true of *Triops cancriformis* the hatching of whose eggs was described by Schaeffer (1756), the animal itself having been first described a few years earlier. Schaeffer gives a minute drawing which shows a nauplius emerging from an egg — the first illustration of a hatching branchiopod. Even in the mid 19th century the ability of the eggs of such animals to remain dormant for long periods and to withstand drought seems not to have been known.

Likewise the true nature of the ephippium of anomopods and its properties were not immediately apparent. Although *Daphnia* was described by Swammerdam as early as 1669, its ephippium was not mentioned until more than 80 years later when the remarkable Schaeffer (1755) gave a very small illustration of an ephippial female of *Daphnia magna*, noting that the individual exhibited a black patch posteriorly above the spine, but was unaware of its nature. In 1785 Müller not only gave what for that time was a reasonable description and illustration of the ephippium of *Daphnia* but bestowed on it the name by which it continues to be known. He had, however, nothing to say about its nature, which was to remain unknown until well into the next century. In 1820 Jurine, an excellent observer, also noticed ephippia but thought that animals which bore them, which he illustrated, were afflicted by a disease, and it was left to Straus (1819–1820) to recognise that the ephippium of *Daphnia* contained resting eggs that enabled the animal to withstand the northern winter, which its producer was unable to do. This appears to be the first genuine observation on diapause in a freshwater crustacean.

That the eggs of certain freshwater ostracods could not only become dormant but could withstand drought was known by the mid 19th century (Baird, 1850). The discovery of diapause in the Copepoda seems to have occurred some time later, and only a few species were known to have this ability until the second half of the present century. While some calanoids and at least one harpacticoid produce resting eggs, the discovery of diapause in cyclopoids and harpacticoids proved to involve stages not known to display the phenomenon in other groups — copepodid stages in a cyclopoid (Birge & Juday, 1908) and adults in a harpacticoid (Wolf, 1905; Lauterborn & Wolf, 1909). In both cases dormant individuals were encased in a gelatinous cyst, apparently with an outer layer of mud particles in the latter though this may not be an essential feature. Harpacticoid cysts were subsequently found by several workers but until the 1950s Birge & Juday's finding of encysted individuals of *Acanthocyclops (Diacyclops) thomasi* (= *A. bicuspidatus thomasi*) or *A. thomasi* remained the only known example of diapause in cyclopoids.

The dramatic nature of diapause in certain cyclopoid copepods

A personal reminiscence shows how dramatic diapause can be in cyclopoid copepods. In January 1953 I had in my room, core samples taken at a depth of c15m from the profundal zone of Esthwaite Water in the English Lake District. After these had stood overnight I noticed that numerous copepods, which proved to be copepodids, almost all stage 5, of *Mesocyclops leuckarti*, were swimming freely in the water overlying the mud. None had been seen at the time of collection. In Britain *M. leuckarti* is a summer species and is not found in the plankton of Esthwaite Water in winter. Where the vast number of individuals that frequent the lake in summer go in winter, was a complete mystery. These cores supplied the answer. Further cores were obtained, some being allowed to warm up to room temperature (c. 16–17 °C) from the 3 °C or so at which they were collected, others being kept in a refrigerator. Copepods soon appeared in the former, not in the latter. Observations were also made at other temperatures, details of which need not concern us here. These simple observations revealed where *M. leuckarti* goes in the winter time. The number of individuals concerned was enormous. From the number emerging in a single core it was easy to calculate the density per square metre of bottom, and this proved to be about 1 000 000. (Far fewer were found in shallow water deposits.) Much clearly depends on just where the corer strikes, and perhaps on the year of investigation, as subsequent work by a colleague (Smyly, 1961) revealed rather smaller, but still enormous, concentrations of dormant copepodids. On the other hand, when resting stages of *Cyclops scutifer* were discovered later by Elgmork (1962), densities as great as 3 800 000 per m^2 were recorded, so we are here dealing with a phenomenon that involves prodigious numbers of individuals.

Naturally the mud from which *M. leuckarti* emerged was searched for dormant individuals. In spite of their abundance these were not easy to find. Less than 0.9 mm long and transparent, each tended to be obscured by adhering mud particles, and not all lay at the same depth in the semi- liquid mud. The interesting feature of these copepods was that they lacked any trace of a mucilaginous coat, being different in this respect from *A. thomasi* and from those of *Microcyclops bicolor* found in shallow water in the English Lake District by my colleague Smyly. The abdomen of each was flexed ventrally approximately at right angles to the cephalothorax, the antennules lay parallel to the sides of the body, the antennae were folded back and the swimming legs directed forward as at the beginning of a propulsive thrust. No movements of the gut, so characteristic of cyclopoids, were to be seen and the animals remained inert, and to all appearances dead, and could be turned over with a needle. Revival is rapid, especially if the temperature is raised quickly, and animals may swim in as little as 5 h. Diapause is thus easily broken in response to a single stimulus — a rise in temperature. Similar diapausing individuals were found also in Windermere at depths of about 65 m. (For further details see Fryer & Smyly, 1954).

Subsequent work on the life cycle of *M. leuckarti* in Esthwaite Water and Windermere and on the behaviour of the copepodids carried out by Smyly (1961, 1962) extended our understanding of the events so briefly recapitulated here. His findings showed, for example, that in nature some copepodids decend to the bottom and enter diapause in autumn while water temperatures are still above 10 °C (whereas in Lago Maggiore *M. leuckarti* remains active all winter even though the temperature falls to c. 7 °C); that copepodids revive in spring while temperatures are still at the winter minimum of c. 4 °C; and that dormancy can not be re-induced in revived copepodids by lowering the temperature to 0–2 °C. As such stimuli as changes in light intensity and disturbance could be ruled out, he concluded that the length of the dormant period is primarily determined by some physiological means and that changes taking place with time ensure that dormancy cannot be prolonged indefinitely even if there is no rise in temperature. The different behaviour in different lakes of populations assigned to this species raises the possibility that different physiological races, or even cryptic species, may be involved.

Curiously, at the same time as the initial work was in progress, Ulomsky (1953) was studying populations attributed to *M. leuckarti* in Northern Siberia and also found that copepodids spend the winter on the bottom, but not, apparently, in diapause. This apparent anomaly seems to have been resolved by Kiefer's (1981) revelation that *M. leuckarti* is confined to Europe and the western parts of northern Asia and that the many records of this species from elsewhere in fact refer to other, closely related, species.

The wider incidence of diapause in freshwater copepods

Dormant copepodids without gelatinous cysts have now been found in other cyclopoids and the phenomenon is clearly widespread: by 1967 Elgmork was able to list 12 species that display this attribute and Smyly (1967) added another. More cases are now known. Elgmork (1959) has given a detailed account of the life cycle of *Cyclops strenuus* and shown, for example, that dormant copepodids sometimes penetrate soft mud to depths of as much as 30 cm. A so far unique situation prevails in some populations of *Cyclops scutifer* where individuals are remarkably long-lived and spend two consecutive winters in a state of diapause, the first as stage II or III copepodids, the second as stage IV or V, and may conceivably sometimes enter diapause a third time (Elgmork, 1981). Remarkably, one-year cycles are known in populations in other lakes with colder water, which again presents intriguing problems for physiologists, and geneticists. Whether such striking biological differences reflect the existence of cryptic species could perhaps be resolved by molecular techniques.

Some non-cyst-forming species use diapause as a means of overwintering, as does the cyst-forming *Microcyclops bicolor*; others employ it as a means of withstanding high summer temperatures and low oxygen concentrations as does the cyst-forming *Acanthocyclops thomasi*.

Various freshwater copepods are now known to be able not merely to undergo diapause but to withstand desiccation. It has been known since the early years of this century that the calanoid *Diaptomus castor* has drought-resistant eggs, whose hatching to produce swarms of nauplii shortly after the filling of the temporary pools in which it often lives, is a characteristic feature of its usual life cycle. More recent work, summarised by Champeau (1970) who made extensive studies in nature, revealed that in the Mediterranean region several calanoids produce drought-resistant resting eggs. Other calanoids which produce such eggs, or eggs that do not have to face this hazard but which tide their producers over periods of adverse conditions in permanent water bodies, have also been found elsewhere, e.g. in North America. Calanoids with such abilities are listed by Hairston & Cáceres (1996). Champeau has also shown that several cyclopoids can undergo diapause as late copepodid stages that can tolerate a considerable degree of drying. Although he found that excessive desiccation kills such diapausing individuals, *Acanthocyclops bicuspidatus odessanus* emerged from dried mud stored for 3 years, and he collected many stage IV copepodids of this species from a water body that had just refilled after being dry for eight years, though rain must have fallen on the area during this time.

Ability to withstand intense desiccation is shown by *Metacyclops minutus* which Rzóska (1961) found encysted in temporary pools in the Sudan where true drought resistance is necessary for survival. Rzóska describes how advanced copepodid stages were found 'struggling out of completely transparent cysts' about 10 hours after a pool filled.

Several freshwater harpacticoid copepods can also enter diapause as adults and some of these can withstand at least some degree of desiccation (see Champeau, 1970 for summary) and at least one species, *Bryocamptus arcticus*, can produce thick-walled resting eggs (Borutzky, 1929).

The widespread occurrence of diapause in freshwater ostracods

That many freshwater ostracods produce drought-resistant eggs often receives casual mention in general accounts of the group, but the ecological and evolutionary significance of this ability has not received the attention it deserves. The importance of the phenomenon in these animals can be emphasised simply by noting that G. O. Sars, who, from the 1880s onwards, reared many freshwater crustaceans from dried mud, hatched no fewer than 48 species of ostracods from such mud received from S. Africa (Sars, 1924). He also records that the annually added water to mud from a sample kept dry from the time of its collection. Each year up to the time of writing, 12 years in all, he hatched species of *Cypridopsis*, whose remarkable ability to remain dormant for long periods in a dry state was thus dramatically demonstrated. Even longer periods of dormancy are now known.

Diapause in the Branchiopoda: its exploitation and its evolutionary significance

Of all crustaceans it is the Branchiopoda that have universally adopted diapause, without which they could not have achieved their present eminence as freshwater animals. Recent discoveries (Walossek, 1993) have confirmed that, as long suspected, branchiopods existed in the Cambrian. Just when they first colonised freshwater is not known, but they probably did so very early in the Palaeozoic, and they have been exclusively freshwater animals for most of their history. The few present-day marine forms are derived from freshwater ancestors. Colonisation of freshwater was almost certainly facilitated by the early-acquired habit of feeding on fine particles of detritus, which must have been one of the few sources of food initially available in freshwater habitats.

Early branchiopods were slow-moving, largely defenceless animals (as are extant representatives of the most primitive forms), but predators were then few. With the advent of active predators, survival of primitive forms — Anostraca, Notostraca and the two 'conchostracan' orders — whose ancient nature is attested by fossil evidence as well as by their morphology, became completely dependent on the possession of drought-resistant eggs which enabled them to live in predator-free temporary ponds and, to a lesser extent, in extremely cold habitats. Such habitats have changed little for vast periods of time and temporary ponds are indeed probably the most permanent of all freshwater habitats (Fryer, 1985). Highly specialised physiological adaptations, again with the possession of resting eggs, have enabled some primitive branchiopods to colonise extreme environments, such as highly saline waters, where predators are also lacking.

But not all branchiopods are conservative types. This group has given rise to some highly successful forms that, by virtue of small size, rapid reproduction, largely by parthenogenesis, and exceedingly efficient feeding mechanisms, have been able to occupy a wide range of freshwater habitats, including those where predation is intense, and a few have secondarily invaded the sea. All these nevertheless retain diapausing eggs, the presence of which, both as a device for countering adverse conditions and as agents of dispersal, has played a key role in their success. The evolution of such eggs has involved considerable diversification.

The Anostraca is probably the most primitive order, not only of the Branchiopoda but of all crustaceans, but the resting eggs are highly specialised. All save a few parthenogenetically reproducing species of *Artemia* produce such eggs. Parthenogenesis is clearly a secondary development, a belief supported by genetic evidence (Abreu- Grobois & Beardmore, 1980).

Anostracan eggs are usually spherical, but flattened or disc- like in *Tanymastix* and tetrahedral in some species of *Streptocephalus*. The embryo is invested in what is often described as a thin, inner, chitinous cuticle, but which is three-layered (Garreau de Loubresse, 1974) and separates into two distinct membranes at the time of hatching. These membranes are apparently produced at different times by the embryo. Outside these there is a much thicker, non-chitinous shell, sometimes called the tertiary shell. Harder than the chitinous cuticle, it consists of an outer cortex and an inner alveolar layer (Kupka, 1940; Gilchrist, 1978). In some species these are separated in places by a sub-cortical space which copes with the irregular nature of the cortex. The alveolar layer, whose thickness differs in different species, is closely associated with the embryonic cuticle which it invests with what can be described as a layer of solidified foam with large air spaces. In *Thamnocephalus platyurus* it is 'double', having a distinct partition that divides it into inner and outer layers (De Walsche *et al.*, 1991). The cortex, often but not always, thinner and denser in texture than the alveolar layer, often has its external surface puckered or drawn out into ridges or ribs. In *Chirocephalus diaphanus* these slender ribs have a honeycomb- like appearance, remarkably similar to the ribs on the surface of the carapace of the so-called honeycomb species of anomopods of the genus *Chydorus*. These surface patterns and egg shapes are of taxonomic value (Mura & Thiery, 1986; Brendonck & Coomans, 1994a, b).

Various workers have studied the histochemistry of the egg shell (e.g. Linder 1960, Fautrey & Fautrez-Firlefyn, 1971) and shown it to be a complex lipo-protein with mucopolysaccharides and other substances and to be impregnated with haemoglobin. Thiery (1987) suggests that the presence of haemoglobin, a polymer of high molecular weight, might be able to explain the formation of the superficial ornamentation of the eggs and draws attention to the remarkably similar patterns that form on inflated and solidified polymers (Tanaka *et al.*, 1987) and to patterns obtained by Galwey *et al.* (1980) by crystallisation under partial vacuum of solutions of chrome alum that bear an astonishing resemblance to the eggs of *C. diaphanus*.

At the time of shedding, eggs appear usually to contain embryos at the gastrula stage of development. Embryos often seem to remain at this stage for some time while the eggs remain fully hydrated. Further development seems to take place during the dry phase. Khalaf & Hall (1975) suggest that in *C. diaphanus* this occurs while the mud on which they lie is still moist and aerobic, and it is hardly feasible for development to take place in dehydrated embryos.

Dried eggs are often truly cryptobiotic. Their contents contract to a cup-like disc that occupies only a small proportion of the egg as bounded by the inner (embryonic) membrane. As such they can remain dormant for long periods, certainly for several years.

Notostracan eggs are spherical and larger than those of the Anostraca but are similar in structure. The embryo is surrounded by what appears to be a single chitinous membrane, but of which there are certainly two at the time of hatching. The tertiary shell consists of a thin external cortex with a scarcely elaborated surface and a thicker robust alveolar layer, bounded internally, according to Tommasini *et al.*, (1989), by a thin layer, so there are three layers in all. There is no sub-cortical space but in at least one species, *Triops granarius*, the alveolar spaces are in communication by a number of fine pores (Thiery, 1985).

Eggs of the Laevicaudata are small and spherical: those of the Spinicaudata are also small and sometimes spherical, but in some species are of complicated shapes or cylindrical. Like those of the Anostraca and Notostraca, most of those investigated have basically similar inner cuticles produced by the embryo (Zaffagnini & Minelli, 1970) and outer, protective shells, but the spinicaudatan *Leptestheria complexi-manus* lacks the thick shell as apparently does *Eocyzicus klunzinger*. The now extensive studies on the structure of the resting eggs of primitive branchiopods are summarised by Martin (1992).*

Eggs of primitive branchiopods are amazingly resistant. In nature those of the notostracan *Lepidurus arcticus* withstand freezing while those of *Triops granarius* tolerate extremely high temperatures. Carlisle (1968) found that, where atmospheric pressure was such that water boiled at 100 °C, eggs remained viable after 16 hours at 98 °C — but failed to hatch after only 15 min. at 102 °C, which would boil out the last trace of water from the eggs. When the pressure was increased to raise the boiling point to 105 °C, eggs remained viable after 16 hours at 103 °C, but were killed by only 30 minutes exposure to 75 °C when reduction of pressure lowered the boiling point to 70 °C. Some anostracan eggs can tolerate temperatures close to absolute zero for several hours, and eggs of *Artemia* have been fixed to the outside of space-craft and retained their viability after a flight (Planel *et al.*, 1980).

The animals long known as the Cladocera in fact comprise a polyphyletic assemblage of four distinct orders separated by many trenchant differences (Fryer, 1987a, b). These include deep-seated differences in the nature of the resting eggs. All four orders have adopted parthenogenesis as a means of rapidly populating environments that are often habitable for only limited periods, but all also employ diapause. The tactic employed for protection of the resting eggs by the Anomopoda is completely different from that of the other orders.

The Haplopoda, consisting only of *Leptodora*, and most of the Onychopoda, reproduce by parthenogenesis in summer but overwinter as sexually produced resting eggs. Some Caspian cercopagids may be obligate parthenogens. Resting eggs are spherical and lack any equivalent of the thick shell of the more primitive orders but the outer of the two embryonic membranes is often thick and confers mechanical protection.

Haplopods and onychopods are open-water, often planktonic, animals whose eggs are shed freely and fall to the bottom. Those of *Polyphemus pediculus*, however, which are sticky, are often attached to vegetation. In nature, especially in planktonic species, they must almost always remain fully hydrated, and indeed

* See note added in proof (p. 14).

there seem to be no reports of drought-resistance except in the littoral *P. pediculus*. Makrushin (1985) says that the resting eggs of neither *Leptodora kindti* nor *Bythotrephes longimanus* can withstand drought and have never been hatched from dried mud. Those of *Leptodora* are indistinguishable on histological criteria from parthenogenetic eggs, having coarse-grained yolk and vacuoles whereas drought-resistant eggs have fine-grained yolk and lack vacuoles.

In the hydrated state resting eggs of *L. kindti* and *B. longimanus* continue to have measurable metabolic rates even at the low temperatures that prevail on the bottoms of north temperate lakes in winter (Andrew & Hertzig, 1984). The metabolic rate also falls with oxygen concentration (Andrew, 1993). This metabolic activity contrasts with the cryptobiotic nature of dried eggs of other branchiopods. Data produced by Andrew & Hertzig (1984) and Hertzig (1985) suggest that most eggs undergo diapause for one winter only, and this is supported by experiments using newly produced resting eggs (Hertzig, 1985), but some eggs collected in nature remained dormant for more than a year, in one case for 17 months. Eggs that sink deeply into sediments may also remain dormant for more than a year, but prolonged dormancy seems not to be possible. Hatching appears to be independent of either photoperiod or temperature.

Resting eggs of the Ctenopoda are also shed freely. These also lack the thick outer shell so characteristic of the Anostraca and other primitive orders, but the outer of the two protective envelopes present is robust, sometimes has a rough surface, and sometimes bears small spines. In some cases (*Sida crystallina*, *Pseudosida variabilis* and some species of *Diaphanosoma*) the outer envelope is sticky and the eggs become attached to vegetation or other objects. Some species (e.g. *Latonopsis australis*) have been reared from dried mud so their eggs are clearly drought- resistant: those of others, such as *S. crystallina*, lose their viability after one day's drying (Makrushin 1981). Resting eggs of the pelagic *Holopedium gibberum* are also unable to withstand drying. (For differences in the nature of the yolk among ctenopods and the relevance of these differences to drought-resistance in these and other branchiopods see Makrushin 1978, 1985. For a general account of the resting eggs of ctenopods see Korovchinsky & Boikova, 1996).

At the time of hatching of ctenopod eggs the inner membrane swells in a distinctive manner, at least in *Holopedium*, *Sida* and *Diaphanosoma* (see below).

Resting eggs of the Anomopoda, usually the product of sexual reproduction, are protected in a manner unique to the order. Ovoid in shape and enveloped only in two embryonic membranes, these are protectyed by the carapace valves or parts thereof which close around them when the animal moults and sheds its clutch of resting eggs. This composite structure takes the name of an ephippium from the saddle-like structure of *Daphnia*, but primitive ephippia ('protoephippia') are much simpler than this. The most primitive situation is found in the Macrothricidae, but some members of the Chydoridae also have primitive ephippia. *Ophryoxus* (Figs 1–6) displays a very primitive condition. Here much of the carapace envelops the eggs (from one to at least 6 in number). Its only modification in connection with storage of the eggs is the development of a thickened rib along the mid-dorsal line. An almost equally primitive ephippium is found in *Acantholeberis* (Figs 7–10) which also holds a variable number of eggs. More specialised modifications of the carapace are seen in other macrothricids, e.g. in *Bunops* that part of the carapace which overlies the eggs is elaborated into closely packed blister-like swellings, and in several cases some means of attachment, sometimes elaborate, has been evolved. (Fryer, 1972; Makrushin, 1972; Makrushin & Markevitch, 1982).

Chydorids have thicker carapaces than macrothricids so their eggs are inevitably better protected. Dark pigmentation covers that part of the carapace which overlies the egg or eggs. Primitive chydorid ephippia, such as those of *Eurycercus* and *Saycia*, contain several eggs: more specialised types have reduced the number to two or, most commonly, a single egg. Descriptions of many of these and of the ephippia of the related family Bosminidae, were given long ago by Scourfield (1899, 1901, 1902). Some chydorid ephippia are attached to substrata by their producers at the time of shedding.

Daphniids, with which I include *Moina*, have the most specialised of anomopod ephippia. Here only part of the carapace is modified, and in very elaborate ways, to form a container for either one or two eggs according to genus. Several descriptions e.g. M. Wolff (1905), Zwack (1905) have long been available and Shultz (1977) provides information on fine structure. Much of the 'packing' that fills the gap between the inner and outer walls of the carapace is of a polygonal lattice nature that recalls the foam-like alveolar layer of anostracan and other primitive branchiopod egg shells, and fulfills a similar protective function. It is worth noting that although daphniids have the most

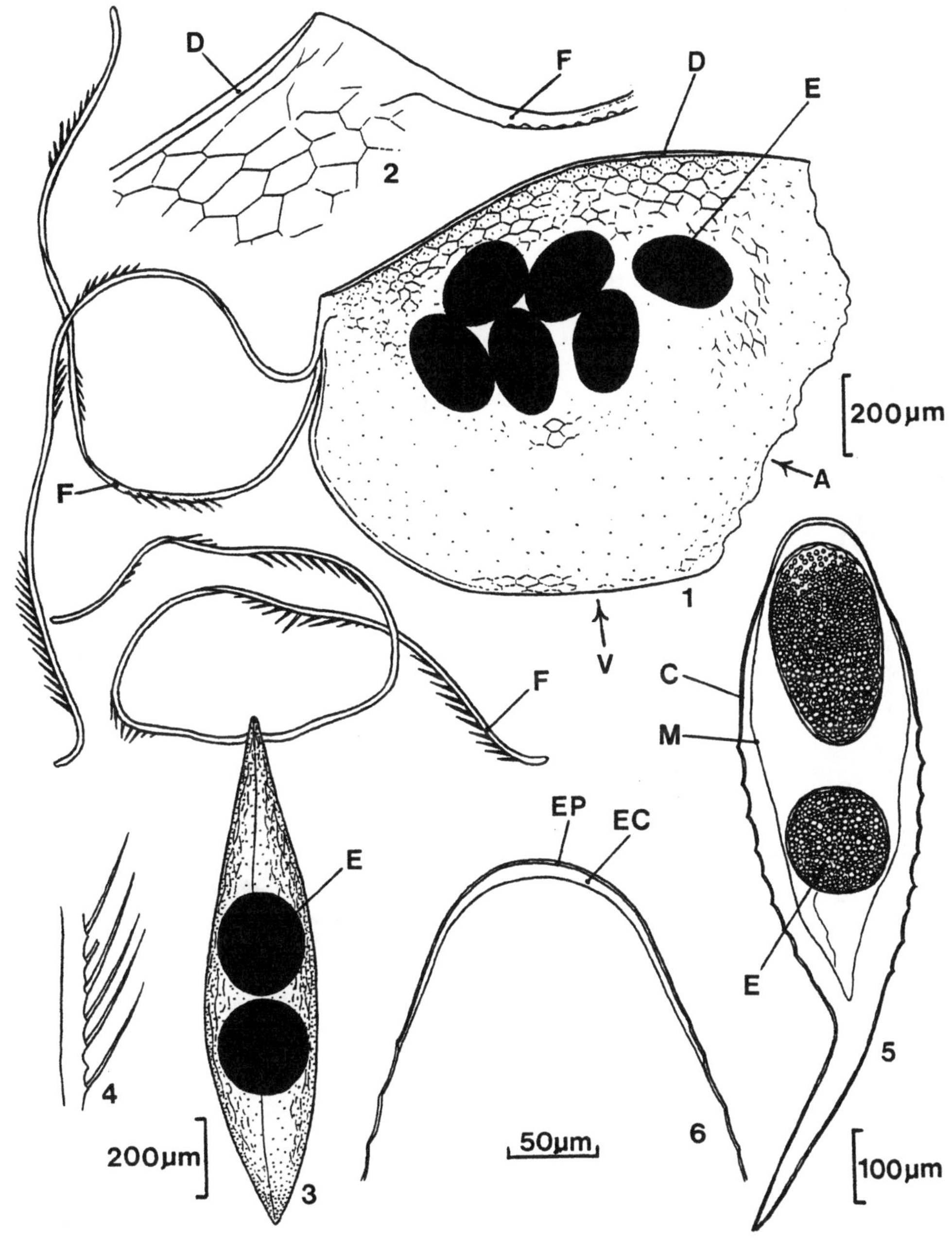

Figs 1–6. The ephippium of *Ophryoxus gracilis* 1. Ephippium, lateral. Note the long filaments (F) which represent the detached rims of the ventral carapace margins. 2. Posterior dorsal corner of the ephippium from the other side showing how the margins here remain attached. 3. Another ephippium containing only two eggs, ventral. 4. Part of one of the filaments. 5. Transverse section showing the very simple nature of the ephippium. The flexure of the delicate ventral portion is not a constant feature of the ephippium and has here probably been accentuated by the embedding process. 6. Transverse section of the dorsal part of the ephippium to show the thickened dorsal plate.

Abbreviations used in Figs 1–14.

A	Broken anterior margin	CT	Cast cuticle	L.	Leaf		from which the rim has become detached
B	Blue-staining* (unsclerotized) endocuticle	D	Dorsal plate	M	Membrane derived from inner cuticular lining of carapace		
		E	Egg			VM	Vitelline membrane
C	Outer cuticular wall	EC	Endocuticle	PC	Moulted cuticle of post-abdominal claws	VR	Ventral region of carapace which usually breaks free from ephippium
CB	Cementing band	EM	Egg membrane				
CC	Cast cuticle and cement	EMB	Embryo	R	Red-staining* (somewhat sclerotized) endocuticle		
CH	Chorion	EP	Epicuticle			Z	Zone of attachment
CM	Cement	F	Filament	V	Ventral margin of carapace	*	With Mallory's triple stain

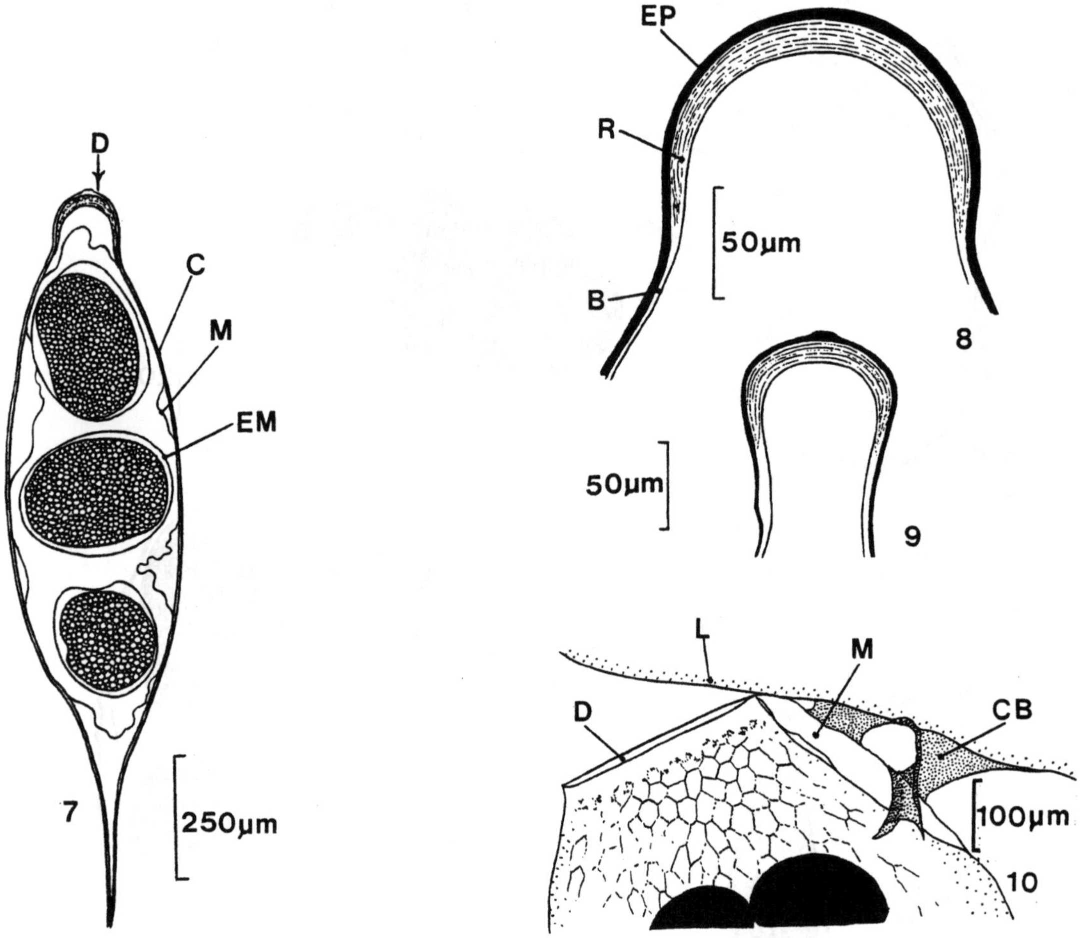

Figs 7–10. The ephippium of *Acantholeberis curvirostris* 7. Transverse section, cut square to the dorsal ridge and therefore somewhat obliquely dorso-laterally across the ephippium. 8. Transverse section through the dorsal ridge near the central portion. 9. The same near to one end. 10. Anterior end of an ephippium attached to a moss leaf, showing the cementing band which has become torn and detached at one point and hangs over the ephippium inclined towards the observer.

advanced ephippia they are nevertheless ancient structures, fossil ephippia assignable even to extant genera, being known from the Lower Cretaceous (Fryer, 1991; Smirnov, 1992).

Even the simplest anomopod ephippia can withstand drought. Although I was aware of this ability in daphniids and certain other anomopods, when I first dried the ephippia of *Eurycercus lamellatus* more than 30 years ago, I saw that the contained eggs simply shrivelled and I assumed that, in this case, they lacked this ability. However, when I immersed the ephippia in water the eggs rapidly hydrated, regained their original shape, and eventually hatched. This is true also of macrothricid eggs that are protected only by thin carapace valves. Resting eggs of the minute *Streblocerus serricaudatus* also shrivel and become grossly distorted when dried but quickly re-hydrate when immersed in water, become indistinguishable from untreated eggs, and often hatch. The same is true of resting eggs of *Acantholeberis curvirostris* that have been either dried or frozen. Though I was not aware of this at the time, eggs of daphniids, hidden in their stout ephippia, also shrivel, as do those of other branchiopods. For example Bishop (1968) found this in *Limnadia stanleyana* (Spinicaudata), and Belk (1982) removed the outer shell of *Eulimnadia antlei* to reveal shrivelled eggs within. Immersed in water, these hydrated and became perfectly spherical in one to three hours and proceeded to develop and hatch. Clearly the thick shell does not prevent desiccation.

Although able to withstand desiccation, resting eggs of *S. serricaudatus* (Figs 11–13) can also develop

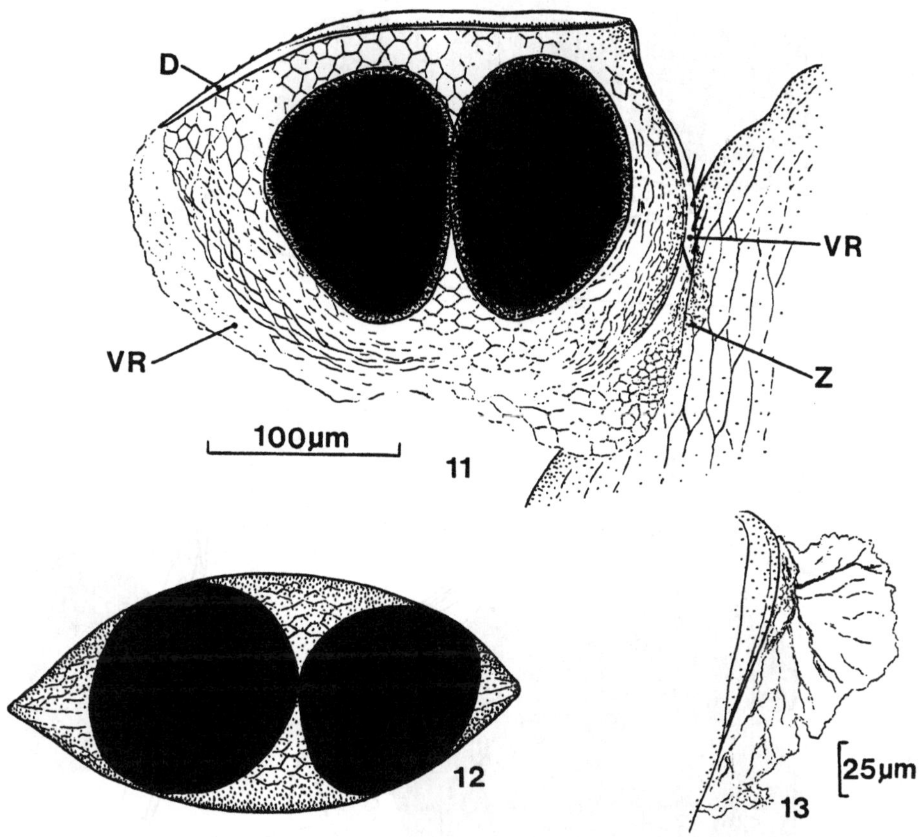

Figs 11–13. The ephippium of *Streblocerus serricaudatus.* 11. Ephippium, lateral, attached to a leaf of *Sphagnum.* 12. Ephippium from a slightly oblique dorsal aspect. Although the dorsal rib is conspicuous in the lateral view it can scarcely be made out from above. 13. The delicate disc by means of which one ephippium was attached to a cover slip.

without being dried. The habits of this species suggest that this is the most usual kind of diapause in nature. The same is true of other species that often deposit ephippia in permanent bodies of water, e.g. *E. lamellatus, O. gracilis* and *Lathonura rectirostris.*

Propagule protection and dispersal in branchiopods

The outer covering of resting eggs and ephippia is obviously protective, but does not prevent the loss of water. It offers protection against abrasion and the attention of small predators. It is also evidently a good protection against the damaging effects of ultra-violet light. Belk (1970) showed that even one day's exposure to full sunlight was sufficient to kill all the eggs of *Eulimnadia antlei* from which the shells had been removed (an operation which does not in itself impair hatchability), but that more than half of those similarly exposed but covered by 1 cm thick glass, opaque to U.V., survived. It is probably significant that anomopod ephippia are often pigmented in the region overlying the eggs, those of daphniids being particularly well protected in this respect.

Drought resistant resting eggs are potential agents of dispersal, and sometimes undoubtedly fulfil this role, but many branchiopods have adopted the tactic of attaching these propagules to firm substrata, which effectively reduces the chances of dispersal. This suggests that it is often more advantageous to ensure persistence of a population in a habitat that has proved favourable than to subject eggs to the hazards involved in chance dispersal to suitable locations elsewhere. Eggs of notostracans are often firmly glued to substrata and some ctenopods and onychopods employ this tactic, but it is the Anomopoda that shows the greatest specialisation in this respect. Several chy-

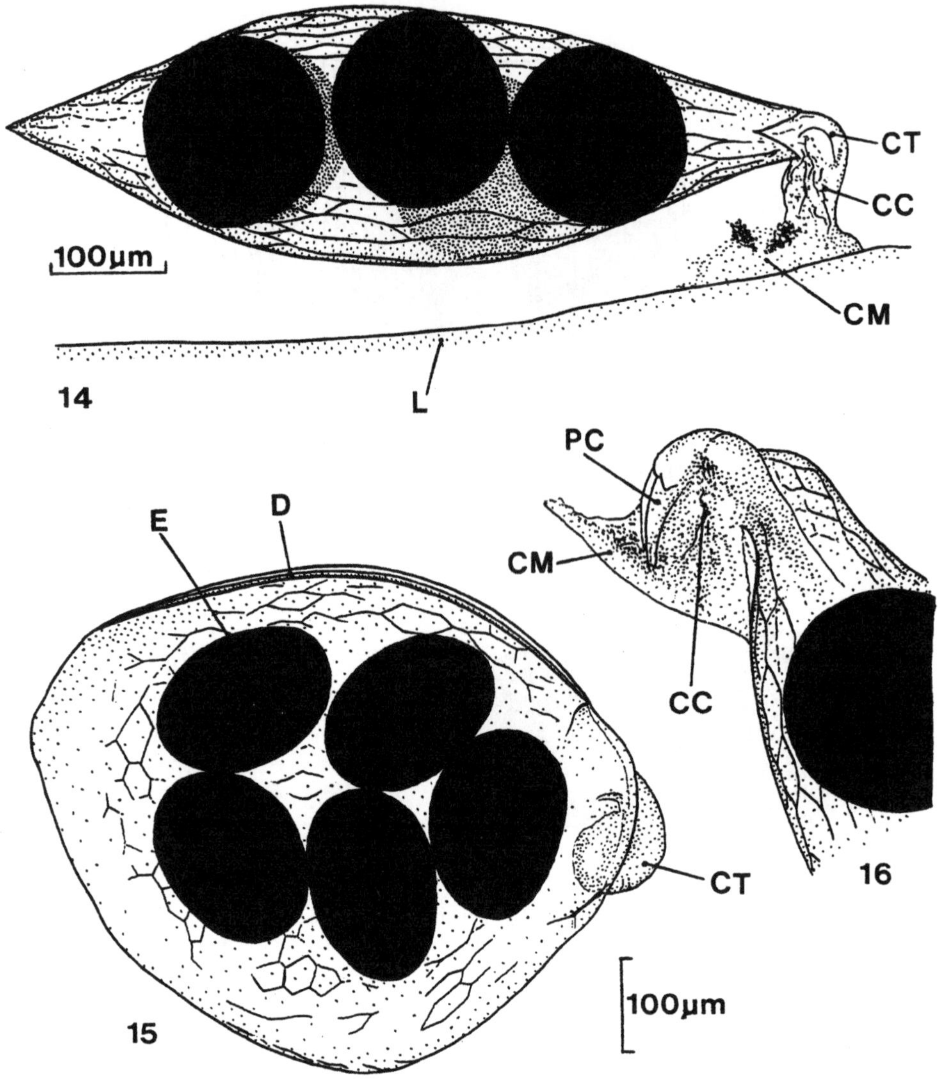

Figs 14–16. The ephippium of *Lathonura rectirostris.* 14. Ephippium, ventral, cemented to a leaf. 15. The same ephippium, lateral: the leaf, to whose surface its lateral faces lie parallel, is omitted for clarity. 16. The cementing structure seen after detachment from the leaf. Note the moulted cuticle of the post-abdominal claws within the cement.

dorids and macrothricids firmly glue their ephippia to suitable objects, (e.g. see Figs 14–16 for *Lathonura rectirostris*) as first noticed by Schödler as long ago as 1846, and later by various workers. Others — the macrothricid *Ophryoxus gracilis* (Figs 1 and 3) and the chydorid *Leydigia acanthocercoides* — use the old carapace margins as hooks which entangle the ephippia in vegetation, a device employed also by certain species of *Daphnia*. Other daphniid ephippia are scattered freely. Free-shedding of eggs or ephippia is employed by planktonic branchiopods. These mostly sink, but some daphniid ephippia float and are sometimes washed up on the shore in large numbers. Eggs of these planktonic forms sometimes accumulate in vast numbers on the bottoms of lakes. In two small English lakes, Smyly (1977) found between 2 and 5 million eggs of *Holopedium gibberum* per m^2, but many of these were apparently infertile and probably represented the accumulation of several years.

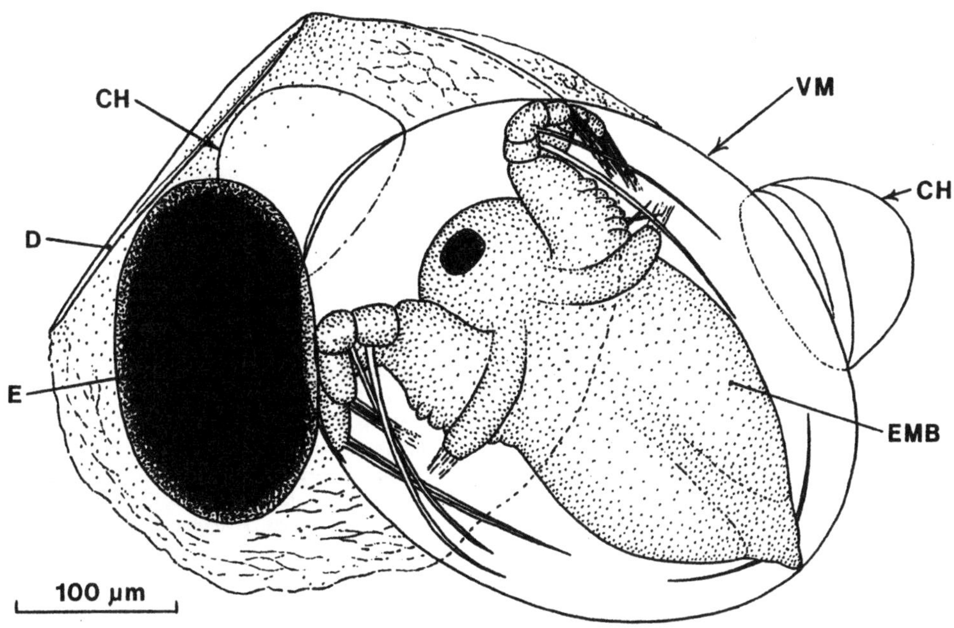

Figs 17. A late embryo of *Streblocerus serricaudatus* which has developed from a resting egg and is near the point of hatching. The embryo is seen somewhat obliquely from the ventral aspect. Note the enormous extent of the osmotic swelling which has distended the vitelline membrane and forced apart the valves of the ephippium. Only the ventral margin of that part of the valve which overlies the embryo is indicated by a dotted line, the embryonated egg being shown as it in fact appears when seen as an optical section.

Induction and breaking of diapause and the hatching process in branchiopods

Although a detailed survey is beyond the scope of this account, brief reference may be made to the problems of what induces and terminates diapause in branchiopods. There are no universal answers. In many temporary pool species all the eggs shed are potentially capable of diapause and if the habitat dries out they enter deep, sometimes prolonged, dormancy, which is usually rapidly terminated when the eggs are again covered by water, though some eggs may not hatch on the first such occasion. Furthermore, eggs, which usually undergo desiccation, may remain hydrated, and enter diapause in this condition, but are easily stimulated to hatch by a rise in temperature. Some resting eggs are inhibited from hatching by darkness, e.g. those of different spinicaudatans (Belk, 1967; Bishop, 1972) and chydorid anomopods (Shan, 1976). In permanent habitats, more subtle stimuli may induce or break diapause though in such planktonic forms as *Leptodora*, *Bythotrephes* and *Holopedium* there may be endogenous rhythms. Thus not all eggs hatch at the same time or temperature and in some lakes where eggs lie in deep water they can scarcely be influenced by changes in light intensity or photoperiod. Another complication is provided by cases such as that of a population of *Holopedium gibberum* which Lampert & Krause (1976) found to produce resting eggs in spring as well as in autumn, which seems to rule out cues provided by temperature or day length. Likewise some populations of *Daphnia pulex* may enter diapause in spring at a time of long day-lengths and rising temperatures, and again in autumn when day lengths are shorter and temperatures falling. In some populations of *Daphnia* however, Stross & Hill (1965) showed a clearly defined relationship between photoperiod and the induction of ephippia-formation, but found that such photo-induction was conditional on population density. Genotypic differences in different populations may also be involved. Such problems are dealt with by Stross in a series of papers (e.g. 1969a, b), part of an extensive literature on the subject.

Hatching follows a basically similar pattern in the resting eggs of all branchiopods. As a result of the osmotic uptake of water, the volume of the space surrounding the nauplius or, in some orders, the juvenile, increases, stretching the inner, elastic embryonic membrane. This ruptures the inelastic outer membrane, often on what appears to be a pre-determined line of weakness (Fig. 17) and, if such is present, the outer egg shell. The ruptured outer embryonic membrane often sits like a pair of caps on the inner membrane but is sometimes readily lost, as for example is the case in

Lepidurus arcticus (Fryer, 1988), which may explain the apparent absence of an outer membrane reported in some species. Final escape by rupture of the inner membrane seems often to be facilitated by movements of the developing nauplius or juvenile.

Differences in details of hatching may be of phyletic interest. For example, the capsule formed by the inner membrane of the ctenopods *Sida*, *Diaphanosoma* and *Holopedium* assumes an elongate, cylindrical shape as it swells prior to hatching as compared with the ovoid shape assumed by its homologue in the Anomopoda. Investigation of other genera is necessary but this difference seems to emphasise the clear distinction of these orders.

Acknowledgments

I am much indebted to Dr N. M. Korovchinsky who kindly translated passages from publications written in Russian. The illustrations were originally prepared for a paper published in 1972. Unknown to the author until they appeared, however, these had been redrawn in a crude and inaccurate manner by a publisher's 'artist'. They are reproduced here from the original, accurate drawings.

References

Abreu-Grobois, F. A. & J. A. Beardmore, 1980. Genetic characterisation of *Artemia* populations — an electrophoretic approach. In The Brine Shrimp *Artemia*. Vol. 1, Eds. G. Persoon, P. Sorgeloos, O. Reoels & E. Jaspers. Wetteren. University Press: 133–146.

Andrew, T. E., 1993. The influence of oxygen concentration on the respiration rate of the resting eggs of *Leptodora kindti* (Focke 1844) and *Bythotrephes longimanus* Leydig 1860. Arch. Hydrobiol. 128: 409–414.

Andrew, T. E. & A. Hertzig, 1984. The respiration rate of the resting eggs of *Leptodora kindti* (Focke 1844) and *Bythotrephes longimanus* Leydig 1860 (Crustacea, Cladocera) at environmentally encountered temperatures. Oecologia (Berlin) 64: 241–244.

Baird, W., 1850. The natural history of the British Entomostraca. London. Ray. Soc.

Belk, D., 1970. Functions of the conchostracan egg shell. Crustaceana, 19: 105–106.

Belk, D., 1972. The biology and ecology of *Eulimnadia antlei* (Conchostraca) Southwest Nat. 16, 297–305.

Belk, D., 1987. Embryonic cuticles of *Artemia* during diapause and hatching: insights for comparison with other Branchiopoda. J Crust. Biol. 7: 691–696.

Birge, E. A. & C. Juday, 1908. A summer resting stage in the development of *Cyclops bicuspidatus* Claus. Trans. Wisc. acad. Sci. Arts & Lett. 16: 1–9.

Bishop, J. A., 1968. Resistance of *Limnadia stanleyana* (Branchiopoda, Conchostaca) to desiccation. Crustaceana 14: 35–38.

Borutzky, E. V., 1929. Zur Frage über den Ruhezustand bei Copepoda Harpacticoida. Dauereier bei *Canthocamptus arcticus* Lilljeborg. Zool. Anz. 83: 225–233.

Carlisle, D. B., 1968. *Triops* (Entomostraca) eggs killed only by boiling. Science 161: 279.

Brendonck, L. & A. Coomans, 1994a. Egg morphology in African Streptocephalidae (Crustacea: Branchiopoda: Anostraca). Part 1 South of Zambezi and Kunene Rivers. Arch. Hydrobiol. Suppl. 99: 313–334.

Brendonck, L. & A. Coomans, 1994b. Ibid. Part 2 North of Zambezi and Kunene Rivers, and Madagascar. Arch. Hydrobiol. Suppl. 99: 335–356.

Champeau, A., 1970. Recherches sur l'écologie et l'adaptation à la vie latente des copépodes des eaux temporaires Provençales et Corses. Thesis. Univ. Aix- Marseille, 360 pp.

De Walsche, C., N. Munuswamy, & H. J. Dumont, 1991. Structural differences between the cyst walls of *Streptocephalus dichotomus* (Baird), *S. torvicornis* (Waga), and *Thamnocephalus platyurus* (Packard) (Crustacea: Anostraca), and a comparison with other genera and species. Hydrobiologia 212 (Dev. Hydrobiol. 64): 195–202.

Elgmork, K., 1959. Seasonal occurrence of *Cyclops strenuus strenuus* in relation to environment in small water bodies in Southern Norway. Folia limnol. Scand. 11: 1–196.

Elgmork, K., 1962. A bottom resting stage in the planktonic freshwater copepod *Cyclops scutifer* Sars. Oikos 13: 306–310.

Elgmork, K., 1967. Ecological aspects of diapause in copepods. Proc. Symp. Crust. Pt. 3: 947–954.

Elgmork, K., 1981. Extraordinary prolongation of the life cycle in a freshwater planktonic copepod, Holarctic Ecol. 4: 278–290.

Fautrez, J. & N. Fautrez-Firlefyn, 1971. Contribution à l'étude des glandes coquilières et des coques de l'oeuf d'*Artemia salina*. Arch. Biol. 82: 41–83.

Fryer, G., 1972. Observations on the ephippia of certain macrothricid cladocerans. Zool. J. Linn. Soc. 51: 76–96.

Fryer, G., 1985. Structure and habits of living branchiopod crustaceans and their bearing on the interpretation of fossil forms. Trans. r. Soc. Edinb. 76: 103–113.

Fryer, G., 1987a. Morphology and the classification of the so-called Cladocera. Hydrobiologia 145 (Dev. Hydrobiol. 35): 19–28.

Fryer, G., 1987b. A new classification of the branchiopod Crustacea. Zool. J. Linn. Soc. 91: 357–383.

Fryer, G., 1988. Studies on the functional morphology and biology of the Notostraca (Crustacea: Branchiopoda). Phil. Trans. r. Soc. B. 321: 27–124.

Fryer, G., 1991. A daphniid ephippium (Branchiopoda: Anomopoda) of Cretaceous age. Zool. J. Linn. Soc. 102: 163–167.

Fryer, G. & W. J. P. Smyly, 1954. Some remarks on the resting stages of some freshwater cyclopoid and harpacticoid copepods. Ann. Mag. Nat. Hist. Ser. 12, 7: 65–72.

Galway, A. K., R. Reed & G. G. T. Guarini, 1980. Observations on internal structures of chrome alum dehydration nuclei. Nature, Lond. 283: 52–54.

Garreau de Loubresse, N., 1974. Étude chronologique de la mise en place des enveloppes de l'oeuf d'un Crustace' phyllopode: *Tanymastix lacunae*. J. Microsc. (Fr) 20: 21–38.

Gilchrist, B. M., 1978. Scanning electron microscope studies of the egg shell in some Anostraca (Crustacea: Branchiopoda). Cell Tiss. Res. 193: 337–351.

Hairston, N. G. Jr. & C. E. Cáceres, 1996. Distribution of Crustacean diapause: micro- and macro-evolutionary pattern and process. Hydrobiologia 320 (Dev. Hydrobiol. 114): 24–44.

Hertzig, A., 1985. Resting eggs — a significant stage in the life cycle of the crustaceans *Leptodora kindti* and *Bythotrephes longimanus*. Verh. int. Ver. Limnol. 22: 3088–3098.

Jurine, L., 1820. Histoire des Monocles, qui se trouvent aux environs de Genève. Paris. Paschaud.

Khalaf, A. N. & R. E. Hall, 1975. Embryonic development and hatching of *Chirocephalus diaphanus* Prévost (Crustacea: Anostraca) in nature. Hydrobiologia 47: 1–11.

Kiefer, F., 1981. Beitrag zur Kenntnis von Morphologie, Taxonomie und Geographischer Verbreitung von *Mesocyclops leuckarti* auctorum. Arch. Hydrobiol. Suppl. 62: 148–190.

Korovchinsky, N. M. & O. S. Boikova, 1996. The resting eggs of the Ctenopoda (Crustacea: Branchiopoda): a review. Hydrobiologia 320 (Dev. Hydrobiol. 114): 131–140.

Kupka, E., 1940. Untersuchungen über die Schalenbildung und Schalenstructur bei den Eiern von *Branchipus schaefferi* (Fischer). Zool. Anz. 132: 130–139.

Lampert, W. & I. Krause, 1976. Zur Biologie der Cladocere *Holopedium gibberum* Zaddach in Windgfällweiher (Schwartzwald). Arch. Hydrobiol. Suppl. 48: 262–286.

Lauterborn, R. & E. Wolf, 1909. Cystenbildung bei *Canthocamptus microstaphylinus*. Zool. Anz., 34: 130–136.

Linder, H. J., 1960. Studies on the freshwater fairy shrimp *Chirocephalopsis bundyi* (Forbes) II Histochemistry of egg-shell formation. J. Morph. 197: 259– 284.

Makrushin, A. V., 1972. Protoephippial glands in Cladocera (Crustacea). Zool. Zhur. 51: 1736–1738 (In Russian).

Makrushin, A. V., 1978. Anhydrobiosis and the structure of Cladoceran egg yolk. Zool. Zhur. 57: 364–373. (In Russian).

Makrushin, A. V., 1981. Resistance of resting eggs of *Sida crystallina* and *Moina macrocopa*. Zool. Zhur. 60: 933–935. (In Russian).

Makrushin, A. V., 1985. Anhydrobiosis of primary invertebrates. Acad. Sci. U.S.S.R. Publ. Leningrad, 101 pp. (In Russian).

Makrushin, A. V. & G. I. Markevich, 1982. On the formation of ephippia in some Cladocera (Crustacea) Zool. Zhur. 61: 1425–1427. (In Russian).

Martin, J. W., 1992. Branchiopoda. In, Microscopical anatomy of invertebrates. Vol. 9 Crustacea. pp. 25–224. Wiley-Liss.

Müller, O. F., 1785. Entomostraca, seu insecta testacea etc. Lipsiae et Harniae.

Mura, G. & A. Thiery, 1986. Taxonomical significance of scanning electron microscopic morphology of the euphyllopods resting eggs from Morocco. Part 1 Anostraca. Vie Milieu 36: 125–131.

Petiver, J., 1709. Gazophylacii Naturae.

Planel, H., Y. Gaubin, R. Kaiser & B. Pianezzi, 1980. Effects of space environment on *Artemia* eggs. In, The Brine Shrimp *Artemia*, Vol. 1 (G. Personne, P. Sargeloos, O. Roels & E. Jaspers) Wetteren Universa Press. 189–198.

Rzóska, J., 1961. Observations on tropical rainpools and general remarks on temporary waters. Hydrobiologia 17: 265–286.

Sars, G. O., 1924. The freshwater Entomostraca of the Cape Province (Union of South Africa), Part II: Ostracoda. Ann. S Afr. Mus. 20: 105–193.

Schaeffer, J. C., 1755. Die grünen Armpolypen, die geschwanzten und ugeschwanzten zackigen Wasserflohe und eine besondere Art kleiner Wasseraale. Regensberg. Weiss.

Schaeffer, J. C., 1756. Der krebsartige Kiefenfuss mit der kurzen und langen Schwanzklappe: 65–200. Ratisbon.

Schödler, J. E., 1846. Ueber *Acanthocercus rigidus*, ein bisher noch un-bekanntes Entomostracon aus der Familie der Cladoceren. Arch. Naturgesch 12: 301–374.

Scourfield, D. J., 1899. The winter egg of a rare water- flea (*Leydigia acanthocercoides* Fischer). J. Quekett Micro. Club 7(2): 171–179.

Scourfield, D. J., 1901. The ephippium of *Bosnina*. J. Quekket Micro. Club. 8: 51–56.

Scourfield, D. J., 1902. The ephippia of the lynceid Entomostraca. J. Quekket Micro. Club 8(2): 217–244.

Schultz, T. W., 1977. Fine structure of the ephippium of *Daphnia pulex* (Crustacea: Cladocera) Trans. am. Micros. Soc. 96: 313–321.

Shan, R. K., 1976. Influence of light on hatching resting eggs of chydorids (Cladocera) Int. Revue Hydrobiol. 55: 295–302.

Smirnov, N. N., 1992. Mesozoic Anomopoda (Crustacea) from Mongolia. Zool. J. Linn. Soc. 104: 97–116.

Smyly, W. J. P., 1961. The life-cycle of the freshwater copepod *Cyclops leuckarti* Claus in Esthwaite Water. J. anim. Ecol. 30: 153–171.

Smyly, W. J. P., 1962. Laboratory experiments with stage V copepodids of the freshwater copepod *Cyclops leuckarti* Claus from Windemere and Esthwaite Water. Crustaceana 4: 273–280.

Smyly, W. J. P., 1967. A resting stage in *Cyclops dybowskii* Lande (Crustacea: Copepoda) Naturalist 92: 125– 126.

Smyly, W. J. P., 1977. A note on the resting egg of *Holopedium gibberum* Zaddach (Crustacea: Cladocera). Microscopy 33: 170–171.

Straus, E. J., 1819–20. Mémoire sur les *Daphnia* de la classe des Crustacés. Mem. Mus. Hist. Nat. Paris 5: 380–425; 6: 149–162.

Stross, R. G., 1969a. Photoperiod control of diapause in *Daphnia* II Induction of winter diapause in the Arctic. Biol. Bull. 136: 264–273.

Stross, R. G., 1969b. Photoperiod control of diapause in *Daphnia* III. Two-stimulus control of long-day, short-day induction. Biol. Bull. 137: 359–374.

Stross, R. G. & G. C. Hill, 1965. Diapause induction in *Daphnia* requires two stimuli. Science 150: 1462–1464.

Tanaka, T., S-T. Sun, Y. Hirokawa, S. Katayama, , J. Kucero, Y. Hirose & T. Amiya, 1987. Mechanical instability of gels at the phase transition. Nature, Lond. 325: 796–798.

Thiery, A., 1985. Ponte et ultrastructure de l'oeuf Chez *Triops granarius* Lucas (Crustacea, Notostraca): adaptations à l'assèchment de l'habitat. Verh. int. Ver. Limnol. 22: 3024–3028.

Thiery, A., 1987. Les crustaces branchiopodes Anostraca, Notostraca et Conchostraca des milieux limnique temporaires (Dayas) au Maroc, Taxonomie, Biogeographie, Ecologie, These Univ. d'Aix-Marseille, 405 pp.

Tommasini, S., F. Scanabissi Sabelli & M. Trentini, 1989. Scanning electron microscope study of eggshell development in *Triops cancriformis* (Bosc.) (Crustacea, Notostraca). Vie Milieu 39: 29–32.

Ulomosky, S. N., 1953. New ecological studies of some species of oar-legged crustaceans (Copepoda), *Mesocyclops leuckarti* (Claus), 1851 C.R. acad. Sci. URSS, 90: 295–297. (In Russian).

Walossek, D., 1993. The Upper Cambrian *Rehbachiella* and the phylogeny of the Branchiopoda and Crustacea. Fossils and Strata No. 32, 1–202.

Wolf, E., 1905. Die Fortpflanzungs-Verhältnisse unsere einheimischen Copepoden. Zool. Jahrb. Syst. 22: 101–280.

Wolff, M., 1905. Das Ephippium von *Daphnia pulex*. ForschBer biol. Stn. Plön. 12: 303–315.

Zwack, A., 1905. Der feiner Bau und die Bildung des Ephippiums von *Daphnia hyalina* Leydig. Z. wiss. Zool. 79: 548–573.

Zaffagnini, F. & G. Minelli, 1970. Origine e natura della membrane che avvologno l'uova di *Limnetis lenticularis* (Crustacea: Conchostraca). Boll. Zool. 37: 139– 149.

Note added in proof

Only recently did I see a paper by Roessler & Sanchez (1986) in a little known journal, of which a revised version by Roessler (1995) has also appeared. The spinicaudatan *Cyclestheria hislopi* produces an 'ephippium'. This resembles anomopod ephippia in several respects but differs in fundamental particulars and is clearly a parallel development of great interest.

References

Roessler, E. W., 1995 Review of Columbian Conchostraca (Crustacea) — ecological aspects and life cycles — family Cyclestheriidae. Hydrobiologia 298 (Dev. Hydrobiol. 103): 113–124.

Roessler, E. W. & Sánchez M., 1986 Estudios sobre los 'Entomostraceos' de Colombia I Contribuciones al conocimiento de la historia natural de *Cyclestheria hislopi* (Baird, 1859), (Arthropoda, Crustacea, Conchostraca), con enfasis en aspectos bioecologicos y del ciclo vital. Caldasia 14: 679–707.

Hydrobiologia **320**: 15–26, 1996.
V. R. Alekseev & G. Fryer (eds), Diapause in the Crustacea.

Types of diapause in Crustacea: definitions, distribution, evolution

Victor R. Alekseev & Yaroslav I. Starobogatov
Experimental Hydrobiology laboratory and Marine Biology laboratory of Zoological Institute of Russian Academy of Sciences, 199034, St. Petersburg, Russia

Key words: Crustacea, diapause, photoperiod responses, evolution

Abstract

Definitions of diapause and of the three types of diapause (embryonal, larval and imaginal) as they apply to the Crustacea are given and the distribution of the types among different orders of crustaceans is discussed. A special role of diapause as a way to escape competition among ecologically similar species is noted and the significance of photoperiod responses as a pace-maker of the biorhythmical function of diapause for a wide range of species is demonstrated. A common unspecific regression for the description of the level of depression of oxygen consumption with duration of diapause in crustaceans of different sizes, different orders and different types of diapause is suggested. A special significance of embryonal diapause for the adaptation of palaeo-, meso- and neolimnic species of crustaceans that migrated from the sea into inland waterbodies is postulated.

Introduction

The term 'diapause' is used widely in entomology but has been applied to Crustacea only recently (Stross, 1965). According to Danilevsky (1961) the definition of 'diapause' can be achieved through the analysis of the annual cycle of an organism that practically always consists of sequences of periods of active growth and reproduction with periods of delay or halting of these processes. This condition of physiological rest of an animal as a result of hormonal activity was named diapause.

There is an opinion that diapause as an adaptation to seasonal changes in the environment appeared independently in different groups of Insecta (Danylevsky, 1961). This leads automatically to the conclusion that this complicated process, consisting of many elements, arose independently in several taxonomically remote groups of invertebrates.

In our opinion there are at least two sets of arguments contrary to this opinion i.e. in favour of the evolutionary transmitted development of diapause mainly as a photoperiodic response.

The first set of arguments is the constancy of temporal and spatial heterogeneity of external conditions such as light and temperature conditioned by the rotation of the globe around the sun from the early days of life on our planet. The necessity of adaptation to the periodicity of these important factors doubtless was a problem for the first inhabitants of water and land. The most successful inventions in this direction, fixed genetically and supported by stabilizing selection related to the acting factor, were the general base from which specific development and specialization in different groups led to diverse forms of diapause of different physiological and ecological significance.

Another set of arguments doubtless necessary for the evaluation of the hypothesis of the origin of the diapause in animals, related to the many elements of internal organization of the system of season-cycle regulation. The so-called neurohormonal mechanism of genetically and phylogenetically remote groups of organisms displays characters that are strikingly similar in different animals. This universality, and the economy of using natural material and the ways in which it is used, are similar in one sense to situations that involve the use of haemoglobin (Alekseev, 1990).

At the same time there is no doubt that the common genetical base does not lead automatically to the common character of properties such as adaptation to specific conditions in evolutionarilly remote group of organisms. A study of similarities and differences in characteristics in closely and distantly related organisms as well as in organisms in different stages of ontogenesis is one of fundamental goals of biology. Therefore resolution of problems related to the degree of similarity or difference in the features of diapause, as well as transformation of the adaptation with the evolution of genetically closed taxa, has importance for general biology.

Diapause and competition in Crustacea

The adaptive function of diapause in crustaceans and other organisms evolves in two directions: biorhythmical adaptation determines the synchronization of an animal's life cycle with the seasonal rhythm of external conditions, and defensive adaptation permits the organism to endure the influence of a complex of conditions that are oppressive or and even lethal for active phases in unfavorable periods.

The defensive function of diapause provides unspecific resistance to a wide complex of unfavourable influences and as its base a lowering (in extreme cases – in anabiosis – lowering to zero) the level of metabolism. The unspecific character of the resistance is discussed below.

The biorhythmical function of diapause is commonly explained by the periodic appearance of conditions unfavourable for active development. This statement is correct but incomplete. Populations of all living organisms persist in a more or less stable state due to licences given to them by the ecosystems of which they are part. The notion of 'licence' was introduced by Günther (1949), but it was defined extremely vaguely as the creation by external and internal media of the ecological conditions, in the wide sense, in which the animal lives.

The distinction between ecological 'niche' and 'licence' must be made. The ecological niche is determined using the notion of species, but the licence using the notion of ecosystem. The fundamental ecological niche is the property of a species but the realized ecological niche is the property of the population of a species. The notion of an 'empty niche' cannot be determined. Licence is that which the ecosystem allows to a species which is present in it or moves in it. Unused licence means that organisms can move in the ecosystem in future.

Many years later, Levchenko (1984) and Starobogatov (1984, 1985) reintroduced this notion defined as: (1) a position in space and time, (2) a role in matter and energy flow and (3) a position in a gradient of external factors permitted by the ecosystem to the population existing in it or to organisms moving into it. It is evident from this definition that one of ways of removing competition (or at least weakening it) between the populations of species is subdivision of licence. It may be in space or in time. In this connection it is more correct to regard the presence of resting phases in the first place as one of the means of subdivision of licences in time. In other words when the population of a species actively takes part in the matter and energy flow, and the population of another species has a resting phase, competition between them weakens or completely disappears, and vice versa.

On other hand for populations of crustaceans living in seasonaly changing condition, the presence of resting stages gives the possibility of surviving when temperature and food availability deteriorate.

To realize the specific goals of diapause it was necessary not only to respond to the selection pressure imposed by competition among species but to acquire an effective mechanism for detecting the time of the year. It must be pace-maker – any external factor alternating periodically and capable of being used as a signal to initiate the beginning or termination of the resting phase. Doubtless the first of such factors is solar radiation and temperature correlated with it. The most interesting factor to serve as a pace-maker is photoperiod. The importance of light periodicity for driving different functions in organisms was assessed for more than 50 marine, freshwater and terrestrial species of crustaceans from 9 families and orders (Table 1). Most of the photoperiodic responses were induction or breaking of diapause. Only five species from the group were neutral to light factors.

Of course, the presence of diapause in the annual cycle of organisms is a very complicated and multivariant adaptation to the seasonal or biotopic heterogeneity in the environment. We concentrate our attention on the competition role of diapause because it seems to us both important and as yet inadequately discussed.

Types of diapause in Crustacea

In the lexicon of zoologists studying diapause in the Crustacea one can find a variety of terms for the phenomenon. The uncertainty makes further analysis and

Table 1. A list of some species of Crustacea examined for photoperiod control of internal functions. (LDP – Long-day photoperiod; SDP – Short-day photoperiod; NPR – No photoperiod response for this function; ID – Induction of diapause; TD – Termination of diapause; BGP – biochemical-histology and other processes; e, l, i – embryonal, larval, imaginal types of diapause)

Order Species	Acting daylength	Photoperiod response	References
ANOSTRACA			
Artemia salina L.	LDP	TD e	1, 2
Streptocephalus torvicornis Waga	SDP	BGP	3
NOTOSTRACA			
Triops granarius (Lucas)	LDP	TD e	4
T. longicaudatus (Le Conte)	LDP	TD e	4
'CLADOCERA' ANOMOPODA			
Alonopsis elongata Sars	SDP	ID e	5
Daphnia magna Straus	SDP	ID e	6, 7, 8, 9, 10
D. magna	SDP	BGP	11
D. pulex Leydig	LDP	TD e	12, 13
D. pulex	SDP	ID e	14, 15, 16
D. middendorffiana Fischer	LDP	TD e	17
D. middendorffiana	SDP	ID e	15
Pleuroxus truncatus (O.F.M.)	SDP	ID e	18
P. procurvus Birge	SDP	ID e	18
P. denticulatus Birge	SDP, LDP	ID e	18
CALANOIDA			
Acartia clausi Gies.	NPR	TD e	19
A. erythreae Gies.	NPR	TD e	19
Diaptomus sanguineus Forbes	LDP	ID e	20, 21
Labidocera aestiva Wilson	SDP	TD e	22
CYCLOPOIDA			
Acanthocyclops americanus (Marsh.)	SDP, LDP	ID l	23
A. vernalis (Fisch.)	SDP, LDP	ID l	23
Cyclops abyssorum sevani Mesh.	LDP	ID l	2
C. kolensis Sars	NPR	ID 1	23
C. strenuus divergens Lindb.	SDP, LDP	ID l	23
C. vicinus Uljanin	LDP	ID l	24, 25, 2
C. vicinus	SDP	TD l	26
Diacyclops bicuspidatus (Claus)	SDP, LDP	ID l	23
D. limnobius Kiefer	SDP	ID l	23
D. navus Herrick	SDP	ID l	27
Diacyclops sp.	LDP	ID l	27
Ectocyclops phaleratus (Koch.)	SDP	ID l	28
Macrocyclops albidus (Jurine)	SDP, LDP	ID l	29
M. fuscus (Jurine)	SDP	ID l	30
Megacyclops viridis (Jurine)	SDP	BGP	23
Mesocyclops leuckarti (Claus)	SDP	ID l	29, 31
M. leuckarti	SDP	BGP	23
Metacyclops minutus (Claus)	SDP	ID l	23, 32

Table 1. Continued.

Order Species	Acting daylength	Photoperiod response	References
Thermocyclops crassus (Fisch.)	SDP	ID I	29
T. dybowskii (Lande)	LDP	ID I	23
T. oithonoides (Sars)	LDP	ID I	23
CIRRIPEDIA			
Balanus balanoides L.	SDP	BGP	33
AMPHIPODA			
Gammarus lacustris (L.)	LDP	TD i	34
G. lawrencianus Bous.	LDP	TD i	35
G. lawrencianus	SDP	ID i	35
Hyalella azteca Sausse	LDP	TD i	36
H. azteca	SDL	ID i	36
Taliturus saltator Montagu	LDP	TD i	37
ISOPODA			
Armadillidium vulgare Latr.	LDP	TD i	38
Asellus aquaticus L.	LDP	TD i	39
A. aquaticus	SDP	ID i	39
Eluma purpurascens Budde-Lund	NPR	TD i	40
DECAPODA			
Astacus astacus L.	NPR	TD i	41
Homarus americanus M.Edw.	LDP	TD i	42
Macrobrachium lammeri (Fabr.)	LDP	BGP	43
M. lanchesteri (De Man)	LDP	BGP	44
Orconectes virilis (Hagen)	LDP	TD i	45, 46
O. virilis	LDP	TD I	45
O. virilis	SDP	TD i	47
Palaemonetes paludosus (Gibles)	SDP	BGP	48
P. pugio (Holthius)	LDP	TD i	49
P. varians Leach	LDP	TD i	50
Penaeus japonicus Bate	LDP	TD i	51
Scylla serrata Fors	LDP	BGP	52
Uca pugax (Smith)	LDP	BGP	53

References 1 – Van der Linden *et al.*, 1985; 2 – Alekseev, 1990; 3 – Bakulin, 1983; 4 – Takahashi, 1977; 5 – Kubertsky, 1977; 6 – Fries, 1964; 7 – Bunner & Halcrow, 1977; 8 – Ferrari & Hebert, 1982; 9 – Carvalho & Hages, 1983; 10 – Alekseev, unpublished; 11 – Korpelainen, 1986; 12 – Pancella & STross, 1963; 13 – Stross, 1971; 14 – Stross, 1965; 15 – Stross, 1969a; 16 – Stimpl, 1971; 17 – Stross & Hill, 1965; 18 – Shan, 1974; 19 – Uye *et al.*, 1979; 20 – Nelson & Olds, 1986; 21 – Hairston & Olds, 1987; 22 – Marcus, 1982; 23 – Alekseev, 1992; 24 – Einsle, 1967; 25 – Spindler, 1971; 26 – Spindler, 1970; 27 – Watson & Smallman, 1971; 28 – Alekseev, 1984a; 29 – Alekseev, 1984b; 30 – Alekseev, 1986; 31 – Alekseev, 1987; 32 – Alekseev, 1988; 33 – Tighe-Ford, 1967; 34 – March, 1982; 35 – Steele, 1981; 36 – de March, 1977; 37 – Williams, 1985; 38 – Juchalt *et al.*, 1980a; 39 – Tadini-Vitagliano *et al.*, 1981; 40 – Juchalt *et al.*, 1980b; 41 – Westin, Gydemo, 1986; 42 – Nelson, 1986; 43 – Sarojini, Gyananth, 1985; 44 – Rao *et al.*, 1984; 45 – Stephens, 1985; 46 – Aiken, 1969a; 47 – Aiken, 1969b; 48 – Paris, Jenner, 1952; 49 – Little, 1968; 50 – Bouchon *et al.*, 1985; 51 – Lauber, Bonichon, 1975; 52 – Nagabhushanam, Farooqui, 1981; 53 – Brown, Stephens, 1951.

an understanding of the essentials of diapause difficult. All the differences of diapause observed in the various orders of the Crustacea may be easily adapted to accord with to the logic pattern advanced by Danylevsky (1961) for Insecta. The foundation of the classification is the instar of development at which an animal goes into hibernation. In Crustacea only embryonal, larval and imaginal types of diapause can be found.

In our further analysis the terms of other authors such as 'resting eggs', 'ephippial eggs', 'winter eggs', 'latent eggs' etc. were replaced by embryonal diapause; 'dormant instars', resting stages' (that are traditionally used for Copepoda) – were replaced by 'larval diapause'. Finally the substitution of imaginal diapause was suggested for terms such as 'pre-embryonal diapause' and 'reproductive diapause' of adults (Alekseev, 1990).

On other hand, sometimes it is important to divide all the types of diapause according to the duration of hibernation. One can suggests oligo-, meso- and superpause as terms for this. The difference among the three pauses is clearly shown if we compare the durations of hibernation and ontogenesis in 'Entomostraca' and shortlived forms of Malacostraca or the season of activity of Decapoda.

Oligopause leads to the halting of vital activities of an animal for an interval much shorter than the duration of ontogenesis of the species (or season length for large-size crustaceans). An example of oligopause may be observed for some small Cyclopoid species in rain pools and other temporary waterbodies of arid regions. The 4th copepodid instars in a population of *Microcyclops varicans* (Lill.) in some studied pools were divided into two parts. One part stopped their development for 7–14 days but the other showed no delay, matured in a short time, and gave rise to the new generation of juveniles. The 4th copepodid instar of the new generation was also divided in the same way and when the first oligopausing animals had broken their arrest of development, the population included a new portion of the diapausing 4th copepodid instars. Oligopause in this species seems to be an adaptation to the temporary state of rain pools (Alekseev, 1978).

Superpause enables an organism to survive unfavourable conditions for years therefore the duration of superpause greatly exceeds the length of ontogenesis of the species. Most examples of superpause are of the embryonal type and are found among the most primitive curstaceans – Branchiopoda (Alekseev, 1990). It is interesting that large species of crustaceans have no superpause at all.

Mesopause occupies an intermediate place and its duration is similar to that of ontogenesis (short-lived animals) or to season length (Decapoda).

The principal classification of diapause on embryonal, larval and imaginal types is based upon both the internal and external differences of the instars. All of the three types may be divided into subtypes according to some essential conditions of the organism.

In Insecta the embryonal type of diapause is subclassified into early, intermediate and late subtypes (Novak, 1972).

In the early embryonal subtype an embryo stops its development at the beginning of cleavage and has no neuro-secretory system. According to Hacker (1902) and Makrushin (1985) most Cladocera and some Diaptomidae arrest cleavage in diapausing embryos at the gastrula stage. It is clear that crustaceans have the early embryonal subtype of diapause.

In the intermediate subtype the neuro-secretory system of an organism is formed but cleavage is not complete. No example of intermediate embryonal diapause is yet known in Crustacea.

Late embryonal diapause is characterized by a ready-to-hatch larva inside the eggshell, but which is waiting for a signal or for favourable conditions before escaping from the egg. Plenty of evidence about the rapid appearance of diaptomid nauplii in temporary pools (6 hours after filling with water) gives us examples of subtype of embryonal diapause in Crustacea (Brewer, 1964).

The larval type in Insecta is subdivided into three stages too but not so clearly as in the embryonal type. For Crustacea we suggested the association of the early subtype with nonsegmented instars (nauplii); the intermediate subtype as classed with the segmented but still not sexually differentiated instars, and late larval instars with segmented instars with visible sexual signs (Alekseev, 1990). All three subtypes of larval diapause are to be seen in Cyclopoida (Elgmork, 1985).

The same subdivision of imaginal diapause in Insecta can be defined according to the copulation and egg laying process (Falkovitch, 1979). The application of the principle to Crustacea gave us similar results, and we can find all the subtypes among the Harpacticoidae (Alekseev, 1990). In the family Canthocamptidae females of many species undergo a period of rest before copulation (Roy, 1979). This can be defined as early imaginal diapause. The female of *Canthocamptus staphylinus* (Jur.), that can dry out among mud after being fertilized – is an example of the intermediate subtype. Finally, in *Arcticocamptus arcticus* (lill.) a female with a sac of eggs can survive being frozen in ice for 10 months (Borutsky, 1952). This more progressive subtype of imaginal diapause, which renders male presence in the population in Spring not important, can be defined as the late one.

Of course, all these types and subtypes of diapause have specific pecularities due to different levels of integration and development of both hormonal and reproductive systems. But is there a principal physiological specification in the suppression of the vital activity and its fundamental characteristic – basal metabolism – among the embryonal, larval and imaginal diapauses?

Danilevsky (1961) predicted for Insecta some correlation between the duration of diapause and its depth. Indeed, to expect from an animal in superpause lower vital activity than it has in oligopause, made sense, but we could find no quantitative data to confirm this prediction. To obtain the data for Crustacea it was necessary to consider a broad spectrum of levels of depression of vital activity in diapausing animals. It is known that the depth of diapause in Insecta depends on a combination of different factors such as temperature, photoperiod, and food availability at the moment of induction of diapause (Danylevsky, 1961). Due to a variation of the diapause producing factors we can expect different levels of depression of vital activity in animals. This can be measured objectively in the form of oxygen consumption by diapausing crustaceans and comparisons can be with the standard data for active animals. Data for Crustacea are shown in Fig. 1. It is clear that for all types of diapause (embryonal, larval and imaginal) for a wide spectrum of animal sizes (cyclops and crayfishes) and for different groups of Crustacea (Anostraca, Cladocera, Copepoda, Decapoda) a common regression can be used.

Table 2. Diapause in Crustacea (Alekseev, 1990 with supplements)

Order/suborder Species	Diapausing instars	Ref.
*ANOSTRACA**		
Streptocephalus torvicornis Waga	Embryo	1
*ANOMOPODA***		
Daphnia middendorffiana Fischer	Embryo	2
OSTRACODA		
Heterocypris incongruens Ramb.	Embryo	3
COPEPODA		
*CALANOIDA****		
Acanthocyclops denticornis (Wier.)	Embryo	4
Labidocera aestiva Wheeler	Embryo	5
Pseudocalanus elongatus (Boeck)	Copepodids 3–4	6
*CYCLOPOIDA*****		
Cyclops scutifer Sars	Copepodids 1–4	7
Mesocyclops leuckarti (Claus)	Copepodids 4–5	8
HARPACTICOIDA		
Canthocamptus staphylinus Jur.	Adults	9
Arctocamptus arcticus Lill.	Female + Embryo	10
CIRRIPEDIA		
Balanus balanoides L.	Adults	11
ISOPODA		
Asellus aquaticus L.	Adults	12
AMPHIPODA		
Gammarus lacustris (L.)	Adults	13
*DECAPODA******		
Scilla serrata Forst	Adults	14
Orconectes virilis (Hagen)	Adults + Juveniles	15
Astacus astacus L.	Adults + Juveniles + Embryo	16

According to our calculation (Alekseev, 1992) diapause is known: * – in 14 species; ** – in 47 species; *** – in 73 species; **** – in 46 species; ***** – in 29 species.
1 – Khmeleva *et al.*, 1981; 2 – Stross, 1969; 3 – Alekseev, 1992; 4 – Hacer, 1902; 5 – Marcus, 1980; 6 – Pertsova, 1981; 7 – Elgmork, 1965; 8 – Ulomsky, 1953; 9 – Donner, 1928; 10 – Borutsky, 1952; 11 – Richard, 1968; 12 – Tadini-Vitagliano *et al.*, 1982; 13 – March, 1982; 14 – Nagabhushanam & Farooqui, 1981; 15 – Aiken, 1969; 16 – Alekseev, 1989

What types of diapause are most common in Crustacea?

Diapause (or a condition with the characters of diapause in earlier papers in crustacean ecology) is found in the life-cycles of the majority of crustacean orders (Table 2). The Cumacea and Mysidacea are not presented in the list, probably reflecting poor study of representatives of these orders. Another explanation of the fact is given later.

Analysis of Tables 1 and 2 demonstrates that most representatives of an order (sometimes suborder) have the same type of diapause and it is interesting in connection with evolutionary changes of types of diapause from obligate embryonal diapause in Branchiopoda, (Anostraca and Anomopoda), Ostracoda and the most archaic of Copepoda – Calanoida, through larval diapause in the Cyclopoida to imaginal diapause in higher and progressive orders such as Isopoda, Amphipoda and Decapoda. As exceptions, the imaginal type

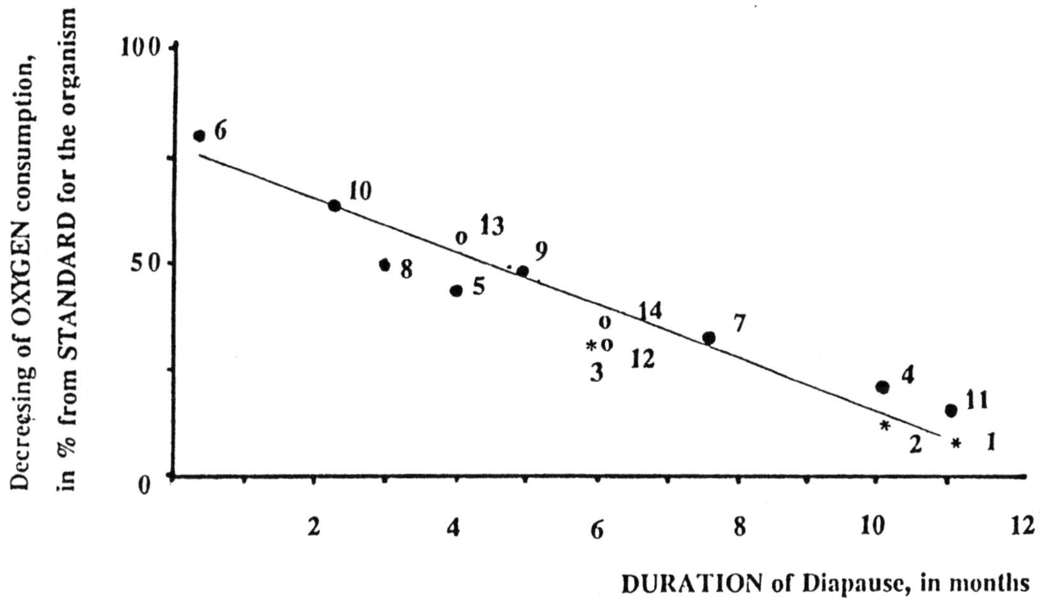

Fig. 1. The correlation between duration of diapause and decrease of oxygen consumption for crustaceans. 1 – *Streptocephalus torvicornis* – Khmeleva *et al.*, 1987; 2 – *Daphnia magna* – Alekseev, 1992; 3 – *Astacus astacus* – the same; 4 – *Cyclops kolensis* – the same; 5 – *C. vicinus* – the same; 6 – *C. abyssorum* – the same; 7 – *Mesocyclops leuckarti* (Day length – 8 hours) – the same; 8 – *M. leuckarti* (Day length – 14 hours) – the same; 9 – *Megacyclops viridis* (Day length – 8 hours) – the same; 10 – *M. viridis* (Day length – 14 hours) – the same; 11 – *Diacyclops navus* – Watson & Smallman, 1971; 12 – *Astacus astacus* (female) – Alekseev, 1989; 13 – *A. astacus* (male) – the same; *A. leptodactillis* – the same. (1–3: species with embryonal diapause; 4–11: species with larval diapause; 12–14: species with imaginal diapause.)

occurs in the Harpacticoida and the embryonal type in the Decapoda.

Some differences in imaginal diapause in lower and higher crustaceans must be mentioned. These relate in the first place to the depth of this condition and, as a consequence of it, independence of the organism from external conditions. Of lower crustaceans the Harpacticoida and of higher forms the Isopoda are capable of true anabiosis, that is why they are able to survive being frozen or dried. The majority of higher crustaceans, Decapoda especially, cannot tolerate complete freezing, or anaerobic conditions, or the full drying of waterbodies.

Species of different orders are very different both in dimensions and in ecology: lower crustaceans as a rule are small animals living in all types of waterbodies including those with an unstable water regime but higher crustaceans are bigger and inhabit in waterbodies that generally have a more constant regime. The capacity for anabiosis among Crustacea changes in the direction from higher crustaceans to lower (Brandt, 1951).

The connection of diapause with different stages of development has its positive and negative sides. Doubtless embryos of dwellers in temporary waterbodies manifest the highest resistance to desiccation, UV – radiation and mechanical abrasion because they are put into special defensive envelopes. The development of such embryos is slowed down at lower levels of integration of vitally important systems (Markushin, 1984). However this high level of defence leads to the necessity of a rather long period of development from the beginning of reactivation to active reproduction because these crustaceans must go through a complete series of stages from egg to female. As a result they cannot begin their reproduction immediately and will lose in competition with another species diapausing in the imaginal state. The presence of significant predation pressure will also decrease survival.

Different groups of lower crustaceans resolve the problem in different ways. 'Cladocera' overcome it by small dimensions, high rate of growth, and parthenogenetic reproduction. Diaptomidae change early embryonal diapause to a later one which permits them to inhabit temporary waterbodies within a few hours after filling with water, and use the small dimensions of their larvae and brisk locomotion. Ostracoda have a bivalved shell which provides good defence, and are small.

But large Branchiopoda which move slowly and have a long period of metamorphosis are defenceless, especially at the beginning of their development. After hatching they are not defended by a dense envelope and are vulnerable to predators. These crustaceans are not capable of remaining for a long period in a waterbody after finishing diapause. They inhabit almost exclusively simplified ecosystems (small temporary pools and peculiar lakes – arctic shallow water or hypersaline in arid zones) where such predators as juvenile fishes, and competitors, develop weakly or are completely absent (Smirnov, 1940).

Imaginal diapause of higher crustaceans is not capable of protecting them from the complex of unfavourable factors to which embryonal diapause give protection, therefore Decapoda and Amphipoda species are absent in true astatic waterbodies, or at least do not form independent self-reproducing populations. However, adults finishing diapause in permanent waterbodies are capable of beginning to reproduce immediately. This is especially important in view of their large size and the longer period needed to reach maturity.

To finish this section the initial question about the most common types of diapause in Crustacea must be answered. The most primitive and ancient Branchiopoda have the embryonal type of diapause only. Among Copepoda (except Cyclopoidae) the embryonal type can be found also, in most ancient Calanoida especially. Finally, in Decapoda we can sometimes observe the same type of diapause too. Therefore the embryonal diapause seems to be the most popular variant of the adaptation among Crustacea, and we think that this is not accidental. The sporadic appearance of embryonal diapause among species with the imaginal type may be serve as indirect evidence of their ancient history.

The evolution of types of diapause in Crustacea

It is important to study the manifestation of diapause in different crustaceans in connection with the age of the taxa and the time of their installation in inland waterbodies.

The system and origin of the Crustacea are currently an object of discussion. Instead of the sole class generally accepted previously, some authors (e.g. Manton, 1977) unite under this name 10–12 classes but other authors (e.g. Starobogatov, 1988) unite them in only 4 classes. We accept the latter system and divide the superclass *Crustacea* into 4 classes: Branchipodiodes (Phylopoda + Cephalocarida + Remipedia), Ascothoracioides (Copepoda + Tantulocarida + Mystacocarida + Scaracarida + Ascothoracida), Halicyniodes (Halicynida + Cirripedia + Thylacocephala + Branchiura + Ostracoda) and Carcinoides (Malacostraca). The most primitive class is Branchipodiodes. Further, it gave origin at different times to the Carcinoides and Ascothoraciodes and the last class to Halicyniodes. There is a series (except Carciniodes): Branchipodiodes – Ascothoraciodes – Halicyniodes. The appearance of these classes in the fossil state does not completely agree with their phylogenetic sequence.

All 4 classes are known from the Cambrian and consequently their divergence was in Pre-Cambrian time. Morphological data give us a possibility of reconstructing the early history of the Crustacea when the fossil evidence is inadequate. This way is connected with a successive diminution of the number of somites and fixation of their number (Starobogatov, 1994). According to the reconstruction, an evolutionary series of Crustacean orders from the most ancient Anostraca (with only embryonal type of diapause) to the relative young Isopoda (with the imaginal type of diapause) may be suggested (Fig. 2).

Another side of the problem must be mentioned. The diapausing embryo in an egg well protected by envelopes is mentioned more often for freshwater than for marine organisms. In this connection it is important to evaluate the time (or at least the sequence) of penetration of the different groups of Crustacea into inland waters.

There is a clear difference: all groups of inhabitants of inland waters are divided, using the time of penetration into freshwater, into four main categories – archaeolimnic, palaeolimnic, mesolimnic and neolimnic (Martinson, 1958, 1967; Starobogatov, 1970, 1986). Archaeolomnic crustaceans appeared in freshwater probably in the Pre-Cambian, palaeolimnic ones – from the Devonian to the Permian, mesolimnic forms – from the Triassic-Jurassic to the Palaeogene and neolimnic forms – from the Neogene to the Quaternary (Fig. 2).

Each of these categories is characterized by its ecological and biogeographical properties. Examining the different groups of the Crustacea we may say that most recent Branchiopoda are palaeolimnic group whereas Polyphemida are mesolimnic animals. The Diaptomidae and Cyclopidae are mesolimnic groups but the freshwater Calanidae are neolimnic. Ostracoda penetrate into inland waters, the Cyprididae being a palaeolimnic group but the remainder (e.g. Candonidae,

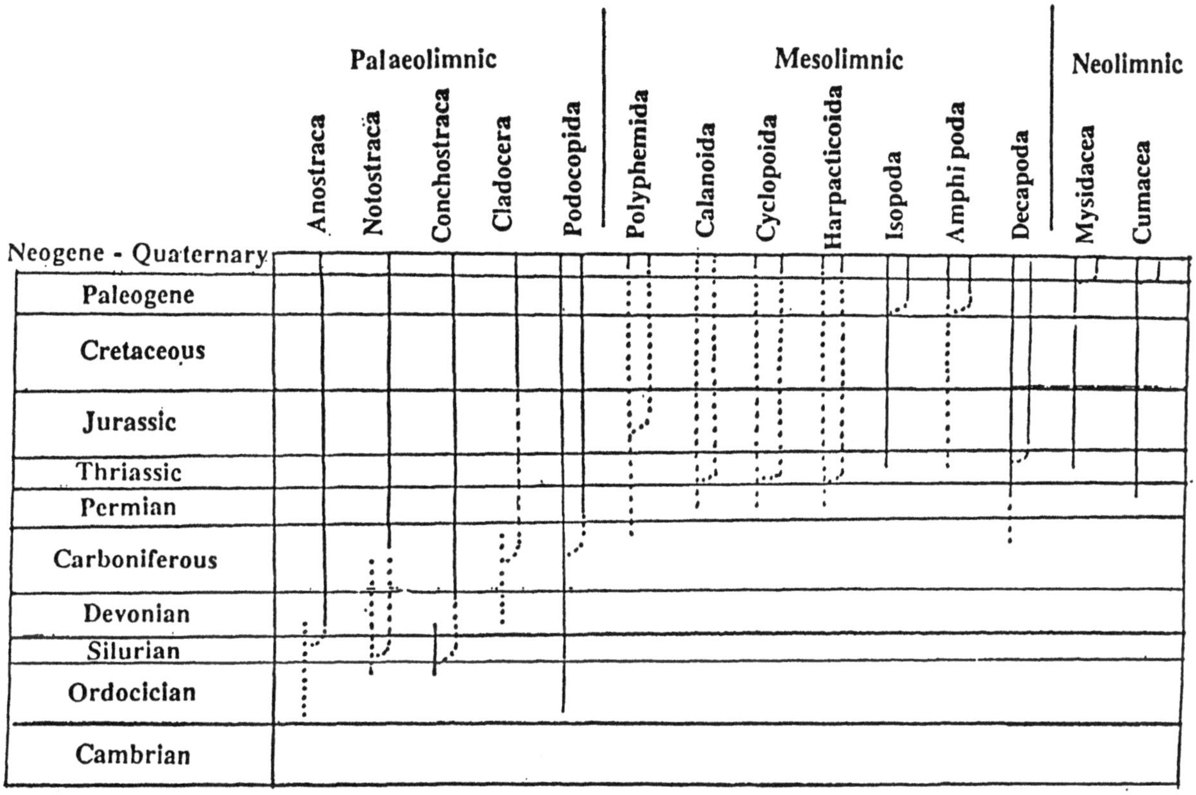

Fig. 2. The history of origin in the sea (left line) and penetration into inland waterbodies (right line) of Palaeolimnic, Mesolimnic and Neolimnic species of Crustacea. (Solid line – palaeontological data; dotted line – morphological data).

Limnocytheridae) are mesolimnic. There are doubtless neolimnic groups, the more so as terrestrial Ostracoda exist (e.g. Terrestricytheridae).

The Isopoda, Amphipoda and Decapoda (except marine forms) belong to mesolimnic group (widely distributed in inland waters) but the Mysidacea and Cumacea belong to neolimnic group (inhabitants of estuaries and 'arctic relicts').

With the kind agreement of our colleagues we have been able to analyze a list of crustacean species presented by Hairston & Caceros (1995, this volume). The contains 161 freshwater and marine species with known duration of diapause or its absence. Of the listed species 43 are palaeolimnic (or secondarily marine, originating from palaeolimnic groups), 43 mesolimnic, 8 neolimnic and 67 marine.

For further analysis we use only three grades of the duration of diapause: one year or more (superpause mainly), less than one year (mesopause) and absence of diapause. Superpause is evidently embryonic, the mesopause may be also embryonic but more frequently it is larval or imaginal.

The result of the analysis is shown in Fig. 3. The duration of diapause of all (100%) palaeolimnic species is one year or more. That of mesolimnic species is a year or more in 44%, less than a year in 40% and with no diapause in 16%. There is no neolimnic species with superpause, 25% with diapause less than a year, and 75% without diapause.

The marine species in the list have diapause of one year and more in 27%, less than a year in 4.5% and no diapause in 68.5%. We may state from the described distribution of types of diapause and its duration that palaeolimnic species have only embryonal diapause of long duration, mesolimnic species have different types of diapause with both long and short duration, and neolimnic species lack diapause or have only short diapause, presumably larval or imaginal.

The analysis presupposes that marine crustaceans have a diapause used mainly for the subdivision of

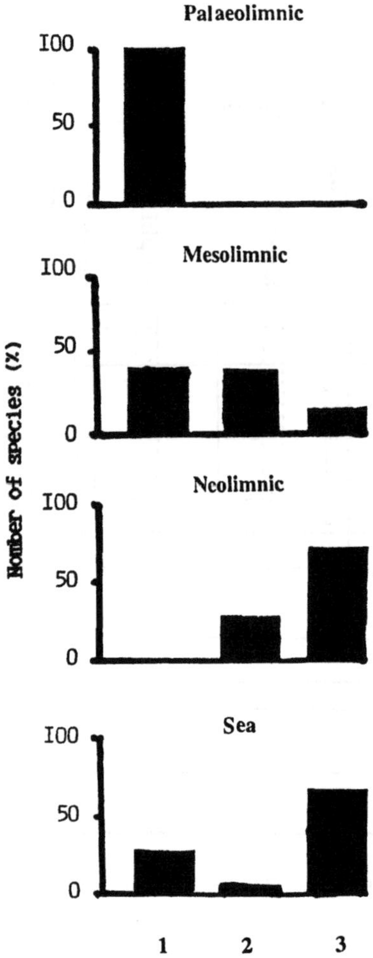

Fig. 3. The distribution of species with superpause (1), mesopause (2) and absence of diapause (3) among Palaeolimnic, Mesolimnic, Neolimnic and Marine crustaceans.

licences. For this purpose oligo- and mesopause of imaginal and larval instars were more suitable.

Embryonal diapause seems likely to have appeared during the penetration into inland waters, the more so as to protect the non-motile egg from unfavourable factors is easier than to protect a motile larva or imago. This process intensified during adaptation to living in temporary waterbodies. Each new wave of curstaceans penetrating inland waters started by adapting to estuarine conditions similar to those in the sea (more or less constant) and moved to temporary waterbodies as the final step of inland inhabitance. The ancient primitive forms of crustaceans could achieve more impressive results in this way.

Acknowledgements

The authors are grateful to Nelson Hairston and Carla Caceres for reading the manuscript and for many fruitful comments. We are especially thankful to Geoffrey Fryer for the very effective help with the improving of the English style in the manuscript.

References

Aiken, D. E., 1969a. Photoperiod, endocrinology and the crustaceans molt cycle. Science, 164: 149–155.

Aiken, D. E., 1969b. Ovarian maturation and egg laying in the crayfish *Orconectes virilis*: influence of temperature and photoperiod. Can. J. Zool. 48: 931–935.

Alekseev, V. R., 1984a. Vlijanie khlorirovania na zooplankton osetrovykh prudov (Influence of treatment by chlorine on the zooplankton of sturgeon farming ponds). Sbornik nauchnykh trudov GosNIORKH. 225: 79–82 (in Russian).

Alekseev, V. R., 1984b. Role photoperioda v formirovanii diapauzy u presnovodnykh cyclopov (Role of photoperiod for forming of diapause in freshwater cyclopoids). Sbornik nauchnykh trudov GosNIORKH. 223: 95–102 (in Russian).

Alekseev, V. R., 1986. Signalnaja role photoperioda v upravlenii sezonnymi cyklami (The signal role of photoperiod for driving by seasonal cycles). Tezisi dokladov V c'ezda VGBO, Tol'atti 15–19 sentjabrja. 1: 130–132.

Alekseev, V. R., 1987a. Role katchestvenykh i kolitchestvenykh photoperioditcheskikh reakcij v gyznedejatelnosti rakoobraznykh (The role of both quality and quantity photoperiod responses for vital activity of crustaceans). Trudy Zool. Inst. Acad. Nauk SSSR, 168: 106–115 (in Russian).

Alekseev, V. R., 1987b. Photoperiodizm i sezonnoe razvitije populjacij *Mesocyclops leuckarti* (Claus) (Crustacea, Copepoda) (Photoperiodism and seasonal development of populations of *Mesocyclops leuckarti* (Claus) (Crustacea, Copepoda)). Hydrobiolog. Journal 23: 22–28 (in Russian).

Alekseev, V. R., 1990. Diapauza u rakoobraznykh: ekologo – phyziologitcheskije aspekty (Diapausa in Crustaceans: ecological – physiological aspects). NAUKA press., Moscow, 144 pp. (in Russian).

Alekseev, V. R., 1992. Ecologo-fisiologitcheskije aspekty diapauzy u racoobraznykh (Ecological and physiological aspects of diapause in crustaceans). Thezisy dissertatciji na soiskanije utchenoj stepeni doctora biologitcheskikh nauk. Moskwa, Moskowskiv Gos. Universitet, 68 pp. (in Russian).

Bakulin, A. M., 1983. Vlijanije dvukh photoperiodov na soderganije lipidov u *Streptocephalus torvicornis* (Crustacea, Anostraca) v period razmnodjenija (Influence of two photoperiods on contents of fats in *Streptocephalus torvicornis* (Crustacea, Anostraca) during breeding period). Vestcy Akad. Nauk Belorus. 6: 106–108 (in Russian).

Borutzky, E. V., 1952. Presnovodije *Harpacticoida*. Fauna SSSR. (Freshwater *Harpacticoida*. Fauna USSR). Academ. izdat, Moskwa – Leningrad, 424 pp. (in Russian).

Bouchon, D., G. Remoissenent & J. P. Mocquard, J. P., 1985. Influence de la temperature sur l'entrée en reproduction de *Palaemonetes varians* Leah. (Crustacea, Decapoda, Natantia). Bull. Soc. Zool. France 110: 439–447.

Brandt, T., 1951. Anaerobioz u bespozvonochnikh (Anaerobiosus in *Invertebrate*). Izdat. inostran. literatury. Moskwa, 335 pp.

Brewer, R. H., 1964. The phenology of *Diatomus stagnalis* (Copepoda, Calanoida); the development and hatching of the egg stage. Physiol. Zool. 37: 1–20.

Brown, F. A. & G. C. Stephens, 1951. Studies of the daily rhythmicity of the feedler carb, *Uca*. Modification by photoperiod. Biol. Bull. 101: 71–83.

Bunner, H. C. & K. Halcrow, 1977. Experimental induction of the production of ephippia by *Daphnia magna* Straus (Cladocera). Crustaceana. 32: 77–86.

Carvalho, G. R. & R. N. Hughes, 1983. The effect of food evailability, female culture density and photoperiod on ephippia production in *Daphnia magna* Straus (Crustacea, Cladocera). Freshwat. Biol. 13: 37–46.

Cherkashina, N. Ya. & I. V. Karnaushenko, 1982. Doembrionalnaja diapauza u raka *Astacus leptodactylus cubanicus Bir. et Win.* (Prae-embrional diapause in the crayfish *Astacus leptodactylus cubanicus Bir. et Win.*). Zhurnal Obshchey Biologii. 43: 687–689 (in Russian).

Cukersis, J., 1989. Presnovodnje raki (Freshwater crayfishes). Mokslas publish., Vilnius, 143 pp. (in Russian).

Danilevsky, A. S., 1961. Photoperiodizm i sezonnoje razvitije nasekomykh (Photoperiodism and seasonal development of insects). Izdat. Leningrad Gos. Universit., Leningrad, 243 pp. (in Russian).

Einsle, U., 1967. Die äusseren Bedingungen der Diapause plankisch lebender *Cyclops* – Arten. Arch. Hydrobiol. 63: 387–403.

Elgmork, K., 1985. Prolonged life cycles in the planktonic copepod *Cyclops scutifer* Sars. Verh. int. Ver. theor. angew. Limnol. 22: 3154–3158.

Falkovitch, M. I., 1979. Sezonnoje razvitie pustynnykh tcheshuekrylykh *(Lepidoptera)* Srednej Azii i ego istoriko – geografitcheskij analiz. (Seasonal development of *Lepidoptera* of antral Asia and its historical and geographical analysis). Entomol. Obozr. 58: 260–281 (in Russian).

Ferrari, D. C. & P. D. N. Hebert, 1982. The induction of sexual reproduction in *Daphnia magna*: genetic differences between arctic and temperate population. Can. J. Zool. 60: 2143–2148.

Fries, G. 1964. Über die Einwirkung der Tagesperiodik und der Temperature auf den Generatiionwechsel, die Weibchengrosse under der Eier von *Daphnia magna* Straus. Ztschr. Morphol. und Ökol. Tiere. 53: 475–516.

Günther, K., 1949. Über Evolutionsfaktoren und die bedeutung des Begriffs der 'ökologische Lizenz' für die Erklarung vom Formenerscheinungen im Tierreich. Festschrift von Erwin Streseman, Heidelberg, Carl Winter Verl.: 23–54.

Häcker, V., 1902. Über die Fortpflanzung der limnetischen Copepoden des Titisees. Ber. naturforsch. Ges. Freiburg 12: 1–18.

Hairston, N. G. & E. J. Olds, 1984. Population differences in the timing of diapause: adaptation in a spatially heterogeneous environment. Oecologia. 61: 42–48.

Hairston, N. G. & E. J. Olds, 1987. Population differences in the timing of diapause: a test of hypotheses. Ibid. 71: 339–344.

Jassem, W., J. P. Mocquard & P. Juchalt, 1981. Senil de l'intensite lumineuse du signal photoperioddique induisant l'entrée en reproduction chez *Armadillidium vulgare* Latr. Bull. Soc. zool. France. 106: 451–455.

Juchault, P., A. Pavese & J. P. Mocquard, 1980a. Determinism de la reproduction saisonniere des femmelles d'*Armadillidium vulgare* Latr. (Crustacea, Isopoda, Oniscicoide) II. Etude phiques differentes. Ann. sci. natur. Zool. biol. anim. 2: 99–108.

Juchault, P., J. P. Mocquard, N. Bougrier & G. Besse, 1980b. Croissomce et cycle reproducteur du Crustace Isopode Oniscoide *Elume purpurascens* Budde-Lund. Etude dans la nature et an Laboratore sous differenntes conditions de temperature et de photoperiode, d'une population du centre-ouest de la France. Vie et milieu. 30: 149–150.

Khmeleva, N. N., R. Z. Klekowski & L. L. Nagorskaya, 1981. Respiration and yolk utilization efficiency during embriogenesis of *Streptocephalus torvicornis* (Waga) (Crustacea, Anostraca). Pol. arch. hydrobiol. 28: 43–53.

Kubersky, E. S., 1977. Worldwide distribution and ecology of *Alonopsis* (Cladocera, Chydoridae) with a description of *Alonopsis americana* sp. nova. Int. Rev ges. Hydrobiol. 62: 649–685.

Laubier-Bonichou, A., 1975. Induction de la maturation sexuelle et ponte ches la overelte *Panaeus japonicus* Bate en milieu controle. C. r. acad. sci. D. 281: 2013–2019.

Levchenko, V. F., 1984. Vnutrennije svijazi i sokhranenije struktury v ecosistemakh (On inner connections and conservativity of structures in ecosystems). Metodologicheskije problemy evolutsionnoj teoriji. Academic publish. Estonii. Tartu: 22–23 (in Russian).

Lipcius, R. N. & W. F. Herrnkind, 1985. Photoperiodic regulation and daily timing of spini lobster mating behavior. J. exp. Mar. Biol. Ecol. 85: 191–204.

Little, G., 1968. Induceed winter breeding and larval development in the shrimp *Palaemonetes pugio* (Holthius). Crustaceana. Suppl. 2: 19–26.

Makrushin, A. V., 1984. O nepolnom sootvetstviji generativnogo cycla morskikh *Cladocera (Crustacea) Podon leuckarti* and *Evadne nordmanni* sezonnym izmenenijam vodnoj sredy (On the incomplete agreement of the cycle of generations of marine *Cladocera (Crustacea) Podon leuckarti* and *Evadne normanni* with seasonal alternations of conditions of media). Ecologia moria. Naukova dumka, Kiev. 18: 59–63 (in Russian).

Manton, S. M., 1977. The Arthropoda: habits, functional morphology and evolution. Oxford Univ. Press, Oxford, 527 pp.

March, B. G. E., 1982. Decreased day length and light intensity as factor inducing reproduction in *Gammarus lacustris* Sars. Can. J. Zool. 60: 2962–2965.

Marcus, N. H., 1980. Photoperiodic control of diapause in the marine calanoid copepod *Labidocera aestiva*. Biol. Bull. 159: 311–318.

Martinson, G. G., 1958. Pzoiskhozhdenie fauny Baicala v svete paleontologicheskikh issledovanij. Origin of the faune of Baical Lake in the light of paleontological study. Doklady Akademii Nauk SSSR. 120: 1155–1158 (in Russian).

Martinson, G. G., 1967. Problema pzoiskhozhdenija fauny Baicala. Problem of origin of Baical Lake fauna. Zool. Zhurn. 46: 1594–1598 (in Russian).

Nagabhushanam, R. & U. M. Farooqui, 1981. Photoperiodic stimulation of ovary and testis maturation inthe immature marine crab *Scylla serrata* Fors. Ind. J. Mar. Sci. 40: 396–398.

Nelson, K., 1986. Photoperiod and reproduction in lobseter *(Homarus)*. An. Zool. 26: 447–457.

Nelson, G. H. J. & E. J. Olds, 1986. Partial photoperiodic control of diapause in three populations of the freshwater copepod *Diaptomus sanguineus*. Biol. Bull. 171: 135–142.

Novak, V. J. A., 1972. Gormonalnye osnovy diapausy u nasekomykh (Hormonal basis of diapause in insects). Izdat. Leningrad. Univer.: Leningrad, 210 pp. (in Russian).

Pancella, J. R. & R. G. Stross, 1963. Light induced hatching of *Daphnia* resting eggs. Chesapkate Sci. 4: 404–425.

Paris, O. H. & G. E. Jenner, 1952. Photoperiodism in the freshwater shrimp, *Palaemonetes paludosus* (Gibbes). J. Elisha Mitchell Sci. Soc. 68: 1–144.

Rao, C., K. Shakuntala & S. R. Reddy, 1984. Effects of environmental factors on gonad development in the prawn *Macrobrachium lanchesteri* (de Man) (Natantia: Palaemonidae). Ind. J. exp. Biol. 22: 471–474.

Roy, J., 1931. Copepodes et Cladoceres de l'ouest de la France. Paris, 224 pp.

Sarojini, R. & G. Gyananth, 1985. Photoperiodic regulation of ovarian maturation in the freshwater prawn *Macrobrachium lammeri*. J. Curr. Biosci. 2: 158–162.

Shan, R. K., 1974. Reproduction in laboratory stocks of *Pleuroxus* (Chydoridae, Cladocera) under influence of photoperiod and light intensity. Int. Rev ges. Hydrobiol. 59: 643–666.

Smirnov, S. S., 1940. Phyllopoda. Zhizn' presnykh vod SSSR. 1. Academic publish. Moscow – Leningrad: 305–303 (in Russian).

Spindeler, K. D., 1969. Experementelle Untersuchungen zur Dormanz bei *Cyclops vicinus*. Naturwissenschaffen. 56: 93–94.

Spindler, K. D., 1970. Die Bedeutung der Photoperiode für die Entwicklung von *Cyclops vicinus*. Zool. Anz. 33: 190–195.

Starobogatov, Ya. I., 1970. Fauna molljuskov i zoogeographitcheskoe rajonirovanije kontinentalnykh vodojemov zemnogo shara (Molluscan fauna and zoogeographic division of inland waters of the Globe). Academ. idat., Leningrad, 372 pp. (in Russian).

Starobogatov, Ya. I., 1984. Evolutsia ecosistem (Evolution of ecosystems). Metodologicheskiye problemy evolutsionnoy tiorii. Academic publish. Estonii. Tartu: 70–72 (in Russian).

Starobogatov, Ya. I., 1985. Problemy vidoobrazovanija (Problem of speciation). Itogi nauki i tekhniki. Obschaya geologiya. VINITI publish. Mosocw. 20, 94 pp. (in Russian).

Starobogatov, Ya. I., 1986. Istorija ozer po dannym izutchenija ikh fauny (The lake history on the data of studying of their fauna). Istoriya ozier SSSR. NAYKA publish.: 33–50 (in Russian).

Starobogatov, Ya. I., 1988. Systematics of Crustacea. J. Crustacean Biol. 8: 300–311.

Starobogatov, Ya. I., 1994. Javlijajutsia li *Polychaeta* edinym klassom ili sbornoji grouppoji neskolkich tipov Zhyvotnykh? (Are *Polychaeta* a single class or a collection of representatives of several animal phyla?). Issledovaniya fauny morey. Academ. izdat., St. Petersburg: 69–75 (in Russian).

Steele, V. J., 1981. The effect of photoperiod on the reproductive cycle of *Gammarus lawrencianus* Bousfield. J. exp. mar. Biol. and Ecol. 53: 1–7.

Stephens, G. C., 1955. Induction of moulting in the crayfish, *Cambarus*, by modification of daily photoperiod. Bull. Bull. 108: 235–241.

Stimpl, K. J., 1980. The effect of four environmental variables and the induction of gamogenesis in *Daphnia pulex* Leydig. Diss. Ind. University, USA, 113 pp.

Stross, R. G., 1969a. Photoperiod control of diapause in *Daphnia*. II. Induction of winter diapause in the Arctic. Biol. Bull. 136: 264–273.

Stross, R. G., 1969b. Photoperiod control of diapause in *Daphnia*. III. Two stimulus control of long-day, short-day induction. Ibid. 137: 359–374.

Stross, R. G. & J. C. Hill, 1968. Photoperiod control of winter diapause in the freshwater Crustacean *Daphnia*. Biol. Bull. 134: 176–198.

Tadini-Vitagliano, G., F. Valentino & L. Migliore, 1982. Is the mechanism present in *Sellus aquaticus* (L.) an evolutionary stage towards an endogenous calendar? Zeitschr. zool. Syst. und Evolutionforsch. 20: 144–155.

Takahashi, F., 1977. Pioneer life of the Tadpole Shrimps, *Triops* spp. (Notosdtraca: Triopsidae). Appl. Entomol. Zool. 12: 104–117.

Tighe-Ford, D. J., 1967. Possible mechanism for the endocrine control of breeding in a cirripede. Nature. 216: 920–921.

Uye, S., S. Kasahara & T. Onbe, 1979. Calanoid copepod eggs in sea-bottom muds. IV. Effects of some environmental factors on the hatching of resting eggs. Mar. Biol. 51: 151–156.

Ulomsky, S. N., 1953. Novosti v ekologii nekotorykh vidov kopepod *Mesocyclops* (News in ecology of certain species of copepods *Mesocyclops*). Doklady Akademii Nauk SSSR. 90: 295–297 (in Russian).

Van der Linden, A., R. Blust & W. Decleir, 1985. The influence of light on the hatching of *Artemia* cysts (Anostraca: Branchiopoda: Crustacea). J. Mar. Biol. Ecol. 92: 207–214.

Watson, N. H. F. & B. N. Smallman, 1971. The role of photoperiod and temperature in the induction and termination of an arrested development in two species of freshwater cyclopid copepods. Can. J. Zool. 49: 855–862.

Westin, L. & R. Gydemo, 1986. Influence of light and temperature on reproduction and moulting frequency of the crayfish *Astacus astacus* L. Aquaculture. 52: 43–50.

Williams, J. A., 1985. The role of photoperiod in the initiation of breeding and brood development in the amphipod *Talitrus saltator* Montagu. J. exp. mar. Biol. Ecol. 86: 59–72.

Hydrobiologia **320**: 27–44, 1996.
V. R. Alekseev & G. Fryer (eds), Diapause in the Crustacea.

Distribution of crustacean diapause: micro- and macroevolutionary pattern and process

Nelson G. Hairston, Jr. & Carla E. Cáceres
Section of Ecology and Systematics, Cornell University, Ithaca, NY 14853, USA

Abstract

Theoretical predictions for the relationships between duration of dormancy, reproductive life span, and dispersal ability developed for plants in temporally varying environments are applied here to crustaceans. Mathematical models suggest that diapause duration should negatively covary with adult life span, and that both diapause and life span should negatively covary with dispersal ability. A survey of 167 crustacean species from 20 orders and three classes confirms that species with prolonged diapause have short adult life spans and those with long adult lives either have diapause lasting less than a year, or do not diapause at all. Prolonged diapause is more common among small or inland water crustaceans than it is among large or marine species, whereas large or marine species have significantly longer adult life spans on average than do those that are small or from inland waters. A greater fraction of species in the Branchiopoda exhibit prolonged diapause than do members of the Maxillopoda which, in turn, are more likely to exhibit prolonged diapause than are the Malacostraca. A greater fraction of malacostracan species have adult life spans exceeding one year than do species in either the Branchiopoda or the Maxillopoda. Cladistic analysis shows that phylogenetic constraint is likely to be at least in part responsible for the expression of diapause among the Crustacea. We conclude that both natural selection and macroevolutionary pattern have influenced the distribution of diapause among modern crustaceans.

Introduction

The diapause trait is by no means evenly distributed within the Crustacea, either among species living in different habitats or among higher taxonomic categories. Crustaceans found in inland bodies of water, especially those that dry occasionally, are more likely to possess a diapausing stage than are those living in the open ocean. The Malacostraca, including the decapods, euphausids, amphipods, isopods, and so on, are markedly depauperate in species exhibiting diapause, whereas the Branchiopoda such as the 'cladocerans', 'conchostracans', anostracans, and notostracans are rich in such species, and the Maxillopoda (e.g., copepods, ostracods) present a mixed picture. Two possible explanations for these uneven ecological and taxonomic distributions of diapause are: (1) the patterns reflect selection pressures exerted on crustacean species by the environments in which they live: diapause is favored in some habitats and not in others, and (2) the patterns are the result of phylogenetic history: the diapause trait is primitive for some groups and not for others, hence its uneven distribution between broad taxonomic categories.

These two possibilities are not necessarily independent since direction and strength of selection are in part a function of the characters an organism already possesses, and these are in part a function of past selection pressures (which are frequently not statistically independent of current selection pressures).

Here we ask how selection might be expected to act on the expression of diapause, by applying predictions of optimization and ESS (Evolutionarily Stable Strategy) models developed principally to understand plant seed dormancy. We then investigate whether the association between diapause and other life-history traits within the Crustacea conforms with theory. We explore the phylogenetic expression of diapause among crustacean taxa, and end by asking how selection at the microevolutionary scale may have interacted with macroevolutionary pattern to produce the distribution of diapause we observe in nature.

Plant ecology for the carcinologist

There is a robust theoretical literature concerning how selection might be expected to act on seed dormancy in plants. Duration of dormancy, expressed as the fraction of seeds expected to germinate in each year, has been related both to environmental features and to other plant life history characteristics. Dormancy response to temporal or spatial variation in reproduction and survival has been investigated by Cohen (1966), Ellner (1984), and Philippi (1993), while response to the life history traits, seed dispersal, adult life span, and offspring survival have been explored by Venable & Lawlor (1980), Levin *et al.* (1984), Cohen & Levin (1985) and Rees (1994). Because the environments in which crustaceans live are also variable, and because prolonged diapause is a prominent feature in the life history of many crustacean taxa, much of this plant dormancy theory has potential relevance for understanding the distribution of the diapause trait among the Crustacea.

Although theory has also been developed specifically for understanding animal diapause in both insects (e.g., Levins, 1969; Cohen, 1970; Taylor, 1980) and crustaceans (Hairston & Munns, 1984; De Stasio & Hairston 1992), these efforts have been restricted largely to predicting seasonal diapause timing. The questions we address here concern instead the evolution of diapause duration, and this theory has been the province of plant ecologists and evolutionary biologists (Livdahl's, 1979, paper on insects is an exception). It is our thesis that the prolonged survival of crustacean diapausing eggs is analogous to seed dormancy that lasts over multiple years (or multiple opportunities to germinate or hatch), and thus subject to many of the same analyses.

Our reason for concentrating on prolonged diapause lies in its relevance to fitness in a temporally varying environment. Most, perhaps even all, environments vary in suitability for reproduction and survival between years. Occasional years may be so poor that no new individuals are recruited to the population, and cannot be survived by an annual organism that saves no eggs for hatching in subsequent years. Although the optimal duration of prolonged dormancy is related to the frequency of low recruitment years, note that it is not necessary that recruitment ever reach zero (Ellner, 1985). Cohen (1966) showed that as the variance in the quality of the environment for recruitment increases, the fraction of seeds that germinate (or diapausing eggs that hatch) should decline, although zero germination or hatching should never occur. Ellner (1985) pointed out that this result depends quantitatively on the probability a seed (egg) has of surviving until the next opportunity to germinate (hatch). The lower the probability, the higher the germination (hatching) fraction. Furthermore, Easterling & Ellner (1994), considering copepod diapausing eggs, have shown that the result also depends on the probability that eggs receive the hatching cue.

In plant ecology, the pool of seeds that do not germinate the year after they are produced is called the 'seed bank'. For crustaceans, the analogous pool of diapausing eggs has been termed the 'dormant propagule pool' (Hairston & De Stasio, 1988) or the 'egg bank' (De Stasio, 1989; Marcus *et al.* 1994). The presence of an egg bank in a population can alter its dynamics by effectively lengthening generation time, which can lead to a slowing of the rate of response to natural selection (Hairston & De Stasio, 1988), maintenance of genetic variation (Ellner & Hairston, 1994), or alterations of the outcome of competition (Ellner, 1984; Shmida & Ellner, 1984; Cáceres, 1994; Hairston *et al.*, 1995a).

Two life-history characters should, in theory, alter expected duration of prolonged diapause in a temporally variable environment: (1) the degree of iteroparity, and (2) dispersal capability. Iteroparity is itself hypothesized to be an adaptation to a temporally varying environment (Murphy, 1968; Stearns, 1976; Roff · 1992). Here we take environmental variation between years to be the most important temporal scale, and so by iteroparity we mean reproduction by one individual in more than one year (our argument holds for other temporal scales as well). If occasional recruitment failures are probable, then iteroparity provides reproduction in future favorable years. In fact, iteroparity and prolonged diapause can be seen as alternative mechanisms to solve the same problem of temporal environmental variation. Rees (1993, 1994) has recently explored the tradeoffs between these two traits for plants. His fundamental conclusion is that when offspring survival varies between years, plants with long adult life span should make seeds with a much higher germination probability (shorter or no prolonged diapause) than plants with a short adult life span. There is no reason to have both iteroparity and prolonged diapause. In a review of the literature for 171 plant species, he found that the general trend predicted was in fact observed (Rees, 1993).

Venable & Lawlor (1980) and Levin *et al.* (1984) explored an alternative tradeoff in plant life histories:

that between duration of seed dormancy and dispersal ability. Their proposal is that if an environment varies both temporally and spatially, so that different patches of the environment are of different quality at any given time, an annual organism may avoid unfavorable conditions by either surviving in dormancy or by dispersing (note that prolonged diapause can be thought of as dispersal in time; Templeton & Levin, 1979). The result depends quantitatively on how successful a dispersing organism is at finding a suitable habitat (Levin *et al.*, 1984; Cohen & Levin, 1985), but the general conclusion is that as dispersal ability increases, the duration of dormancy should decrease. Venable & Lawlor (1980) reviewed data for 28 plant species with seed polymorphisms for dispersal near-to and far-from the parent plant. Consistent with theory, they found a repeated pattern of near dispersal being associated with delayed germination and far dispersal with rapid germination. Similarly, Rees (1993) found that species with less seed dormancy are significantly more likely to posses efficient dispersal strategies (defined by seed morphology).

Do crustaceans show similar patterns? To answer this question, we reviewed the literature for 167 species (Appendix 1), taken from 20 orders in 3 classes (Table 1). For our analysis we used a definition of diapause taken from Tauber *et al.* (1986): 'a neurohormonally mediated, dynamic state of low metabolic activity'. Because few crustacean species have been sufficiently studied to know their neurohormonal state at different seasons or life history stages, in practice we used instead the list of associated characteristics that Tauber *et al.* (1986) present: 'reduced morphogenesis, increased resistance to environmental extremes, and altered or reduced behavioral activity', with the added stipulation that 'once diapause has begun, metabolic activity is suppressed even if conditions favorable for development prevail'. For phylogenetic comparisons, we adopted Schram's (1986) systematic terminology for the Crustacea with some modification taken from Brusca & Brusca (1990) and Fryer (1987). For each species, we recorded duration of diapause, duration of adult reproductive life, dispersal ability, habitat, and adult body mass. These data were gathered using the following guidelines.

1. Duration of diapause and duration of life span. In general, maximum values are used where only one or two durations were found in the literature to compensate for the fact that long durations are less likely to be discovered in the short-term studies typical of ecological investigations. Where more complete data were obtained, we used mean durations or threshold values beyond which survival of dormant stages or adults drops substantially. Where specific durations were not given, we based our judgement on literature discussions, attempting to err on the side of short duration for the sake of being conservative.

2. Dispersal ability. The data available for quantifying this trait are scanty. For some species long distance movement of pelagic larvae far from parent populations have been documented (e.g., Rothlisberg & Church, 1994, for *Penaeus*), while for other species dispersal is clearly limited (e.g., some pond-dwelling species). One might suppose that molecular genetic similarities between populations would give an indication of dispersal ability, but these measures are subject not only to dispersal but also assume equilibrium conditions with selectively neutral underlying loci (Endler, 1977; Hedgcock *et al.*, 1982). In general, attempts to establish a close fit between gene flow and dispersal

Table 1. The average duration of diapause for the 167 species in 20 crustacean orders reviewed in this paper.

Class	Order	Number of species	Average duration of diapause (years)	Standard deviation
Branchiopoda	Anomopoda	18	16.0	20.7
	Anostraca	8	3.9	3.2
	Ctenopoda	1	1.0	0.0
	Haplopoda	1	1.5	0.0
	Notostraca	1	15.0	0.0
	Onychopoda	2	8.8	10.3
	Spinicaudata	6	4.3	1.6
Maxillopoda	Calanoida	34	6.5	13.6
	Cyclopoida	11	0.9	0.7
	Harpacticoida	5	0.6	0.2
	Podocopida	15	3.5	6.2
	Thoracica	3	0.0	0.0
Malacostraca	Amphipoda	3	0.0	0.0
	Anaspidacea	1	0.6	0.0
	Cumacea	1	0.0	0.0
	Decapoda	18	0.04	0.1
	Euphausiacea	9	0.0	0.0
	Isopoda	19	0.04	0.1
	Mysida	8	0.0	0.0
	Stomatopoda	3	0.0	0.0

Table 2. A list of groupings of species from Fig. 1 that have < 1 year diapause duration and < 1 year adult life span. The groupings are explained in the text.

	No. spp.	Dispersal	Notes
1. High dispersal (≥ 2.5)			
Marine calanoid copepods	14	2.5	Diapause duration > 1 yr.?
Marine ostracods, open water	3	2.5	Diapause not studied
Marine amphipods	3	3	
Marine decapod shrimps	6	2.5	
Marine euphausids	4	3	
Marine mysids	3	3	
	Total 33	Mean 2.7	
2. Misclassified ?			
a. Higher dispersal?			
Inland water cyclopoid copepods	10	1	First to enter new ponds
Inland w. harpacticoid copepods	4	1, 2	Late stage or egg diapause
Marine harpacticoid copepod	1	2	
Marine isopods	8	2	
b. Longer diapause?			
Inland water Branchiopoda	4	2	Diap. duration really unkwown
Inland water ostracod	1	2	Diap. dur. conservative estimate
Tide-pool ostracods	4	1	Diap. dur. conservative estimate
	Total 32	Mean 1.5	
3. Relict species			
Inland water mysids	5	1	Glacial relicts w/ low dispersal
Inland water calanoid copepod	1	1	*Limnocalanus macrurus*
	Total 6	Mean 1	
4. No plausible explanation			
Inland w. and terrestrial isopods	7	1	
	Total 7	Mean 1	
	Overall Total 78	Mean 1.9	

in natural populations, including aquatic invertebrates, have failed (Endler, 1977; Burton, 1983). For our purposes, we adopted the following qualitative ranking: 1 – lowest dispersal: pond and lake dwelling species without resistant diapausing eggs. 2 – moderate dispersal: pond and lake dwelling species with resistant diapausing eggs. The logic here is that resistant eggs are often depicted in the literature as adaptations for dispersal (Thienemann, 1950; Maguire, 1963) in addition to being a means of surviving seasonally harsh conditions. We note, however, that Fryer (1972) and Fryer & Frey (1981) emphasized that many littoral 'cladocerans' glue their diapausing eggs to plants or other substrata suggesting that dispersal may be low. Also assigned a score of '2' are the subset of estuarine and near-shore marine species whose larvae are known to spend only a short time in the plankton. 3 – highest dispersal: benthic marine species with long-lived pelagic larvae, pelagic estuarine and marine species. We have also included intermediate scores (1.5 or 2.5) where such scoring seemed appropriate.

3. Habitat. This simply indicates whether the species is freshwater, marine (includes estuarine and intertidal), or terrestrial as an adult.

Relationships between diapause, life span, and dispersal

Diapause duration plotted as a function of adult life span (Fig. 1) shows a striking relationship consistent with Rees' (1994) theory. Crustaceans either have long adult life, or they have prolonged diapause. None of the crustaceans for which we obtained data have both. The result is significantly non-random (Spearman Rank Correlation, $r_S = -0.73$, $n = 167$, $P<0.001$). The relationship holds as well in a log-log plot, even when the many species with a diapause duration of zero are omitted from the analysis ($r^2 = 0.23$, $n = 103$, $P \ll 0.0001$).

There is a large group of crustaceans, however, that has neither long life nor prolonged diapause (Fig. 1, inset). One explanation for how these species survive environmental variation is that they should have high dispersal ability. Theory suggests that dispersal ability will be higher in groups without prolonged diapause than in those that have this trait (Cohen & Levin, 1985), and we presume that the same is true for species without multi-year iteroparity. Is this the case? Of the 78 species that have both diapause durations and adult life spans of ≤ 1 year, we ranked only 33 (Table 2) as having especially high dispersal ability (2.5 or 3). Dispersal scores for this 'corner' group are not significantly greater than those for the remainder of the species that have either diapause duration >1 year or adult life span >1 year (Mann-Whitney $U = 3720$, $P = 0.4$). One interpretation of this result is that the theory is flawed. Another is that temporal and spatial variation are not important components of crustacean environments, which can be rejected. A third alternative, however, is that some corner group species are misclassified for one or more of the categories:

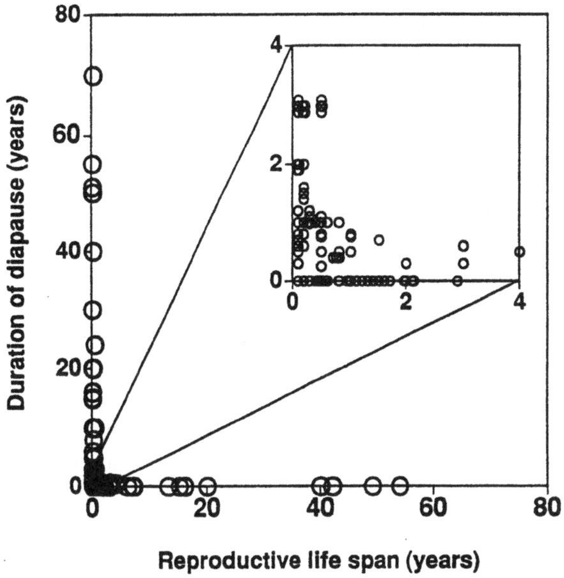

Fig. 1. The relationship between duration of diapause and reproductive life span for 167 crustacean species (see Appendix 1). Inset shows the distribution of data points near the origin.

dispersal, diapause, and life span. For example, our dispersal rankings assume that without a resistant diapausing egg, lake and pond species must have a very weak ability to move between water bodies. Included in this category are the cyclopoid copepods which encyst in a late instar. Yet we know that cyclopoids are among the first species to colonize newly established ponds. Jenkins and Buikema (1994) recently reported that reproductive *Eucyclops agilis* females were found in 12 new ponds dug in an old field within one week after filling from a chlorinated municipal water supply. Two other cyclopoid species were found in many of the ponds within a year of filling. In contrast, no calanoid copepods (many of which make diapausing eggs) were seen during this period. Furthermore, the late copepodid diapause of cyclopoids may be more resistant than is commonly believed: Rylov (1963) recounted an example of encysted cyclopoids surviving in dried mud for 3 years, and more recently Wyngaard *et al.* (1991) reported that several cyclopoid species reappeared in a temporary pond after two or more years of failed reproduction. These latter observations also raise the question of whether the short-term (0.5–0.7 years) diapause duration typically ascribed to cyclopoids is correct.

As with the cyclopoids, harpacticoid copepods also possess a late stage diapause, and it is possible that they are similarly misclassified (Table 2). Other species that we may have incorrectly placed in the corner are benthic marine isopods which may have greater dispersal than we have credited to them, and a variety of inland water and marine Branchiopoda and Ostracoda that we classified as having diapause durations of ≤ 1 year because we had no information to the contrary. Given the diapause durations reported for many of the species in the latter two groups, misclassification is certainly a plausible explanation for their inclusion in the corner group. Altogether there are 32 species in the corner for which we can make a case for misclassification.

There remain 13 species for which we have no plausible rationale for exclusion from the corner. Included in this group are six crustacean species (Table 2) whose natural distribution dates from the last glaciation. The freshwater mysids, and the large copepod *Limnocalanus macrurus*, are all thought to have been stranded in large, deep inland basins as the continental ice sheet receded at the end of the Pleistocene (Segerstrale, 1962; Holmquist, 1972; Williamson, 1991). Recently, Katajisto (1994) has reported finding diapausing eggs of *L. macrurus* in Baltic sediments, though their presence in lakes remains uncertain (Roff, 1972). The mysids at least must have poor dispersal ability since they have recently been successfully introduced into a large number of lakes from which they were previously absent (Lasenby *et al.*, 1986). Given that these six species also have short adult life spans and apparently lack a diapause stage, we conclude that they truly would be vulnerable to short-term recruitment failures. They apparently survive only where they do because the lakes in which they live are relatively constant environments that have existed for a geologically brief period.

In addition to species without prolonged diapause, long reproductive life span or high dispersal, there are some species that have both high dispersal ability and either long adult life (e.g., many marine decapods) or prolonged diapause (e.g., some marine calanoid copepods). Although Venable & Lawlor (1980) predicted that these traits should not be associated, Cohen & Levin (1985) found in models for annual plants that either variation in environmental variability or simultaneous variation in both dispersal success and seed survival would cause dormancy duration and dispersal ability to increase in concert. Presumably the same also applies to dispersal and iteroparity.

Covariates of diapause and life span

A number of interesting associations are apparent, even when we include all of the crustaceans classified as having both short diapause duration and short reproductive life span in our analysis. These relate the form of temporal bet-hedging (diapause or iteroparity) to the body sizes, habitats and phylogenetic affinities of the crustaceans surveyed.

Body size – Animals with long adult lives are relatively large, whereas those with prolonged diapause are small. The relationship between body weight and reproductive life span (Fig. 2) is highly significant ($r^2 = 0.53$, $P \ll 0.0001$, $n = 145$) with a slope on a log-log plot [log(reproductive life span) = −0.33 + 0.17 log (mg body weight)] very similar to that reported by others for a variety of animals (e.g., Blueweiss *et al.*, 1978; Peters, 1983). If we take a life span of greater than one year to represent a minimal level of multi-year bet-hedging, then the data in Fig. 2 show that it is principally crustaceans greater than 10 mg adult body weight that achieve this. The relationship between adult life span and body weight is presumably driven at least in part by the fact that it takes longer to reach a large than a small size, though some relatively large animals are semelparous (e.g., octopods and some salmonid fishes). The largest crustacean in our survey with a life span of <1 year is the pelagic shrimp *Acetes japonicus* with an adult body weight of just 26 g.

The relationship between duration of diapause and body weight (Fig. 3) is also highly significant (Spearman Rank $r_S = -0.61$, $n=145$, $P<0.001$), with smaller crustaceans more likely to exhibit prolonged diapause. The nonparametric relationship is principally the result of animals larger than 10 mg having no diapause (with the exception of notostracans). Thus, crustaceans larger than 10 mg tend to have long adult lives, whereas those smaller than 10 mg frequently have a prolonged diapause.

We can think of no *a priori* physiological reason why a large or a long-lived animal should not be able to produce diapausing eggs. A negative correlation has been noted, however, for marine invertebrates between body size and brood care, in which small animals invest more resources per offspring than do large-bodied animals (Strathmann & Strathmann, 1982). The correspondence we observe between small body size and presence of diapause may, then, be an extension of this correlation since it costs more to make diapausing eggs than it does to make eggs that hatch immediately. All of the hypotheses advanced by Strathmann &

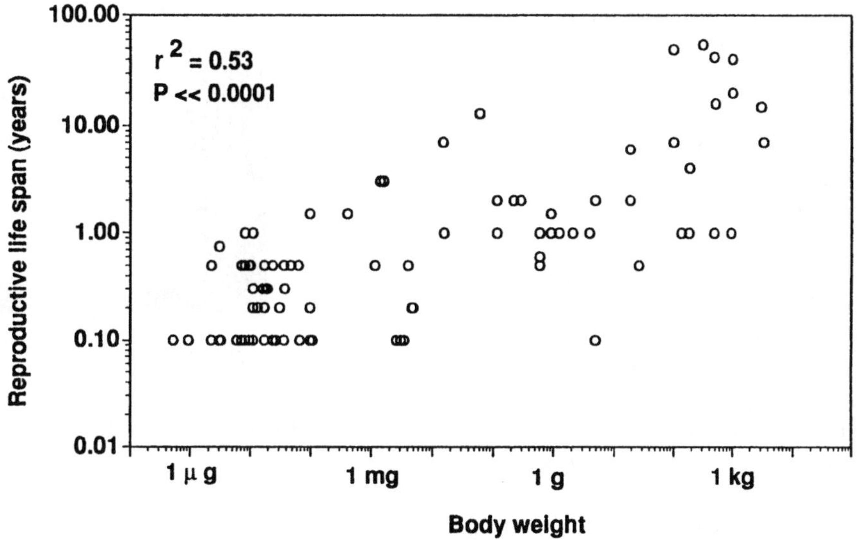

Fig. 2. The relationship between reproductive life span and body weight for 145 crustacean species. Regression equation given in text.

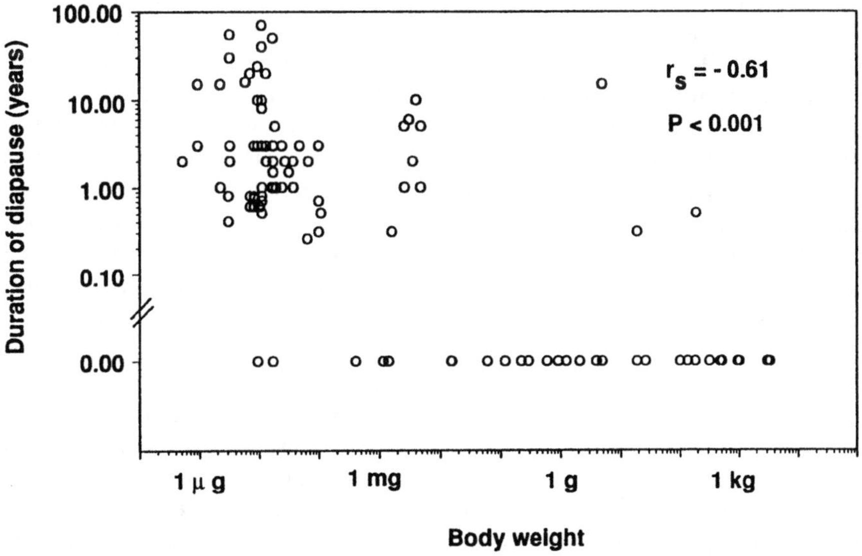

Fig. 3. The relationship between duration of diapause and body weight for 145 crustacean species. Results of nonparametric Spearman Rank correlation are shown.

Strathmann (1982) to explain the brooding-allometry relationship they observed rely upon the greater ability of large animals with numerous offspring to sample a varying environment through spatial dispersal. If it is true, however, that large animals tend also to be iteroparous, then temporal sampling of a variable environment must also be considered. It is also true that diapause at later stages in a large animal might be difficult due to the requirement of protecting such a large mass from attack by predators, microbes, and physical processes. Body size itself may be constrained by the environment in several ways. If it takes longer for a large animal to reach reproductive size than a small one, large crustaceans in highly seasonal environments (i.e., with 'catastrophic' periods that can only be survived in diapause) may simply be unable to survive. It does appear to be true that only small crustaceans live in highly seasonal environments, with the exception of species in which individuals can migrate away from the harsh conditions or can find microhabitats that are not uninhabitable (e.g., crayfish and the other crustaceans that survive drought in burrows that descend to the water table, Wiggins *et al.*, 1980).

Habitat – Crustaceans living in inland waters are significantly more likely to have a diapause stage that can survive longer than one year than are those living in marine environments (Fig. 4; $X^2 = 40.9$, $n = 160$, $P \ll 0.0001$). At the same time, marine crustaceans are more likely than those in inland waters to have a reproductive life span that exceeds one year ($X^2 = 12.5$, $n = 160$, $P < 0.0005$). Although exceptions exist, the general trends in Fig. 1 are that the species lying along the axis of 'reproductive life span' are typically marine and large bodied, while those lying along the axis of 'duration of diapause' are typically relatively small and inland water taxa.

Phylogeny – In the introduction we asserted that the diapause trait is not evenly distributed among crustacean taxa. For the species we reviewed, significantly more members of the Branchiopoda have a diapause exceeding one year than do members of the Maxillopoda (Fig. 5; $X^2 = 25.3$, $n = 105$, $P \ll 0.0001$) which in turn show a higher incidence of prolonged diapause than do the Malacostraca ($X^2 = 29.6$, $n = 130$, $P \ll 0.0001$). In direct contrast, species within the Malacostraca are more likely to have a long adult life span than are members of either the Maxillopoda or the Branchiopoda ($X^2 = 20.6$, $n = 130$, $P \ll 0.0001$ and $X^2 = 18.9$, $n = 99$, $P < 0.0002$ respectively). Life spans of Maxillopoda and Branchiopoda do not differ significantly ($X^2 = 2.3$, $n = 105$, $P = 0.13$).

Discussion

There are two alternative interpretations for our principal result that duration of diapause is negatively associated with reproductive life span: one is microevolutionary, the other macroevolutionary. From a adaptationist perspective, Fig. 1 may be taken as support for the hypothesis that animals with long adult life have not responded to selection for producing a long-lived diapausing stage as Rees' (1994) theoretical result suggests. Similarly, animals with prolonged diapause apparently have not responded to selection for long adult life.

An alternative, macroevolutionary, interpretation is that phylogenetic constraint contributes to the pattern we observe. If the ancestors of some taxa did not evolve diapause, their descendants may also lack the trait whatever the selection pressures may have been.

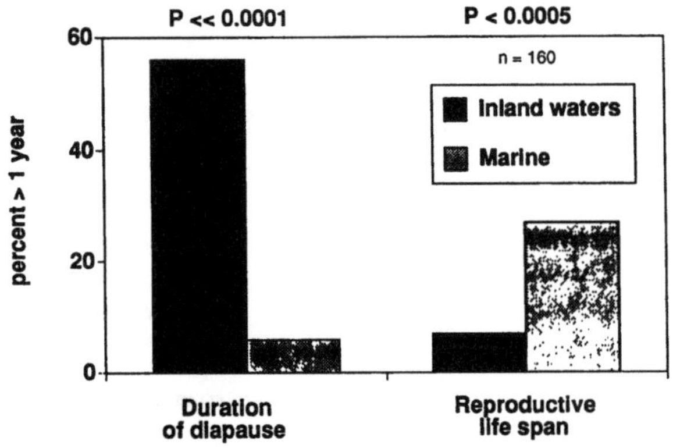

Fig. 4. Percentages of crustacean species from inland aquatic and marine habitats that have either diapause durations or reproductive life spans exceeding one year. P-values are for X^2 tests (see text).

If these same taxa tend to have long adult life spans, then fortuitous associations may appear that have little directly to do with adaptation. When we illustrate the presence or absence of diapause on Schram's (1986) cladogram for the crustacean orders, several striking phylogenetic patterns emerge (Fig. 6). The first is the observation that distinct types of diapause occur in the different groups, supporting independent evolution of diapause in different major taxa. Diapausing eggs are limited to the Branchiopoda, and to some of the Maxillopoda (i.e., some ostracods, some calanoid copepods, and one harpacticoid species). Analysis of the structure of Branchiopod egg coats suggests that production of diapausing eggs in this group is a primitive trait (Fryer, 1994; personal communication). Its phylogenetic relationship to the diapausing eggs in the Maxillopoda is, however, less certain. Two possibilities exist: either diapause is primitive for the Branchiopoda and Maxillopoda and there has been selection against diapause in more derived members of the Maxillopoda, or diapause has evolved independently in these two groups.

Late-instar, encysted diapause is limited to the cyclopoid and harpacticoid copepods. The late instar non-encysted diapause of some marine calanoids is almost certainly an independent phenomenon. With the exception of short-duration (i.e., <1 year), overwintering, 'reproductive diapause' in a few members of the Malacostraca (Hickman, 1937; Tandini-Vitagliano *et al.*, 1982; Alekseev, 1989), none of the other groups exhibit the diapause trait. Clearly this is a nonrandom pattern (cf. Fig. 5). Although there are short-lived members of the Malacostraca, they have not evolved prolonged diapause even though some of them (e.g.,

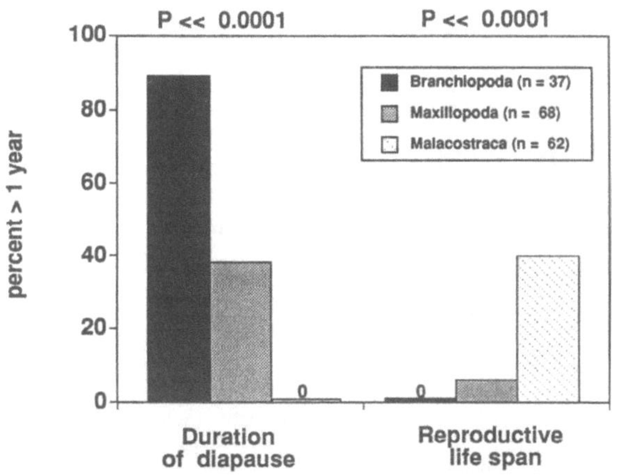

Fig. 5. Percentages of crustacean species from each of the three classes represented in our survey that have either diapause durations or reproductive life spans exceeding one year. P-values are for comparisons of all three classes at once (diapause duration $X^2 = 81.1$, $n = 167$, $P \ll 0.0001$; adult life span $X^2 = 34.6$, $n = 167$, $P \ll 0.0001$). See text for pairwise comparisons.

amphipods and isopods) may occasionally occur in the same temporally variable habitats where diapausing-egg producing Branchiopoda thrive (Wiggins *et al.*, 1980). It is still possible that taxa like amphipods have evolved (or already possessed) high dispersal capability as an alternative to diapause, but as pointed out earlier, we were unable to obtain a convincing assessment of dispersal for many species.

The fact that the harpacticoid and cyclopoid copepods express a very different type of diapause from the other groups, and must represent at least one independent evolutionary event (cladistic analyses typically place harpacticoids and cyclopoids more closely together than either is to the calanoids; Huys & Boxshall, 1991), suggests that the Maxillopoda have evolved diapause independently of the Branchiopoda. As suggested earlier, it may be that possessing either long adult life or prolonged diapause reduces selection for the alternative life history pattern (e.g., long-lived crustaceans do not diapause in part because there is no selection pressure to do so). Finally, it may be that the association between habitat and diapause or habitat and life span facilitates the invasion of a particular habitat. Lindley (1992) has proposed that calanoid copepods of the superfamily Centropagoidea are preadapted to invade freshwater by their ability to produce diapausing eggs in near-shore marine habitats. He ascribed diapause in nearshore environments to adaptation to avoid abrasion and predation on eggs at the sea bottom, and proposed that the diapausing egg coat facilitated invasion of freshwater by serving as a protection against osmotic stress. His interpretation has the problem that the egg coat does not protect nauplii newly hatching into fresh water from osmotic stress. An alternative interpretation is that the ability to produce diapausing eggs facilitates survival in highly variable lake and pond environments. Certainly the diaptomids (many, perhaps all, of which produce diapausing eggs, Hairston & Van Brunt, 1994) have experienced enormous diversification in inland waters. Although we do not attempt here to categorize the temporal (especially interannual) or spatial variability of habitats, most freshwater lakes and ponds are generally considered to be more variable than most marine environments.

Conclusions

Three more or less distinct crustacean groups fall out of our analysis that are both largely consistent with expected patterns of adaptation and with underlying phylogeny. (1) Species with prolonged diapause tend to be small and tend to be found in inland waters. If the variability of inland water habitats is such that bad years for recruitment are also often bad for adult survival, then temporal bet-hedging is best accomplished by rapidly-maturing, small species capable of diapause. Taxa with these characteristics will be the ones likely to invade and proliferate in such inland habitats. Selection for long adult life span is weak or nonexistent, and species possessing this trait are not likely to be successful invaders. (2) Species with long adult life spans tend to be large and tend to be found in marine habitats. Where temporal variability in recruitment is not accompanied by equal or greater variability in adult mortality, iteroparity is a viable bet-hedging strategy. Taxa with these characteristics will be the ones likely to populate marine habitats. (3) A third group of species has small body size and neither long adult life nor prolonged diapause. These animals are difficult to categorize based on the information we have available. They may have high dispersal ability to cope with environmental variation, they may reside in habitats that are relatively invariant, or they may be doomed to a high likelihood of local extinction. It seems likely that this final group must be subject to the most intense selection for some means of spatial or temporal bet-hedging. They may represent the species for which we would most expect novel exceptions to the underlying phylogenetic patterns to appear. It is

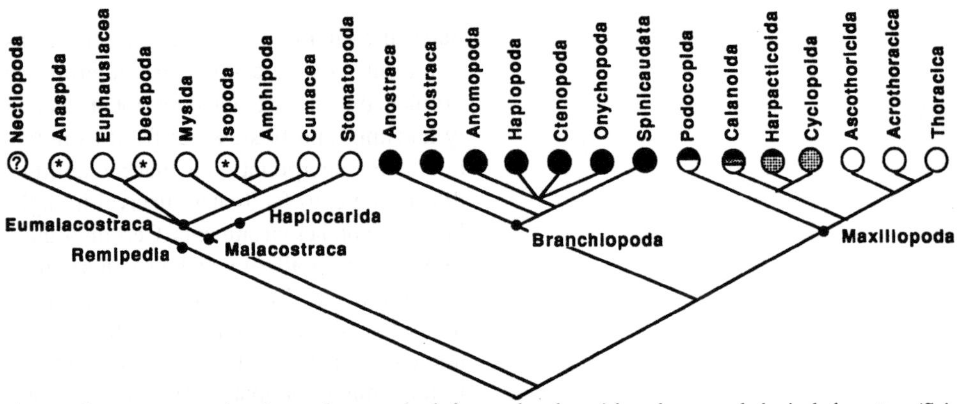

Fig. 6. A cladogram for crustacean orders (some classes and subclasses also shown) based on morphological characters (Schram, 1986) with diapause trait expression represented by the symbols: Open – no diapause; Open with asterisk – most species have no diapause; a few exhibit short duration reproductive diapause; Open with question mark – presence or absence of diapause unknown; Black – diapausing eggs produced; Shaded – late instar diapause. Mixed symbols represent orders in which species express a variety of diapause traits as depicted by the solid symbols. Some taxonomic categories have been modified according to Brusca & Brusca (1990) and Fryer (1987). The orders Leptostraca and Brachypoda, which Schram (1986) included in the clade 'Phyllopoda', have been omitted here because their affinity with the Branchiopoda is controversial, and because we were unable to obtain any information about diapause expression in either of these minor orders.

tempting to speculate that this is what the diversity of diapause patterns in the Maxillopoda represents.

Acknowledgments

We thank A. Bohonak, F. Ferrari, R. Harrison, C. Holtmeier, and C. Kearns for useful discussions. L. Yampolsky, A. Kondrashov, and H. Schwanninger helped with translations, and V. R. Alekseev, G. Fryer, C. D. Harvell, A. R. McCune, Y. I. Starobogatov and A. Ghilarov provided contructive criticism of the manuscript. This research was funded by U.S. National Science Foundation grants DEB-9118894 to NGH and S. Ellner, and DEB-9410966 to CEC. Carla Cáceres was supported by an NSF Predoctoral Fellowship.

References

Aiken, D. E., 1969. Ovarian maturation and egg laying in the crayfish *Orconectes virilis*: Influence of temperature and photoperiod. Can. J. Zool. 47: 931–935.

Alekseev, V., 1989. Influence of diapause upon level of oxygen consumption in *Astacus* (Decapoda). State Research Institute for Lake and River Fisheries, Research Reports. 300: 80–90. (In Russian).

Andrew, T. E. & A. Herzig, 1984. The respiration rate of the resting eggs of *Leptodora kindti* and *Bythotrephes longimanus* Leydig 1860 (Crustacea, Cladocera) at environmentally encountered temperatures. Oecologia 64: 241–244.

Angell, R. W. & J. W. Hancock, 1989. Response of eggs of *Heterocypris incongruens* (Ostracoda) to experimental stress. J. crustacean Biol. 9: 381–386.

Arashkevich, E. G., A. V. Drits & A. G. Timonin, 1994. Diapause in the life cycle of the copepod *Calanoides carinatus*. International Symposium: Diapause in Crustacea, Abstracts p. 7.

Ban, S., 1992. Seasonal distribution, abundance and viability of diapause eggs of *Eurytemora affinis* (Copepoda: Calanoida) in the sediment of Lake Ohnuma, Hokkaido. Bull. Plankton Soc. Japan 39: 41–48.

Blueweiss, L., H. Fox, V. Kudzma, D. Nakashima, P. Peters & S. Sams, 1978. Relationships between body size and some life history parameters. Oecologia 37: 257–272.

Borowsky, B., 1980. Reproductive patterns of three intertidal saltmarsh gammaridean amphipods. Mar. Biol. 55: 327–334.

Borutzky, E.W., 1929. Zur Frage über den Ruhezustand bei Copepoda-Harpacticoida. Dauereier bei *Canthocamptus arcticus* Lilljeborg. Zool. Anz. 83: 225–233.

Brendonck, L., D. M. Centeno & G. Persoone, 1993. Fecundity and resting egg characteristics of some subtropical fairy shrimp and clam shrimp species (Crustacea: Branchiopoda) reared under laboratory conditions. Arch. f. Hydrobiol. 126: 445–459

Brusca, R. C. & G. J. Brusca, 1991. Invertebrates. Sinauer, Sunderland. 922 pp.

Burton, R. S., 1983. Protein polymorphisms and genetic differentiation of marine invertebrate populations. Mar. Biol. Lett. 4: 193–206.

Cáceres, C. E., 1994. Egg bank dynamics and daphnid species diversity. Bull. Ecol. Soc. Amer./Supp. 75: 30.

Carefoot, T. H., 1973. Studies on the growth, reproduction, and life cycle of the supralittoral isopod *Ligia pallasii*. Mar. Biol. 18: 302–311.

Carvalho, G. R. & H. G. Wolf, 1989. Resting eggs of lake-*Daphnia* I. Distribution, abundance and hatching of eggs collected from various depths in lake sediments. Freshwat. Biol. 22: 459–470.

Castro, M., 1992. A methodology for obtaining information on the age structure and growth rates of the Norway Lobster, *Nephrops norvegicus* (L.) (Decapoda, Nephropoidea). Crustaceana 63: 29–43.

Champeau, A., 1970. Etude de la vie latente chez les Calanoïdes (Copépodes) caractéristiques des eaux temporaires de Basse-Provence. Ann. Fac. Sci. Marseille 44: 155–189.

Cohen, A. C. & J. G. Morin, 1990. Patterns of reproduction in ostracods: A review. J. crustacean Biol. 10: 184–211.

Cohen, D., 1966. Optimizing reproduction in a randomly varying environment. J. theor. Biol. 12: 119–129.

Cohen, D., 1970. A theoretical model for the optimal timing of diapause. Am. Nat. 104: 389–400.

Cohen, D. & S. A. Levin, 1985. The interaction between dispersal and dormancy strategies in varying and heterogeneous environments. In: E. Teramoto and M. Yamaguti (eds), Mathematical topics in population biology, morphogenesis and neurosciences, Proceedings. Springer-Verlag, Heidelberg: 110–122.

Cooper, R. A. & J. R. Uzmann, 1977. Ecology of juvenile and adult clawed lobsters. In: B. F. Phillips & J. S. Cobb (eds), Workshop on lobster and rock lobster ecology and physiology: 187–208.

Cukerzis, J., 1970. La biologie de l'écrevisse (Translation to French 1984). Institut Nationale de la Recherche Agronomique, Paris, 313 pp.

Daly, K. L. & D. M. Damkaer, 1986. Population dynamics and distribution of *Neomysis mercedis* and *Alienacanthomysis macropsis* (Crustacea: Mysidacea) in relation to the parasitic copepod *Hansenulus trebax* in the Columbia River estuary. J. crustacean Biol. 6: 840–857.

Dana, G. L., 1984. *Artemia* in temporary alkaline ponds near Fallon, Nevada with a comparison of its life history strategies in temporary and permanent habitats. In: S. Jain & P. Moyle (eds), Vernal pools and intermittent streams. Institute of Ecology Publication No. 28, University of California, Davis: 115–125.

Delorme, L., 1991. Ostracoda. In: J. H. Thorp and A. P. Covich (eds) Ecology and classification of North American freshwater invertebrates. Academic, San Diego: 691–722.

De Stasio, B. T., Jr., 1989. The seed bank of a freshwater crustacean: Copepodology for the plant ecologist. Ecology 70: 1377–1389.

De Stasio, B. T., Jr., 1990. The role of dormancy and emergence patterns in the dynamics of a freshwater zooplankton community. Limnol. Oceanogr. 35: 1079–1090.

De Stasio, B. T., Jr. & N. G. Hairston, Jr., 1992. Environmental variability and the persistence of multiple emergence strategies. Bull. math. Biol. 54: 313–334.

Dugan, J. E., A. M. Wenner & D. M. Hubbard, 1991. Geographic variation in the reproductive biology of the sand crab *Emerita analoga* (Stimpson) on the California coast. J. exp. Mar. Biol. Ecol. 150: 63–81.

Easterling, M. R. & S. Ellner, 1994. Hatching strategies in a random environment: A structured model. Bull. Ecol. Soc. Amer./Supp. 75: 59.

Edmondson, W. T., 1955. The seasonal life history of *Daphnia* in an arctic lake. Ecology 36: 439–455.

Elgmork, K., 1955. A resting stage without encystment in the annual cycle of the freshwater copepod *Cyclops strenuus strenuus*. Ecology 36: 739–743.

Elgmork, K., G. Halvorsen, J. A. Eie & A. Langeland, 1990. Coexistence with similar life cycles in two species of freshwater copepods (Crustacea). Hydrobiologia 208: 187–200.

Ellner, S., 1984. Asymptotic behavior of some stochastic difference equation population models. J. math. Biol. 19: 169–200.

Ellner, S., 1985. ESS germination strategies in randomly varying environments I. Logistic-type models. Theor. Pop. Biol. 28: 50–79.

Ellner, S. & N. G. Hairston, Jr., 1994. Role of overlapping generations in maintaining genetic variation in a fluctuating environment. Am. Nat. 143: 403–417.

Endler, J. A., 1977. Geographic variation, speciation, and clines. Princeton Univ. Press, Princeton, NJ., 246 pp.

Farmer, A. S., 1973. Age and growth in *Nephrops norvegicus* (Decapoda: Nephropidae). Mar. Biol. 23: 315–329.

Fryer, G., 1972. Observations on the ephippia of certain macrothricid cladocerans. Zool. J. Linn. Soc. 51: 79–86.

Fryer, G., 1987. A new classification of the branchiopod Crustacea. Zool. J. Linn. Soc. 91: 357–383.

Fryer, G. & D. G. Frey, 1981. Two-egged ephippia in the chydorid Cladocera. Freshwat. Biol. 11: 391–394.

Fryer, G., 1994. Diapause, a potent force in the evolution of freshwater crustaceans. International Symposium: Diapause in Crustacea. Abstracts, p. 11.

Grice, G. D. & N. H. Marcus, 1981. Dormant eggs of marine copepods. Oceanogr. Mar. Biol. annu. Rev. 19: 125–140.

Hairston, N. G., Jr. & B. T. De Stasio, Jr., 1988. Rate of evolution slowed by a dormant propagule pool. Nature 336: 239–242.

Hairston, N. G., Jr. & W. R. Munns, Jr., 1984. The timing of copepod diapause as an evolutionary stable strategy. Am. Nat. 123: 733–751.

Hairston, N. G., Jr. & R. A. Van Brunt, 1994. Diapause dynamics of two diaptomid copepod species in a large lake. Hydrobiologia 292/293 (Dev. Hydrobiol. 102): 209–218.

Hairston, N. G., Jr., S. Ellner & C. M. Kearns, 1995a. Overlapping generations: the storage effect and the maintenance of biotic diversity. In: O. E. Rhodes, R. K. Chesser & M. H. Smith (eds), Population dynamics in ecological space and time. Univ. Chicago Press, Chicago (in press).

Hairston, N. G., Jr, R. A. Van Brunt, C. M. Kearns & D. R. Engstrom, 1995b. Age and survivorship of diapausing eggs in a sediment egg bank. Ecology 1706–1711.

Hartland-Rowe, R. C. B., 1972. The limnology of temporary waters and the ecology of Euphyllopoda. In: R. B. Clark and R. J. Wooton (eds), Essays in hydrobiology. University of Exeter, Exeter: 15–32.

Hedgecock, D., M. L. Tracey & K. Nelson, 1982. Genetics. In: L. G. Abele (ed.), The biology of crustacea, Vol. 2. Academic, New York: 283–403.

Hickman, V. V., 1937. The embryology of the syncarid crustacean, *Anaspides tasmaniae*. Pap. & Proc. r. Soc. Tasm. 1936: 1–35.

Hildrew, A. G., 1985. A quantitative study of the life history of a fairy shrimp (Branchiopoda: Anostraca) in relation to the temporary nature of its habitat, a Kenyan rainpool. J. anim. Ecol. 54: 99–110.

Hines, A. H., 1979. The comparative reproductive ecology of three species of intertidal barnacles. In: S. E. Stancyk (ed.), Reproductive ecology of marine invertebrates. University of South Carolina Press, Columbia, SC: 213–234.

Holmquist, C., 1972. Das Zooplankton der Binnengewässer V. Mysidacea. Binnengewässer 26: 247–256.

Huys, R. & G. A. Boxshall, 1991. Copepod evolution. The Ray Society, London, 468 pp.

Jenkins, D. G. & A. L. Buikema, in prep. Structure and function of zooplankton communities in twelve new experimental ponds.

Johannsson, O., 1992. Life history and productivity of *Mysis relicta* in Lake Ontario. J. Great Lakes Res. 18: 154–168.

Juchault, P., W. Jassem & J.-P. Mocquard, 1982. Déterminisme de la reproduction saisonnière de femelles d'*Armadillidium vulgare* Latr. (Crustacé, Isopoda, Oniscoïda). VI. Mise en évidence d'une photopériode critique permettant l'entrée en reproduction; modatités du maintien en reproduction. Ann. Sci. Nat. Zool. 4: 203–210.

Katajisto, T., 1996. Copepod eggs survive a decade in the sediments of the Baltic Sea. Hydrobiologia 320 (Dev. Hydrobiol. 114): 153–159.

Khmeleva, N. M. & V. M. Baichorov, 1987. Comparative estimate of reproductive cycles in two relict, ecologically different mysid species. Polsk. Arch. Hydrobiol. 34: 321–329.

Korpelainen, H., 1986. The effect of diapause on the genetic structure of *Daphnia magna* populations. Z. zool. Syst. Evolut. -forsch. 24: 291–299.

Kulikov, A. S., 1980. On the ecology of two gammarid and one mysid species from the cryopelagic biocenose of the central arctic basin. Biol. Cent. Ark. Bass: 111–118. (In Russian).

Lampert, W. & I. Krause, 1976. Zur Biologie der Cladocere *Holopedium gibberum* Zaddach im Windgfällweiher (Schwarzwald). Arch. f. Hydrobiol./Supp. 48: 262–286.

Lasenby, D. C., T. G. Northcote & M. Fürst, 1986. Theory, practice, and effects of *Mysis relicta* introductions to North American and Scandinavian lakes. Can. J. Fish. aquat. Sci. 43: 1277–1284.

Levin, S. A., D. Cohen & A. Hastings, 1984. Dispersal strategies in patchy environments. Theor. Pop. Biol. 26: 165–191.

Levins, R., 1969. Dormancy as an adaptive strategy. Symp. Soc. exp. Biol. 23: 1–10.

Lindley, J. A., 1992. Resistant eggs of the Centropagoidea (Copepoda: Calanoida): A possible preadaptation to colonization of inland waters. J. crustacean Biol. 12: 368–371.

Livdahl, T. P., 1979. Environmental uncertainty and selection for life cycle delays in opportunistic species. Am. Nat. 113: 835–842.

Maguire, B., Jr., 1963. The passive dispersal of small aquatic organisms and their colonization of isolated bodies of water. Ecol. Monogr. 33: 161–185.

Maier, G., 1989. Variable life cycles in the freshwater copepod *Cyclops vicinus* (Uljanin 1875): Support for the predator avoidance hypothesis? Arch. f. Hydrobiol. 45: 203–219.

Marcus, N. H., R. Lutz, W. Burnett & P. Cable, 1994. Age, viability, and vertical distribution of zooplankton resting eggs from an anoxic basin: Evidence of an egg bank. Limnol. Oceanogr. 39: 154–158.

Mattox, N. T. & J. T. Verlardo, 1950. Effect of temperature on the development of the eggs of a conchostracan phyllopod, *Caenestheriella gynecia*. Ecology 31: 497–506.

Mauchline, J. & L. R. Fisher, 1969. The biology of euphausids. Adv. Mar. Biol. 7: 1–454.

McCaughran, D. A. & G. C. Powell, 1977. Growth model for Alaska king crab (*Paralithodes camtschatica*) J. Fish. Res. Bd. Can. 34: 989–995.

Miller, C. B. & K. S. Tande, 1993. Stage duration estimation for *Calanus* populations: a modeling study. Mar. Ecol. Prog. Ser. 102: 15–34.

Moghraby, A. el., 1977. A study of diapause of zooplankton in a tropical river – The Blue Nile. Freshwat. Biol. 7: 207–212.

Momot, W. T., 1986. Production and exploitation of the crayfish, *Orconectes virilis*, in northern climates. Can. Special Pub. Fish. aquat. Sci. 92: 154–167.

Momot, W. T., 1992. Further range extensions of the crayfish, *Orconectes virilis*, in the Lake Superior Basin of northwestern Ontario. Can. Field-Nat. 106: 397–399.

Moritz, C., 1987. A note on the hatching and viability of *Ceriodaphnia* ephippia collected from lake sediment. Hydrobiologia 145 (Dev. Hydrobiol. 35): 309–314.

Moritz, C., 1988. Die Verteilung der Ephippien von *Bosmina longirostris* und *Ceriodaphnia pulchella* im Sediment des Piburger Sees (Ôtztal, Tirol). Ber. nat.-med. Verein Innsbruck 75: 91–107.

Murphy, G. I., 1968. Pattern in life history and the environment. Am. Nat. 102: 391–403.

Naess, T. & J. P. Nilssen, 1991. Diapausing fertilized adults: a new pattern of copepod life cycle. Oecologia 86: 368–371.

Ng, P. K. L., 1988. The freshwater crabs of peninsular Malaysia and Singapore. S. Lee Press, Kent Ridge, Singapore, 156 pp.

Omori, M., 1971. The biology of pelagic shrimps in the ocean. Adv. Mar. Biol. 12: 233–324.

Otto, R. S., 1986. Management and assessment of Eastern Bering Sea king crab stocks. Can. Special Pub. Fish. aquat. Sci. 92: 83–106.

Payne, J. F. & J. O. Price, 1983. Studies of the life history and ecology of *Orconectes palmeri palmeri* (Faxon). In: C. R. Goldman (ed.), Freshwater crayfish V. AVI Publishing Company, Westport, Connecticut: 183–191.

Pearce, A. F. & B. F. Phillips, 1994. Oceanic processes, Puerulus settlement and recruitment of the Western Rock Lobster, *Panulirus cygnus*. In: P. W. Sammarco & M. L. Heron (eds), The bio-physics of marine larval dispersal. Amer. Geophys. Union, Washington, D.C.: 279–303.

Peters, R. H., 1983. The ecological implications of body size. Cambridge University Press, Cambridge, UK, 329 pp.

Philippi, T., 1993. Bet-hedging germination of desert annuals: beyond the first year. Am. Nat. 142: 474–487.

Reaka, M. L., 1979. The evolutionary ecology of life history patterns in stomatopod crustacea. In: S. E. Stancyk (ed.), Reproductive ecology of marine invertebrates. University of South Carolina Press, Columbia, SC: 235–260.

Rees, M., 1993. Trade-offs among dispersal strategies in British plants. Nature 366: 150–152.

Rees, M., 1994. Delayed germination of seeds: A look at the effects of adult longevity, the timing of reproduction, and population age/stage structure. Am. Nat. 144: 43–64.

Roff, D. A., 1992. The evolution of life histories. Chapman & Hall, New York, 535 pp.

Roff, J. C., 1972. Aspects of the reproductive biology of the planktonic copepod *Limnocalanus macrurus* Sars, 1863. Crustaceana 22: 155–160.

Rothlisberg, P. C. & J. A. Church, 1994. Processes controlling the larval dispersal and postlarval recruitment of penaeid prawns. In: P. W. Sammarco & M. L. Heron (eds), The bio-physics of marine larval dispersal. Amer. Geophys. Union, Washington, DC: 235–252.

Rylov, V. M., 1963. Freshwater cyclopoida. In: E. N. Pavlovskii & A. A. Shtakel'berg (eds), Fauna of the USSR, Vol 3. Translated from Russian by the Israel Program for Scientific Translations, Jerusalem, 314 pp.

Sarvala, J., 1979a. Benthic resting periods of pelagic cyclopoids in an oligotrophic lake. Holarct. Ecol. 2: 88–100.

Sarvala, J., 1979b. A parthenogenetic life cycle in a population of *Canthocamptus staphylinus* (Copepoda, Harpacticoida). Hydrobiologia 62: 113–129.

Sazhina, L. I.,1987. Reproduction, growth and production of marine copepoda. Institute of Biology of South Seas, Ukrainian Academy of Science, 155 pp.

Schmitt, W. L., 1965. Crustaceans. Univ. of Michigan Press, Ann Arbor, 204 pp.

Schram, F. R., 1986. Crustacea. Oxford, New York, 606 pp.

Segerstrale, S. G., 1962. The immigration and prehistory of the glacial relicts of Eurasia and North America. A survey and discussion of modern views. Int. Revue ges. Hydrobiol. 47: 1–25.

Shmida, A. & S. Ellner, 1984. Coexistence of plant species with similar niches. Vegetatio 58: 29–55.

Stearns, S. C., 1976. Life-history tactics: a review of the ideas. Quart. Rev. Biol. 51: 3–47.

Strathmann, R. R. & M. F. Strathmann, 1982. The relationship between adult size and brooding in marine invertebrates. Am. Nat. 119: 91–101.

Swain, R. & C. I. Reid, 1983. Observations on the life history and ecology of *Anaspides tasmaniae* (Thomson) (Syncarida: Anaspididae). J. crustacean Biol. 3: 163–172.

Tadini-Vitagliano, G., F. Valentino & L. Migliore, 1982. Is the mechanism present in *Asellus aquaticus* (L.) an evolutionary stage towards an endogenous calendar? Z. zool. Syst. Evolut. -forsch. 20: 144–155.

Taylor, B. E. & D. L. Mahoney, 1990. Zooplankton in Rainbow Bay, a Carolina Bay pond: population dynamics in a temporary habitat. Freshwat. Biol. 24: 597–612.

Taylor, B. E., G. A. Wyngaard & D. L. Mahoney, 1990. Hatching of *Diaptomus stagnalis* eggs from a temporary pond after a prolonged dry period. Arch. f. Hydrobiol. 117: 271–278.

Taylor, F., 1980. Optimal switching to diapause in relation to the onset of winter. Theor. Pop. Biol. 18: 125–133.

Templeton, A. R. & D. A. Levin, 1979. Evolutionary consequences of seed pools. Am. Nat. 114: 232–249.

Theisen, B. F., 1966. The life history of seven species of ostracodes from a Danish brackish–water locality. Meddelelser fra Danmarks Fiskeri – og Havundersøgelser, n.s. 4: 215–270.

Thienemann, A., 1950. Verbreitungsgeschichte der Süsswassertierwelt Europas. Die Binnengewässer 18: 1–809.

Toda, H., M. Takahashi & S. Ichimura, 1982. Abundance and life history of *Neomysis intermedia* Czerniawsky in Lake Kasumigaura. Hydrobiologia 93 (Dev. Hydrobiol. 10): 31–39.

Venable, D. L. & L. Lawlor, 1980. Delayed germination and dispersal in desert annuals: Escape in space and time. Oecologia 46: 272–282.

Watson, N. H. F. & B. N. Smallman, 1971. The role of photoperiod and temperature on the induction and termination of an arrested development in two species of freshwater cyclopoid copepods. Can. J. Zool. 49: 855–862.

Weaver, C. R., 1943. Observation on the life cycle of the fairy shrimp *Eubranchipus vernalis*. Ecology 24: 500–502.

Weslawski, J. M., 1989. Occurrence and life cycle of *Mysis oculata* (Crustacea) from Spitsbergen. Pol. Arch. Hydrobiol. 36: 207–215.

Westin, L. & R. Gydemo, 1986. Influence of light and temperature on reproduction and moulting frequency of the crayfish, *Astacus astacus* L. Aquaculture 52: 43–50.

Wiggins, G. B., R. J. Mackay & I. M. Smith, 1980. Evolutionary and ecological strategies of animals in annual temporary pools. Arch. f. Hydrobiol./Supp. 58: 97–206.

Williams-Howze, J. & B. C. Coull, 1992. Are temperature and photoperiod necessary cues for encystment in the marine benthic harpacticoid copepod *Heteropsyllus nunni* Coull? Biol. Bull. 182: 109–116.

Williamson, C. E., 1991. Copepoda. In: J. H. Thorp and A. P. Covich (eds), Ecology and classification of North American freshwater invertebrates. Academic, San Diego: 787–822.

Wolcott, T. G., 1988. Ecology. In: W. W. Burggren & B. R. McMahon (eds), Biology of the land crabs. Cambridge University Press, Cambridge: 55–96.

Wyngaard, G. A., 1988. Geographical variation in dormancy in a copepod: evidence from population crosses. Hydrobiologia 167/168 (Dev. Hydrobiol. 47): 367–374.

Wyngaard, G. A., B. E. Taylor & D. L. Mahoney, 1991. Emergence and dynamics of cyclopoid copepods in an unpredictable environment. Freshwat. Biol. 25: 219–232.

Appendix 1. List of the crustacean species used in this paper and their associated reproductive life spans, relative dispersal abilities, diapause durations, and habitats (see text for explanations). References are numbered and listed in the footnote.

Class Order Species	Reproductive life span (years)	Relative Dispersal	Duration of diapause (years)	Habitat	References
Branchiopoda					
Anomopoda					
Bosmina longirostris	0.1	2	15	Inland waters	1
Ceriodaphnia pulchella	0.1	2	15	Inland waters	1,2
Ceriodaphnia quadrangula	0.1	2	2	Inland waters	3
Ceriodaphnia reticulata	0.1	2	3	Inland waters	4
Chydorus sphaericus	0.1	2	2	Inland waters	3
Daphnia cucullata	0.1	2	50	Inland waters	5
Daphnia galeata	0.1	2	50	Inland waters	5,6
Daphnia laevis	0.1	2	2	Inland waters	7,8
Daphnia magna	0.1	2	2	Inland waters	9
Daphnia middendorffiana	0.1	2	2	Inland waters	3
Daphnia pulicaria	0.1	2	50	Inland waters	6
Daphnia pulex	0.1	2	2	Inland waters	10
Daphnia retrocurva	0.1	2	30	Inland waters	6
Eubosmina longispina	0.1	2	3	Inland waters	4
Macrothrix sp.	0.1	2	55	Inland waters	11,12
Scapholeberis sp.	0.1	2	2	Inland waters	7,8
Simocephalus exspinosus	0.1	2	2	Inland waters	7,8
Simocephalus serrulatus	0.1	2	2	Inland waters	7,8
Anostraca					
Artemia monica	0.2	2	1	Inland waters	13
Artemia salina	0.5	2	10	Inland waters	13
Branchinecta coloradensis	0.1	2	2	Inland waters	3
Eubranchipus vernalis	0.8	2	1	Inland waters	14
Streptocephalus probocideus	0.2	2	5	Inland waters	15
Streptocephalus seali	0.1	2	6	Inland waters	16
Streptocephalus sudanicus	0.2	2	5	Inland waters	14
Streptocephalus vitreus	0.2	2	1.5	Inland waters	17
Ctenopoda					
Holopedium gibberum	0.1	2	1	Inland waters	18
Haplopoda					
Leptodora kindti	0.2	2	1.5	Inland waters	19
Notostraca					
Triops longicaudatus	0.1	2	15	Inland waters	16,20
Onychopoda					
Bythotrephes longimanus	0.2	2	1.5	Inland waters	19
Podon polyphemoides	0.1	2.5	16	Marine	21

Appendix 1. Continued.

Class Order Species	Reproductive life span (years)	Relative Dispersal	Duration of diapause (years)	Habitat	References
Spinicaudata					
Caenestheriella australis	0.2	2	5	Inland waters	15
Caenestheriella gynecia	0.2	2	5	Inland waters	22
Eocyzicus klunzingeri	0.1	2	5	Inland waters	15
Eulimnadia africana	0.1	2	5	Inland waters	15
Leptestheria aegyptiaca	0.1	2	5	Inland waters	15
Limnadia stanleyana	0.1	2	1	Inland waters	16
Maxillopoda					
Calanoida					
Acartia bifilosa	0.2	2.5	8	Marine	75
Acartia californiensis	0.3	2.5	1	Marine	23
Acartia erythraea	0.3	2.5	1	Marine	23
Acartia hudsonica	0.2	2.5	10	Marine	21
Acartia tonsa	0.3	2.5	1	Marine	23
Aglaodiaptomus spatulocrenatus	0.2	2	3	Inland waters	24
Aglaodiaptomus stagnalis	0.2	2	3	Inland waters	7
Arctodiaptomus wierzejskii	0.2	2	3	Inland waters	25
Calanoides carinatus	0.1	3	0.3	Marine	76
Calanopia thompsoni	0.3	2.5	1	Marine	23
Calanus finmarchicus	1.5	3	0.7	Marine	26
Centropages abdominalis	0.3	2.5	1	Marine	23
Centropages hamatus	0.3	2.5	1	Marine	23
Centropages ponticus	0.3	2.5	1	Marine	23,27
Centropages yamadai	0.3	2.5	1	Marine	23
Diaptomus castor	0.2	2	3	Inland waters	25
Diaptomus cyaneus	0.2	2	3	Inland waters	25
Epilabidocera longipedata	0.3	2.5	1	Marine	23
Epischura nordenskilodi	0.2	2	3	Inland waters	24
Eudiaptomus vulgaris	0.2	2	3	Inland waters	25
Eurytemora affinis	0.2	2	40	Inland waters	21,28,75
Hemidiaptomus ingens	0.2	2	3	Inland waters	25
Labidocera aestiva	0.3	2.5	1	Marine	23
Labidocera bipinnata	0.3	2.5	1	Marine	23
Leptodiaptomus coloradensis	0.2	2	2	Inland waters	3
Leptodiaptomus minutus	0.1	2	20	Inland waters	29
Limnocalanus macrurus	0.5	1	0.25	Inland waters	30
Mixodiaptomus kupelwieseri	0.2	2	3	Inland waters	25
Mixodiaptomus laciniatus	0.2	2	3	Inland waters	25
Onychodiaptomus sanguineus	0.2	2	70	Inland waters	31
Pontella maedi	0.3	2.5	1	Marine	23
Pontella mediterranea	0.3	2.5	1	Marine	23
Skistodiaptomus oregonensis	0.2	2	20	Inland waters	29
Tortanus forcipatus	0.3	2.5	5	Marine	21

Appendix 1. Continued.

Class Order Species	Reproductive life span (years)	Relative Dispersal	Duration of diapause (years)	Habitat	References
Cyclopoida					
Acanthocyclops vernalis	0.1	1	3	Inland waters	32,33
Cyclops kolensis	0.1	1	0.8	Inland waters	34
Cyclops scutifer	1	1	0.75	Inland waters	35
Cyclops strenuus	1	1	0.5	Inland waters	36
Cyclops strenuus strenuus	1	1	0.8	Inland waters	37
Cyclops vicinus	0.1	1	0.7	Inland waters	38
Diacyclops navus	0.1	1	0.6	Inland waters	39
Diacyclops thomasi	0.1	1	0.6	Inland waters	40
Mesocyclops edax	0.5	1	0.6	Inland waters	41
Mesocyclops leuckarti	0.5	1	0.75	Inland waters	34,35
Thermocyclops oithonoides	0.2	1	0.8	Inland waters	34
Harpacticoida					
Canthocamptus arcticus	0.1	2	0.8	Inland waters	77
Canthocamptus staphylinoides	0.7	1	0.4	Inland waters	42
Canthocamptus staphylinus	0.7	1	0.4	Inland waters	42
Canthocamptus microstaphylinus	0.7	1	0.4	Inland waters	42
Heteropsyllus nunni	0.5	2	0.8	Marine	43
Podocopida					
Candona decora	0.5	2	0	Inland waters	44,45
Cypricercus reticulatus	0.1	2	10	Inland waters	45,46
Cypridopsis aculeata	0.5	2.5	1	Marine	44,45
Cyprinotus carobineusis	0.5	2	3	Inland waters	44,45
Cyprois marginata	0.5	2	3	Inland waters	44,45
Cypris sp.	0.5	2	24	Inland waters	44
Cytherura gibba	0.5	1	0	Marine	44,47
Cytherois arenicola	0.5	1	0	Marine	44,47
Cytherois fischeri	0.5	1	0	Marine	44,47
Elofsonia baltica	0.5	1	1	Marine	44,47
Eucypris fuscata	0.5	2	3	Inland waters	44,45
Herpetocypris reptans	0.5	2	3	Inland waters	44,45
Heterocypris incongurens	0.5	2.5	1	Marine	44,45
Heterocypris salinus	0.5	2.5	1	Marine	44,45
Notodromas monacha	0.5	2	3	Inland waters	44,45
Thoracica					
Balanus glandula	7	3	0	Marine	48
Chthamalus fiscus	3	3	0	Marine	48
Tetraclita squamosa	13	3	0	Marine	48
Malacostraca					
Amphipoda					

Appendix 1. Continued.

Class / Order / Species	Reproductive life span (years)	Relative Dispersal	Duration of diapause (years)	Habitat	References
Gammarus mucronatus	0.5	2.5	0	Marine	49
Gammarus palustris	0.5	2.5	0	Marine	49
Melita nitida	0.5	2.5	0	Marine	49
Anaspidacea					
Anaspides tasmaniae	3	1.5	0.6	Inland waters	83,84
Cumacea					
Diastylis goodsiri	7	1	0	Inland waters	50
Decapoda					
Acanthephyra pelagica	1	2.5	0	Marine	62
Acanthephyra quadrispinosa	1	2.5	0	Marine	62
Acetes japonicus	0.5	2.5	0	Marine	62
Astacus astacus	4	1.5	0.5	Inland waters	51,78,79
Cardisoma guanhumi	16	1.5	0	Terrestrial	52,53
Emerita analoga	2	3	0	Marine	54
Homarus americanus	40	2.5	0	Marine	55
Lucifer chacei	1	2.5	0	Marine	62
Nephrops norvegicus	7	3	0	Marine	56,57
Orconectes palmeri	2	1.5	0	Inland waters	58
Orconectes rusticus	6	2.5	0	Inland waters	59
Orconectes virilis	2	1.5	0.3	Inland waters	59,80
Panaeus merguiesis	1	3	0	Marine	63
Penulirus cygnus	20	3	0	Marine	60
Paralithodes camtschatica	15	2.5	0	Marine	61
Paralithodes platypus	7	2.5	0	Marine	61
Sergestes similis	1	2.5	0	Marine	62
Sergia lucens	1	2.5	0	Marine	62
Euphausiacea					
Euphausia pacifica	1	3	0	Marine	64
Euphausia superba	1	3	0	Marine	64
Euphausia triacantha	2	3	0	Marine	64
Meganyctiphanes norvegica	2	3	0	Marine	64
Thysanoessa longicaudata	2	3	0	Marine	64
Thysanoessa longipes	2	3	0	Marine	64
Thysanoessa raschii	2	3	0	Marine	64
Thysanoessa spineffera	1	3	0	Marine	64
Thysaropoda acutifrons	1	3	0	Marine	64
Isopoda					
Armadillidium vulgare	3	1	0.3	Terrestrial	65,81
Asellus aquaticus	0.8	1	0.5	Inland waters	65,82
Asellus intermedius	0.8	1	0	Inland waters	65

Appendix 1. Continued.

Class Order Species	Reproductive life span (years)	Relative Dispersal	Duration of diapause (years)	Habitat	References
Asellus meridianus	0.8	1	0	Inland waters	65
Asellus tomalensis	0.1	1	0	Inland waters	65
Cylisticus convexus	1	1	0	Terrestrial	65
Dynamene bidentata	0.1	2	0	Marine	65
Idotes emarginata	0.3	2	0	Marine	65
Jaera albifrons	0.5	2	0	Marine	65
Jaera forsmani	0.5	2	0	Marine	65
Jaera ishiosetosa	0.5	2	0	Marine	65
Jaera praehisuta	0.5	2	0	Marine	65
Ligia oceanica	1.5	2	0	Marine	65
Ligia pallasii	1	2	0	Marine	65
Ligidium japonica	1	1	0	Terrestrial	65
Mancasellus macrourus	0.5	1	0	Inland waters	65
Oniscus asellus	1.5	1	0	Terrestrial	65
Porcellio scaber	1.5	1	0	Terrestrial	65
Triconiscus pusillus	2	1	0	Terrestrial	65
Mysida					
Alienacanthomysis macropsis	0.5	1	0	Inland waters	66
Gnathophausia ingens	1	3	0	Marine	50
Mysis oculata	0.5	3	0	Marine	67
Mysis polaris	0.5	3	0	Marine	68
Mysis relicta	1	1	0	Inland waters	69,70
Neomysis intermedia	0.5	1	0	Inland waters	71
Neomysis mercedes	0.5	1	0	Inland waters	66
Paramysis lacustris	0.6	1	0	Inland waters	70
Stomatopoda					
Gonodactylus chiragra	54	2	0	Marine	72
Gonodactylus falcatus	42	2	0	Marine	72
Gonodactylus graphurus	49	2	0	Marine	72

*1: Moritz, 1988. 2: Moritz, 1987. 3: A. J. Bohonak, pers. comm. 4: De Stasio, 1990. 5: Carvalho & Wolf, 1989. 6: C. E. Cáceres, unpublished data. 7: Taylor *et al.*, 1990. 8: Taylor & Mahoney, 1990. 9: Korpelainen, 1986. 10: Edmondson, 1955. 11: Moghraby, 1977. 12: Fryer, 1972. 13: Dana, 1984. 14: Weaver, 1943. 15: Brendonck *et al.*, 1993. 16: Hartland-Rowe, 1972. 17: Hildrew, 1985. 18: Lampert & Krause, 1976. 19: Andrew & Herzig, 1984. 20: Schmitt, 1965. 21: Marcus *et al.*, 1994. 22: Mattox & Verlardo, 1950. 23: Grice and Marcus, 1981. 24: N. G. Hairston, Jr., unpublished data. 25: Champeau, 1970. 26: Miller and Tande, 1993. 27: Sazhina, 1987. 28: Ban, 1992. 29: Hairston and Van Brunt, 1994. 30: Roff, 1972. 31: Hairston *et al.*, 1995b. 32: Wyngaard *et al.*, 1991. 33: Rylov, 1963. 34: Sarvala, 1979a. 35: Elgmork *et al.*, 1990. 36: Naess & Nilssen, 1991. 37: Elgmork, 1955. 38: Maier, 1989. 39: Watson & Smallman, 1971. 40: A. M. Hansen, unpublished data. 41: Wyngaard, 1988. 42: Sarvala, 1979b. 43: Williams-Howze & Coull, 1992. 44: Cohen & Morin, 1990. 45: Angell & Hancock, 1989. 46: Delorme, 1991. 47: Theisen, 1966. 48: Hines, 1979. 49: Borowsky, 1980. 50: Schram, 1986. 51 : Cukerzis, 1970. 52: Wolcott, 1988. 53: Ng, 1988. 54: Dugan *et al.*, 1991. 55: Cooper & Uzmann, 1977. 56: Farmer, 1973. 57: Castro, 1992. 58: Payne & Price, 1983. 59: Momot, 1986. 60: Pearce & Phillips, 1994. 61: Otto, 1986. 62: Omori, 1971. 63: Rothlisberg & Church, 1994. 64: Mauchline and Fisher, 1969. 65: Carefoot, 1973. 66: Daly & Damkaer, 1986. 67: Weslawski, 1989. 68: Kulikov, 1980. 69: Johannsson, 1992. 70: Khmeleva & Baichorov, 1987. 71: Toda *et al.*, 1982. 72: Reaka, 1979. 73: McCaughran & Powell, 1977. 74: Momot, 1992. 75: Katajisto, 1995. 76: Arashkevich *et al.*, 1994. 77: Borutzky, 1929. 78: Westin & Gydemo, 1986. 79: Alekseev, 1989. 80: Aiken, 1969. 81: Juchault *et al.*, 1982. 82: Tandini-Vitagliano *et al.*, 1982. 83: Swain & Reid, 1983. 84: Hickman, 1937.

Hydrobiologia **320**: 45–54, 1996.
V. R. Alekseev & G. Fryer (eds), Diapause in the Crustacea.

Egg diapause and clonal structure in parthenogenetic populations of *Heterocypris incongruens* (Ostracoda)

V. Rossi, A. Gandolfi & P. Menozzi
Instituto di Ecologia, Universita' di Parma, Viale delle Sienze, 43100 Parma, Italy

Key words: diapausing egg, electrophoretic genotypes, seasonal cycle, ephemeral habitats

Abstract

A study of the clonal structure of parthenogenetic populations of *Heterocypris incongruens* from rice-fields in Northern Italy carried out over two-year period is summarized. Significantly different levels of genetic polymorphism were found among populations. The coexistence of different electrophoretic clones and similar patterns of clonal seasonal succession have been observed in at least two different rice fields. The clone or group of clones present in fall, winter and early spring is substituted by other multilocus genotypes in late spring and summer. Different egg diapause induction mechanisms drive the clonal substitution in different clones. Here we report a new laboratory experiment designed to test the effect of temperature and photoperiod found in winter (12 °C 8:16 L:D), spring (24 °C 12:12 L:D) and summer (28 °C 16:8 L:D) conditions on the deposition and hatching of diapausal eggs in different multilocus genotypes. Clones respond in a way compatible with their sustained presence in the field in different seasons.

Introduction

Many animals living in permanent or ephemeral habitats produce diapausal stages (e.g. cyclopoids and bdelloid rotifers) or dormant eggs (e.g. calanoids, cladocerans, ostracods) that can survive for a long period in otherwise life-prohibiting conditions (Hutchinson, 1967; Wiggins *et al.*, 1980; Hairston & De Stasio, 1988; Angell & Hancock, 1989; De Stasio, 1990; Marcus *et al.*, 1994). Tauber & Tauber (1981) in their vast review of literature on insect diapause underline the different meaning of dormancy (a term with both ecological and physiological implications that involves overt expressions that can be studied by observation only) and diapause (a term with a physiological connotation; a hormonally mediated state that can be studied by a physiological approach). To our knowledge, no study of the physiological mediators implied in the setting of diapause in small crustaceans has been carried out. Given the ecological nature of this study, the two terms will be used interchangeably.

The importance of diapause in determining the temporal distribution of small crustaceans and the similarity of its ecological role to that of seed banks in plants has been stressed (De Stasio, 1989). Seed banks are widely recognised as reservoirs of potentially large amounts of biological diversity and genetic variability in plant populations (Silvertown, 1987). Hairston & De Stasio (1988) underlined the role of dormant propagules in modulating the rate of genetic change since they remove a fraction of the gene pool from the influence of micro evolutionary forces in each generation.

Few studies exist on the genetic variability of dormant eggs of crustaceans and its relationship to the genetic structure of adult populations (Mort, 1991). A large interpopulation variability for timing of diapause has been found in *Diaptomus sanguineus* as well as a substantial heritability of traits related to the production and timing of diapausing eggs (Hairston & Dillon, 1990). Hairston & Olds (1987) suggested that *D. sanguineus* is locally adapted to specific ponds, although genetic variation in sensitivity to environmental cues between individuals from different habitats exists. In Cladocera, ephippium production is proximally determined by environmental cues. There is evidence, however that the response is, at least in part, under genetic control. Loaring & Hebert (1981) described thermal differences in ephippia production of *D. pulex* clones and suggested that the fitness of a clone depends on the number of ephippial eggs produced. In seven *Daphnia* species and clones of *D. pulex* and *D. obtusa* Schwartz & Hebert (1987) reported variation in resting egg activation factors such as length and temperature of dark incubation and hatching temperature. Korpelainen (1987) described differences in reproductive characteristics among *Daphnia magna* genotypes in natural populations. For different clones of the same species, evidence of genetic differences in ephippium production (Yampolsky, 1992) and in photoperiod-induced resting egg production (Carvalho & Hughes, 1983) have been reported. Larsson (1991) found variability among *D. pulex* clones in resting egg production and in sensitivity to ephippia-inducing stimuli (i.e. food levels).

According to De Stasio (1989) the duration of dormancy should reflect a trade-off between gains in short term fitness due to immediate reproductive success and long term survival in the diapausal stage. In particular, differential hatching of genetically different individuals could create a sort of resource partitioning in time and allow the coexistence and the succession of different clones in parthenogenetic species. Here we review our previous work on differences among clones of the freshwater ostracod *Heterocypris incongruens* in diapausing egg production and hatching in different environmental conditions and report new data on the relationship between clonal structure and egg diapause strategies in this species.

Material and methods

The organism: reproductive strategies of Heterocypris incongruens clones

Clones (defined as multilocus genotypes) were identified using starch gel electrophoresis of enzymes carried out using homogenates of individuals.

The natural populations of *Heterocypris incongruens* that we have been studying for almost a decade in Northern Italy reproduce parthenogenetically (Rossi *et al.*, 1993). In repeated tests over the same time span, the comparison of electrophoretic patterns in subsequent generations produced no evidence of recombination. No difference, in this respect, was found in individuals born from subitaneous or resting eggs.

During its life time the same individual can produce diapausing eggs and eggs that hatch within a few days. The two types of eggs do not present any apparent structural difference and can be distinguished only *a posteriori* by their hatching time. Angell & Hancock (1989) classify the eggs that hatch within 10 days of laying as subitaneous. In our experience the proportion of resting eggs depends on genotype, temperature and photoperiod (Rossi *et al.*, 1991; Rozzi *et al.*, 1991).

Studying a total of 850 individuals, using 3 enzyme loci (GPI, MPI and PGM) over a 3 year period, we described the clonal structure of a parthenogenetic population of *H. incongruens* in La Bigliana, a rice-field located in the Po plain in Northern Italy (Rossi & Menozzi, 1990). A total of 5 clones were found. We reported the seasonal cycle and the thermal tolerance of three of these multilocus genotypes. A winter clone (W), present in the fall, winter and early spring, is replaced by a summer clone (S) from the late spring onwards. An experiment was carried out to test the effect of temperature and photoperiod on different clones. The relatively rare clone I, that showed no clear seasonal trend in nature, reached the highest densities in laboratory conditions. Its resting egg production did not respond to temperature or photoperiod. The percentage of dormant eggs increases with temperature and daylight duration in the winter clone W and decreases in the summer one S (Rossi *et al.*, 1991; Rozzi *et al.*, 1991). Hatching decreases at higher temperature (but is not affected by photoperiod) in clone W and increases with temperature and daylight duration in clone S. Temperature and photoperiod seem the most relevant environmental factors driving the observed clonal substitution of clones W and S in the field.

In laboratory experiments differences among clones have been detected in resting egg activation. The highest hatching percentage was obtained by keeping resting eggs in water, at low temperature (4 ° C) in the case of clone S and in water but at high temperature (32 °C) in the case of clone W (Rossi *et al.*, 1991; Rozzi *et al.*, 1991).

The study from October 1990 to July 1992 of *H. incongruens* in S. Cristina rice-field, 150 Km west of La Bigliana, confirmed the picture found in our previous study and added new interesting details to the complex life-history strategy of the different clones of this species. A less intensive sampling effort revealed a much richer clonal structure: from the analysis of 608 individuals collected in the field in 14 sampling dates a total of 35 different multilocus genotypes were described (Rossi & Menozzi, 1994). Out of 35 clones,

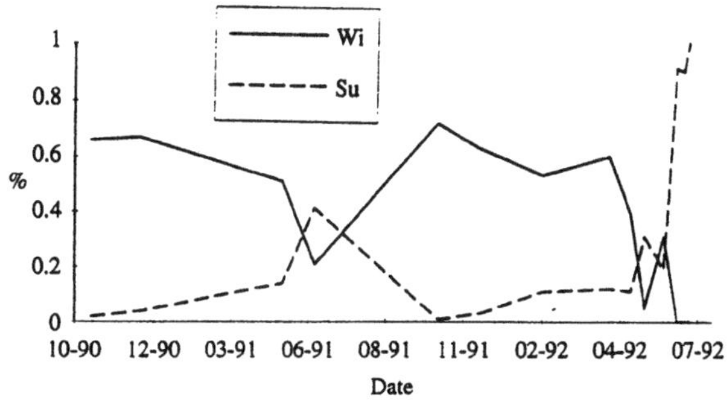

Fig. 1. Seasonal distribution of Wi and Su indexes which summarise the frequencies of winter-like and summer-like clones in S. Cristina as reported by Rossi & Menozzi (1994).

21 were observed in 1991 and 1992. The other 14, 7 observed in 1991 and 7 in 1992, were represented, by only few individuals. Clones can be grouped by their genetic similarity with clones S and W from La Bigliana. Clones similar to the winter clone W are present in relevant numbers in winter and early spring. Clones similar to the summer clone S (fewer in number) are present in late spring or summer. An index summarising the frequency of clones with Summer-like (Su) and Winter-like (Wi) overall genetic characteristics show a clear seasonal cycle (Fig. 1).

A new experiment to investigate the relationship between genetic make-up, diapausing egg production and hatching strategy

Six multilocus genotypes representative of the spectrum of genetic variation found in S. Cristina were used. In Fig. 2 the zymograms of the 6 clones are reported along with their relative abundance observed in the field. Using their abundance in different seasons and their genetic similarity to the typical winter and summer clones (W and S) from la Bigliana we can characterise the clones as follows:

– Clones W1 and W2 are not found in samples collected in the summer. They can be considered typical winter and spring clones. Clone W2 is electrophoretically identical to clone W found in La Bigliana. Clone W1 is similar to clone W (83% alleles in common) and is homozygous for all the markers considered; W2 is heterozygous for PGM.

– Clone I1 is genetically not very similar to clone W from La Bigliana (67% alleles in common). Its abundance in different seasons is similar to clones W1 and W2. It has been labelled as intermediate from the results of the experiment here reported.

– Clone I2, found in samples from any season, with the exception of the collections made in late June and July 1992, can be considered more of a generalist in terms of seasonality. In samples taken in May and December 1991 and in February and May 1992 clone I2 was the most frequent multilocus genotype. The clone is intermediate also in terms of genetic make-up as it shares all alleles at PGM and GPI loci with the winter clones and 50% of its alleles at MPI locus with summer clones.

– Clones S1 and S2 can be considered typical summer clones. They were found in late spring and summer only. In 1992 S1 abundance shows two peaks indicative of more than one generation. Clone S2 was found over a shorter time interval in both years, a fact suggesting a univoltine life cycle. Clone S1 is electrophoretically identical to clone S from La Bigliana and shares 67% of its alleles with clone S2. The two clones share no allele with the winter multilocus genotypes.

To evaluate the role of different seasonal conditions on diapausing egg production and hatching by different multilocus genotypes we set up 3 experimental situations simulating temperature and photoperiod at the winter solstice (12 °C and 8:16 L:D), spring equinox (24 °C and 12:12 L:D) and summer solstice (28 °C and 16:8 L:D) at intermediate latitudes.

Twelve neonates for each of the six clones and for each of the three experimental conditions were reared in a separate well of tissue culture plates using Evian spring water as a medium and an ad libitum diet of the mat-forming alga *Tolypothrix tenuis* (Cyanobacteria) as food.

Using a stereomicroscope, neonates were counted and removed every day and hatching time was recorded. Experimental observations were carried out for 6 months, a period encompassing two seasons. In field conditions this is a time long enough to hatch even for eggs meant to diapause through unfavourable conditions. Bimodality in hatching time distribution allowed a distinction between subitaneous and diapausing eggs. At the end of the experiment, eggs that did not hatch were counted using a microscope. The number of days needed for 50% of non diapausing eggs to hatch was used as an estimate of egg mean developmental time. For an actual measurement, eggs should have been removed every day and maintained in separate wells until hatched. We knew that eggs are laid in successive days next to previously laid eggs and that egg manipulation can affect hatching (see Results). The

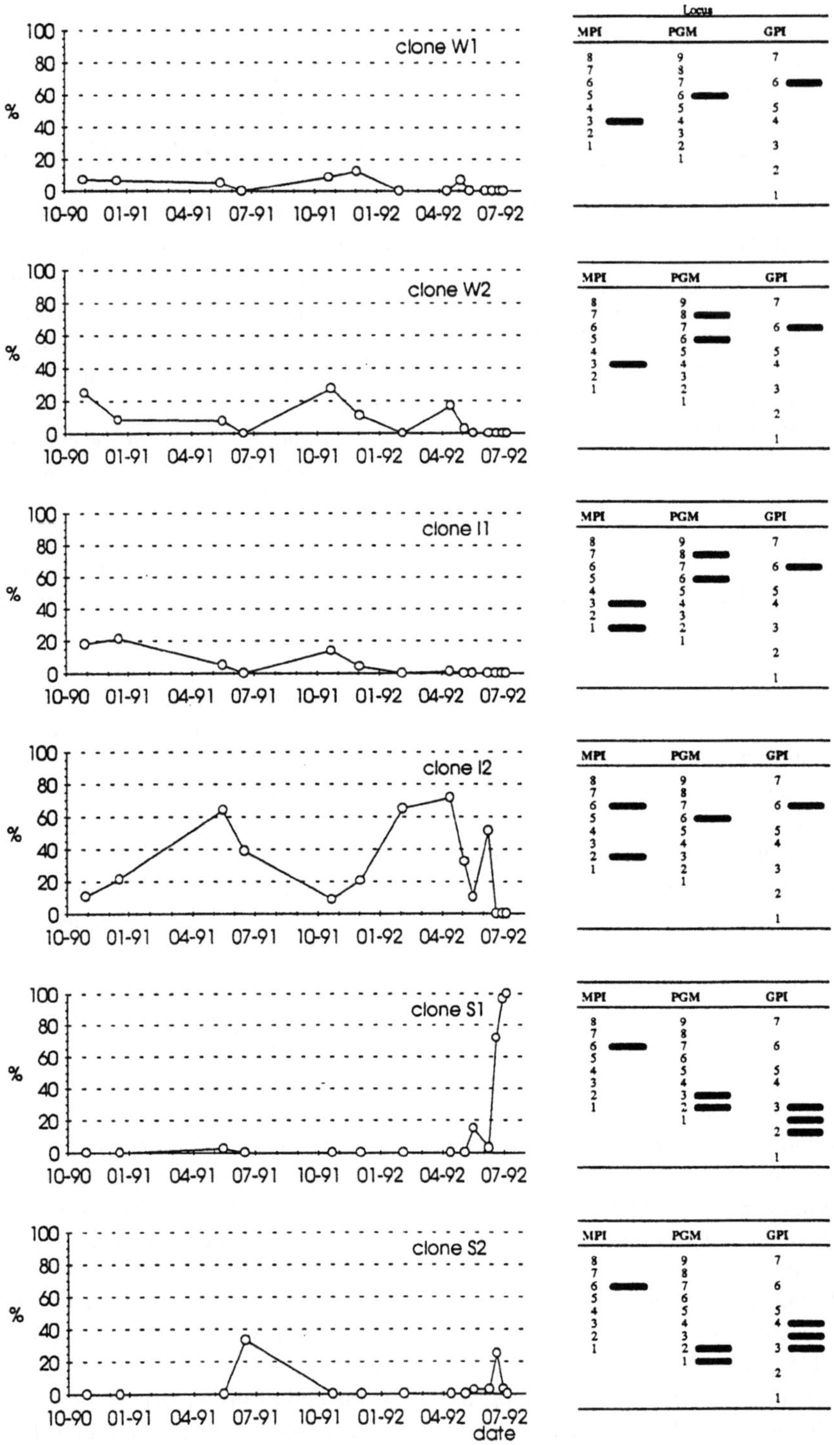

Fig. 2. Seasonal distribution of relative abundance and zymograms for 3 loci (MPI, PGM and GPI) of each clone used in the experiment.

removal could have affected the deposition rate and hatching time, an undesirable side effect in an experiment aimed at explaining events observed in nature. To improve estimates used in the computations we subtracted from the number of days to hatching the age at first deposition of each mother. Days to 50% hatching seemed way of reporting the dormancy of eggs in a quantitative manner. A qualitative definition was

Table 1. Number of females that laid eggs, total number of eggs produced, number of eggs hatched, percentage of resting eggs and days to 50% hatching at 12 °C 8:16 L:D. Due to the small number of replicates statistical tests were not performed.

Clone	N	Eggs	Hatched	%	Days
W1	1	200	13	93	26
W2	0	0	0	0	–
I1	2	84	11	87	79
I2	3	108	26	76	51
S1	0	0	0	0	–
S2	0	0	0	0	–

Table 2. Number of females that laid eggs, total number of eggs produced, number of eggs hatched, percentage of resting eggs and days to 50% hatching at 24 °C 12:12 L:D. Letters indicate statistically different Scheffe pair-wise comparison test groupings.

Clone	N	Eggs	Hatched	%	Days
W1	12	810	520	36 B	104 B
W2	11	1211	754	38 B	14 A
I1	11	576	574	0 A	8 A
I2	10	931	921	1 A	19 A
S1	12	2531	2504	1 A	13 A
S2	6	489	44	91 C	25 A

afforded by considering the orange coloured eggs (an indication of viability in our experience) that had not hatched by the end of the experiment, to be in a state of diapause.

Chi square tests were computed to evaluate differences in percentage of diapausing egg production among clones and experimental conditions. One way Analysis of Variance and Scheffe pair-wise comparison test between pairs of means were used to compare times to 50% of egg hatching. All computations were carried out using the SAS statistical software (SAS, 1985).

Results

The experimental conditions simulating winter solstice (12 °C 8:16 L:D) turned out to be close to the tolerance limits for all clones (Table 1). Only 1 individual of clone W1, 2 of clone I1 and 3 of clone I2 laid

Table 3. Number of females that laid eggs, total number of eggs produced, number of eggs hatched, percentage of resting eggs and days to 50% hatching at 28 °C 16:8 L:D. Letters indicate statistically different Scheffe pair-wise comparison test groupings.

Clone	N	Eggs	Hatched	%	Days
W1	10	665	75	89 D	80 AB
W2	12	880	734	17 B	112 B
I1	12	1169	1166	0 C	53 A
I2	9	224	221	1 A	43 A
S1	12	2104	2035	3 A	8 C
S2	3	121	9	93 D	34 A

eggs. The low temperature slowed down the life cycle of all clones. It seems impossible to distinguish the effect of a slower metabolism from a true diapausing egg production strategy. The very high proportion of diapausing eggs is not comparable with the percentages observed in other experimental conditions. The evidence from this experiment will be used only qualitatively for evaluating diapausing egg production and hatching of these clones.

Figure 3 shows the distribution of hatching time (measured in days since the beginning of the experiment) of different clones at spring and summer conditions. In each frequency distribution arrows indicate the mean age at first deposition and mean age at death computed for mothers that laid eggs. With the exception of clone I1 at 28 °C 16:8 L:D, the observed distributions seem unimodal although we often observe a low frequency of hatching far away from the central values. On the time scale of the experiment the variation in hatching time due to the unmeasured variation in deposition time (roughly estimated by the time interval between the arrows indicating average date at first deposition and death) does not blur the picture of a rather restricted time of hatching for different clones at different conditions. In clone W1 at 24 °C 12:12 L:D and in clone W2 at 28 °C 16:8 L:D the late hatching observed occurred after the eggs were manipulated for counting close to the end of the experiment.

Tables 2 and 3 report data on egg production and hatching for different clones at conditions simulating spring equinox (24 °C 12:12 L:D) and summer solstice (28 °C 16:8 L:D).

In spring conditions clones seem to show essentially 3 strategies in diapausing egg production and hatching as suggested by the statistical significance of

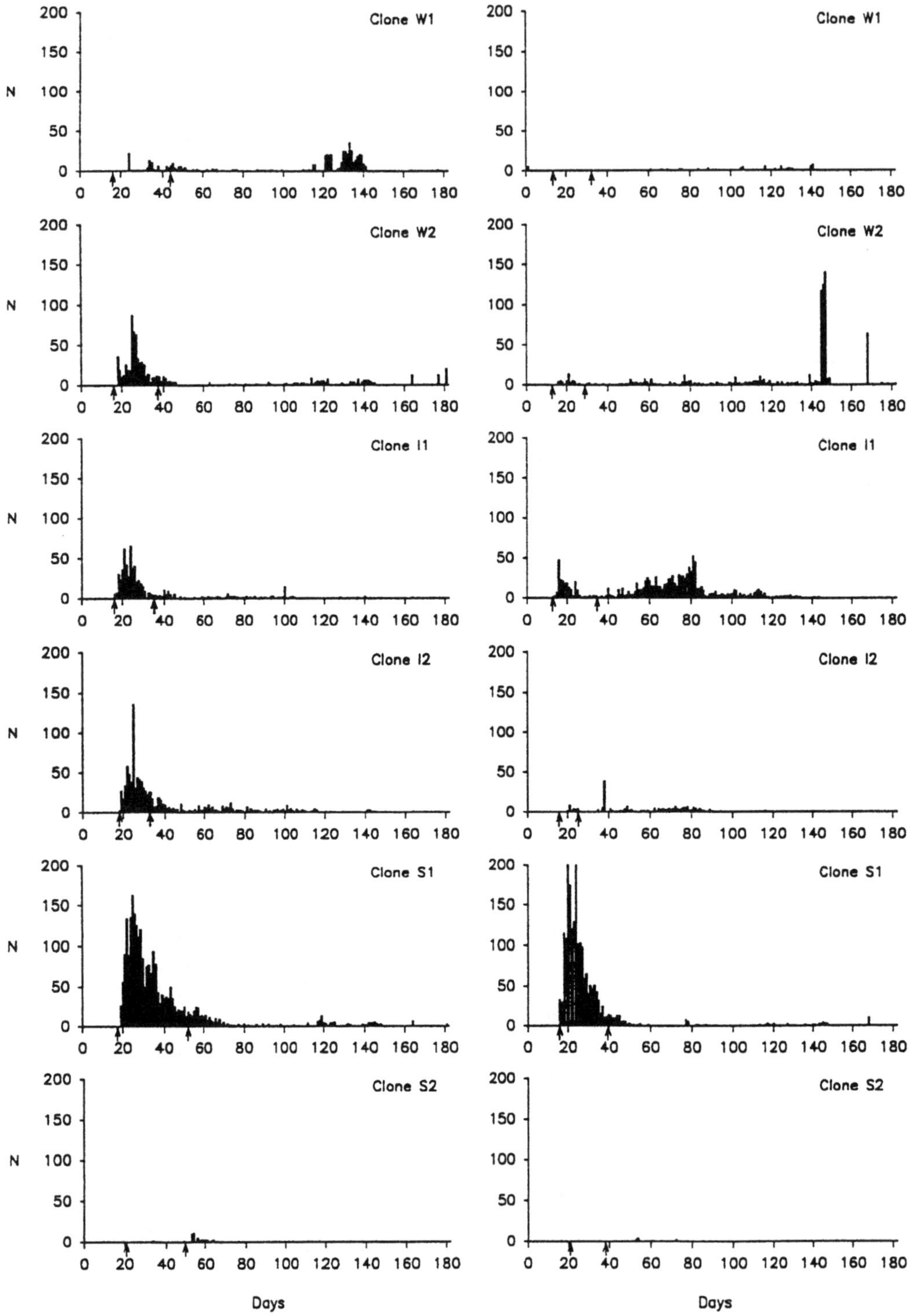

Fig. 3. The distributions of the number of hatchings for each clone at 24 °C 12:12 L:D, on the left, and 28 °C 16:8 L:D, on the right, are reported. Days are measured since the beginning of the experiment. The arrows indicate mothers' mean age at first deposition and at death.

pair-wise comparisons (Table 2). Almost all eggs laid by clone S2 were diapausing. At the other extreme, clone I1, I2 and S1 produced almost no diapausing eggs with most eggs hatching within a short time. An intermediate strategy is shown by clones W1 and W2 that produced a little more than a third of diapausing eggs. This tendency to a certain level of dormancy is more clear in clone W1 whose eggs took a much longer time (104 days) to 50% hatching than clone W2 (14 days to 50% hatching).

In summer conditions the picture is qualitatively similar to that just described, although a less clear classification of the clones' diapausal strategies emerges by the analysis of pair-wise comparisons with a larger number of groups of non significantly different pairs. Clone S2 produced almost all resting eggs: only 7% of the eggs produced by only 3 females hatched after a relatively long period (33 days for 50% hatching). Clone S1 still represents the other extreme with almost no resting eggs laid and a short 50% time to hatching. Almost all eggs produced by clones I1 and I2 hatched, although in a significantly longer time than clone S1. Clones W1 and W2 clearly show a tendency to produce eggs that do not hatch: clone W1 with a high proportion of diapausing eggs (89%) and clone W2 with a delayed 50% hatching time (112 days).

Comparing the results obtained in spring and summer conditions we can gain further insight into the diapausal strategies of different clones. With the exception of clone I1, we found a significant decrease in total egg production in all clones, a possible indication of a general worsening of environmental conditions in this season for the species as a whole.

Some clones maintain essentially the same strategy under spring and summer conditions. Clone S2 lays resting eggs only. Other clones modify their strategies in quantitative terms only, although the observed differences are statistically significant. Clone S1 produces almost no resting eggs under both conditions, although a significant increase in resting eggs ($\chi^2 = 27.7$, $p = 0.0001$) and a significant decrease in time to 50% hatching ($F = 7.6$, $p = 0.0100$) is observed in summer conditions. Clones I1 and 12 do not change their proportion of resting eggs but significantly lengthen their time to 50% hatching at 28 °C 16:8 L:D ($F = 96.6$, $p = 0.0001$ and $F = 10.9$, $p = 0.0040$ respectively). Some clones show clear changes in their reproductive patterns. Clone W1 and clone W2 switch their reproductive allocation to the production of resting eggs by more than doubling the proportion of eggs that do not hatch in clone W1 and increasing one order of magnitude the time to 50% hatching in clone W2.

Discussion

The use of hatching time distribution allowed an objective definition and quantification of resting egg production. It is interesting that the description of diapausing egg production and hatching would be quite different if we had used a fixed time to distinguish between subitaneous and resting eggs. If, for example, we consider as resting the eggs that do not hatch in 20 days (a time double that suggested in the literature to allow for the variability of our estimates (Angell & Hancock, 1989), at 28 °C 16:8 L:D most of the eggs for all clones but S1 should be considered diapausing. While for clone S1 90% of non dormant eggs hatch in 20 days, only 5%, 6%, 13% and 17% do so for clones W1, W2, I2 and I1 respectively.

We did not carry out systematic observations of the possible effect of manipulation as a stimulus for hatching. The surprisingly high number of neonates observed at the end of the experiment for clone W2 at summer conditions and for clone W1 at spring conditions after removal of the algal mat covering clusters of eggs and their manipulation for counting, suggests a role of bioturbation in the hatching of resting eggs in field conditions.

Our previous work in field and laboratory and the results of the experiment on the 6 clones here reported clearly indicate that the clonal structure of *H. incongruens* is made up of essentially 3 groups of clones that can be classified as Winter–Spring, Intermediate and Summer multi-locus genotypes. A key role in the maintenance of the clonal structure observed in the field is played by diapausing egg production and hatching strategies developed by the different clones to respond to the alteration of their preferred condition produced by the seasonal cycle. The experiment here reported increased our understanding of these strategies for the six clones representative of the multi-locus genotypes found in the field. The complexity of the strategies developed by different clones is revealed by considering the clones' classification in the 3 proposed categories that is reached by combining evidence from all experimental conditions and different life history traits.

In winter conditions, clones W1, W2, I1 and I2 all seem to belong to the same group: they grow, moult and, except for clone W2, reach sexual maturity. In spring and summer conditions, clones W1 and W2 produce a significant percentage of diapausing eggs, while clones I1 and I2 lay almost no resting eggs: this clearly separates the first two as winter spring multi-locus genotypes and points to the intermediate nature of the other two clones. This distinction is confirmed by differences in time to 50% hatching, particularly clear in summer conditions. The summer clones (S 1 and S2) can be qualitatively distinguished from the intermediate ones by their inability to grow in winter conditions. A set of life strategy traits that allows the species a fine tuning of its adaptation to a variety of environmental conditions seems associated with different multi-locus genotypes. This seems confirmed by the differences, quite strong in some cases, that exist between clones classified in one of the 3 groups. One of the 2 summer clones, S2 produces almost all diapausing eggs suggesting a univoltine lite cycle, in contrast to the low percentage produced by clone S1. One of the 2 winter clones, W1 responds in a much clearer way to the approaching of unfavourable conditions represented by the summer conditions of our experiment by switching to producing a very high proportion of diapausing eggs.

In an adaptational perspective such diversity in life history strategies is not surprising given the unpredictability of the ephemeral habitats where the species lives, to which is to be added the variability due to the seasonal cycle. In future work it might be interesting to investigate a possible relationship between time to 50% hatching of a clone and length of minimal conditions for its survival and reproduction in the field. Very useful for this kind of investigation would be the theoretical work developed to predict the optimal timing of onset and duration of resting phases in different organisms from plant seed banks (Cohen, 1966, 1967) to insects (Dingle, 1978) to zooplankton (Hutchinson, 1967; Stross, 1966). A vast amount of observational and experimental information has accumulated on the induction of diapause or dormancy by photoperiod and temperature in insects, vertebrates, plants and crustaceans (reviewed in Hairston & Mumms, 1984).

Relatively few studies exist on the relationship between egg diapause and clonal dynamics in small crustaceans populations, although many investigations on the change of clonal structure in time have been carried out (Mort & Wolf, 1986; Carvalho & Crisp, 1987; Jacobs, 1990; Muller & Seitz, 1994). The relationship of diapause and the genetic structure of *Daphnia magna* has been studied revealing large effects but no definite pattern (Korpelainen, 1986). Genotypic variation accounts for differences in hatching time and in minimum dormancy period in different *Daphnia* species but evidence was not considered sufficient to explain the genetic composition of the different species (Carvalho & Wolf, 1989; Wolf & Carvalho, 1989). The comparative study of the response to short-term cues that allow the tracking of the environment revealed differences in genetically different clones of *D. magna* (Carvalho & Hughes, 1983).

Our multi-annual study provides a satisfactory explanation of the variation in time of genotype frequencies observed in field in populations of *H. incongruens* living in the Po plain. Photoperiod and temperature regulate diapause differently in different clones of *H. incongruens*. When cues of unfavourable conditions for a multilocus genotype are perceived, diapausing eggs are produced and added to the diapausing pool. The length of eggs' diapause affects the genetic structure of the population. We observed a regular seasonal succession of specialised clones (Winter–Spring and Summer clones) that respond strongly to seasonal environmental cues inducing resting egg production and hatching. Generalist clones (Intermediate clones) respond less clearly to seasonal cues or perhaps are sensitive to other environmental factors such as wet-dry cycles, changes in water chemistry, crowding etc. This is a possible explanation of their appearance in different proportions in different years. Clone I1 and clone I2, for instance, probably produce resting eggs only in response to high population density: we observed no resting egg production in the experiment here reported in which each individual is maintained in a separate well, while we have obtained them in clonal laboratory mass cultures in which density can be very high (Rossi & Menozzi, 1990).

The resting egg pool consists of genotypes which are favoured in different seasons. If the eggs hatch over several years they will provide a homeostatic mechanism for the clonal structure and ultimately for species permanence. Of special interest will be the study of factors controlling the duration of dormancy of diapausing eggs produced by clone S2 which seems to have developed a univoltine life cycle. It seems important to emphasize that an obligate parthenogenetic population relies on clonal differentiation for maintaining differences in life strategy traits which are under genetic control. A situation quite different from cyclical parthenogens in which regular recombination and segregation of variants occurs.

The experiment here reported, our previous laboratory work and field observations seem to indicate as optimal for the community of clones that make up the species populations in the Po plain (Italy) the climate conditions usually found in spring and very early summer. Even clones that show clear signs of adaptation to winter and summer conditions seem to perform better at more intermediate situations. This poses the interesting hypothesis that the species, in spite of its subdivision into clones with distinct ecological requirements, still maintains its ecological optima as a species. Studying the ecological requirements of clones of this cosmopolitan species from localities at different latitudes and longitudes will allow a test of this hypothesis.

Acknowledgments

We would like to thank Dr Fryer and an anonymous referee for helpful comments on a previous version of the manuscript. This research was partially supported by the EU Network 'Evolutionary ecology of reproductive modes in non marine Ostracodes' of the Human Capital and Mobility Program (contract no. CHRX CT93 0253) and by a grant from the Italian Ministero Pubblica Istruzione.

References

Angell, R. W. & J. W. Hancock, 1989. Response of eggs of *Heterocypris incongruens* (Ostracoda) to experimental stress. J. Crust. Bio. 9: 381–386.

Carvalho, G. R. & D. J.Crisp, 1987. The clonal ecology of *Daphnia magna* (Crustacea: Cladocera) I. Temporal changes in the clonal structure of a natural population. J. anim. Ecol. 56: 453–468.

Carvalho, G. R. & H. G. Wolf, 1989. Resting eggs of lake-*Daphnia* I. Distribution, abundance and hatching of eggs collected from various depths in lake sediments. Freshwat. Biol. 22: 459–470.

Carvalho, G. R. & R. N. Hughes, 1983. The effect of food availability, female culture-density and photoperiod on ephippia production in *Daphnia magna* Straus (Crustacea: Cladocera). Freshwat. Biol. 13: 37–46.

Cohen, D., 1966. Optimizing reproduction in a randomly varying environment. J. theor. Biol. 12: 119–129.

Cohen, D., 1967. Optimizing reproduction in a randomly varying environment when a correlation may exist between the conditions at the time a choise has to be made and the subsequent outcome. J. theor. Biol. 16: 1–14.

De Stasio, B. T., Jr, 1989. The seed bank of a freshwater crustacean: copepodology for the plant ecologist. Ecology 70: 1377–1389.

De Stasio, B. T., Jr, 1990. The role of dormancy and emergence patterns in the dynamics of freshwater zooplankton community. Limnol. Oceanogr. 35: 1079–1090.

Dingle, H., 1978. Evolution of insect migration and diapause. Springer, NY.

Hairston, N. G. Jr & B. T. De Stasio Jr, 1988. Rate of evolution slowed by dormant propagule pool. Nature 336: 239–242.

Hairston, N. G. Jr & W. R. Munns Jr, 1984. The timing of copepod diapause as an evolutionary stable strategy. Am. Nat. 123: 733–751.

Hairston, N. G. Jr & E. J. Olds, 1987. Population differences in the timing of diapause: a test of hypotheses. Oecologia 71: 339–344.

Hairston, N. G. Jr & T. A. Dillon, 1990. Fluctuating selection and response in a population of freshwater copepods. Evolution 44: 1796–1805.

Hutchinson, G. E., 1967. A treatise on limnology. Vol. 2. Wiley.

Jacobs, J., 1990. Microevolution in predominantly clonal populations of pelagic *Daphnia* (Crustacea: Phyllopoda): selection, exchange and sex. J. Evol. Biol. 3: 257–282.

Korpelainen, H., 1986. The effect of diapause on the genetic structure of *Daphnia magna* populations. Z. zool. Syst. Evolut.-forsch. 24: 291–299.

Korpelainen, H., 1987. Reproductive strategies of *Daphnia magna* genotypes. Hereditas 106: 181–188.

Larsson, P., 1991. Intraspecific variability in response to stimuli for male and ephippia formation in *Daphnia pulex*. Hydrobiologia 225: 281–290.

Loaring, J. M. & P. D. H. Hebert, 1981. Ecological differences among clones of *Daphnia pulex* Leydig. Oecologia 51: 162–168.

Marcus, N. H., R. Lutz, W. Burnett & P. Cable, 1994. Age, variability, and vertical distribution of zooplancton resting eggs from an anoxic basin: evidence of an egg bank. Limnol. Oceanogr. 39: 154–158.

Mort, M. A., 1991. Bridging the gap between ecology and genetics: the case of freshwater zooplankton. TREE 6: 41–45.

Mort, M. A. & H. G. Wolf, 1986. The genetic structure of large lake *Daphnia* populations. Evolution 40: 756–766.

Muller, J. & A. Seitz, 1994. Influence of differential natality and mortality on temporal fluctuations of *Daphnia* genotypes in natural populations. In A. R. Beaumont (ed.), Genetics and evolution of aquatic organisms. Chapman & Hall, London: 342–350.

Rossi, V. & P. Menozzi, 1990. The clonal ecology of *Heterocypris incongruens* (Ostracoda). Oikos 57: 388–398.

Rossi, V. & P. Menozzi, 1993. The clonal ecology of *Heterocypris incongruens* (Ostracoda): life-history traits and photoperiod. Funct. Ecol. 7: 177–182.

Rossi, V. & P. Menozzi, 1994. Enzyme and DNA polymorphism in ostracod evolutionary ecology. In D. J. Horne & K. Martens (eds), 'The evolutionary ecology of reproductive mode in non marine ostracoda'. Greenwich University Press: 43–54.

Rossi, V., P. Giordano & P. Menozzi, 1993. Genetic variability in parthenogenetic populations of *Heterocypris incongruens* (Crustacea, Ostracoda). In K. G. McKenzie & P. J. Jones (eds), Ostracoda in the earth and life sciences. A. A. Balkema, Rotterdam: 369–383.

Rossi, V., M. C. Rozzi & P. Menozzi, 1991. Life strategy differences among electrophoretic clones of *Heterocypris incongruens* (Crustacea, Ostracoda). Verh. int. Ver. Limnol. 24: 2816–2819.

Rozzi, M. C., V. Rossi, G. Benassi & P. Menozzi, 1991. Effetto della temperatura e del fotoperiodo sull'attivazione di uova durature di cloni elettroforetici di *Heterocypris incongruens* (Ostracoda). S It E. Atti 12: 591–594.

SAS Institute Inc., 1985. SAS Users' guide: statistical version. edn 5. SAS Institute Inc., Cary, North Carolina.

Schwartz, S. S. & P. D. N. Hebert, 1987. Methods for the activation of the resting eggs of *Daphnia*. Freshwat. Biol. 17: 373–379.

Silvertown, J., 1987. Introduction to plant population ecology. Longman Scientific & Technical, NY.

Stross, R. G., 1966. Light and temperature requirements for diapause development and release in *Daphnia*. Ecology 47: 368–374.

Tauber, C. A. & M. J. Tauber, 1981. Insect seasonal cycles: genetic and evolution. Annu. Revue Ecol. Syst. 12: 281–308.

Wiggins, G. B., R. J. Mackay & I. M. Smith, 1980. Evolutionary and ecological strategies of animals in annual temporary pools. Arch. Hydrobiol./ Suppl. 58: 97–206.

Wolf, H. G. & G. R. Carvalho, 1989. Resting eggs of lake-*Daphnia* II. In situ observations on the hatching of eggs and their contribution to population and community structure. Freshwat. Biol. 22: 471–478.

Yampolsky, L. Y., 1992. Genetic variation in the sexual reproduction rate within a population of cyclic parthenogen, *Daphnia magna*. Evolution 46: 833–873.

Hydrobiologia **320**: 55–61, 1996.
V. R. Alekseev & G. Fryer (eds), Diapause in the Crustacea.

Influence of parthenogenetic reproduction on the genotypic constitution and evolutionary success of populations and species

Sergey D. Grebelnyi
Zoological Institute, Russian Ac. Sci., St. Petersburg, Russia

Key words: variability, life cycles, gene recombination stoppage, parthenogenesis, resting eggs, *Daphnia*

Abstract

The life cycle of *Daphnia* commonly includes parthenogenesis, sexual reproduction, and diapause. Although there are no genotypic transformations during diapause, it separates the 'clonal' and 'pseudopanmictic' states of the population. During parthenogenetic reproduction primarily polymorphic natural populations gradually degenerate into a mixture of a few clones. As resting eggs are usually produced sexually (= by means of sexual reproduction), after the diapause the vast diversity of individual genotypes normal for panmictic populations is observed. It is the competitive interactions between parthenogenetic clones which again eventually decrease the genotypic polymorphism. Forms in which the sexual process and diapause are rare (or species which are represented by parthenogenetic populations in one part of their area and by bisexual ones in another) demonstrate the most significant differences between clonal and panmictic populations.

Parthenogenesis, apomixis and other kinds of reproduction without genetic recombination are widely spread in branchiopod crustaceans (Notostraca, Anostraca, 'Conchostraca', 'Cladocera') and in many other animals and plants. Moreover, reproduction without recombination plays an important role in evolution of faunas and floras. It is emphasized that non-recombinating races and species having high heterozygosity but a low level of genetic variation, enjoy short term advantages, but die out after change in the environment.

Changes in the polymorphism of animal populations in connection with cyclic parthenogenetic reproduction

To date many investigations on variation in natural populations of *Daphnia* and related genera of Anomopoda have been carried out. *Daphnia* is now the most popular object of population structure studies and the subject of some interesting discussions. Though numerous papers on this problem contain contradictory facts and opinions, we surely can use some reliable findings.

The extent of variation within a population depends not only on the array of alleles involved but also on the mode of reproduction of the population at the moment of observation. The life cycle of *Daphnia* and other Anomopoda usually includes a more or less long parthenogenetic phase in which several all-female generations follow one another. Then, as a rule, at the approach of unfavourable conditions, males and sexual females appear. After copulation the females produce fertilized resting eggs, which undergo diapause.

The essential peculiarity of unisexual propagation is that in parthenogenetic females of *Daphnia* so called 'ameiotic' ovogenesis takes place (Kühn, 1908; Mortimer, 1936; Ojima, 1958, Zaffagnini & Sabelli, 1972; Zaffagnini, 1987). Offspring produced parthenogenetically are always of the same genotype as their mother. Enzyme polymorphism in hundreds of clones was examined electrophoretically and no exceptions to this rule were found (Hebert & Ward, 1972).

There is a considerable difference between large and small ponds, permanent and temporary habitats. In spring, when small ponds and other temporary habitats are inundated the *Daphnia* populations consist of individuals which have developed from fertilized resting eggs. In such just re-established populations allele frequencies at polymorphic gene loci are in good correspondence with Hardy-Weinberg expectations as has to be the case in common bisexual animals, and as it is in birds and mammals. But from the start of parthenogenetic reproduction competitive exclusion favours some clones in comparison to others that drop behind in viability or fecundity. In small waterbodies which are frequently subjected to desiccation or freezing, interclonal competition is periodically interrupted by sexual reproduction which, after a period of diapause, restores the vast extent of individual genetic polymorphism. In permanent large ponds where environmental conditions are more stable, and therefore parthenogenetic reproduction holds for a longer time, electrophoretic analysis shows superior survival of highly heterozygous clones. Such a phenomenon is an inescapable result of sexual genetic recombination stoppage (= blockage) (Here and below I use the term 'recombination' to designate the whole complex of processes leading to the creation of new allele combinations). As it was summarized by Young (1979) 'heterozygote excesses appear to be a general feature of permanent *Daphnia* populations'. In connection with that, Young mentioned the opinion of Bęrger (1976) who has suggested that heterozygosity at some enzyme loci 'confers increased fitness'. Nevertheless Young also refers to alternative observations: in some populations of *Daphnia* and *Simocephalus* homozygous clones predominate (Young, 1979: p. 960). It is, however, neither known nor maintained whether *Simocephalus* has ameiotic ovogenesis as is the case in *Daphnia*.

Therefore I draw attention to the fact that the predominance of homozygosity in *Simocephalus* could be explained if automictic modification of meiosis, which is well known in many animals (see in Zaffagnini & Sabelli, 1972; Vasetskii, 1977), also prevails in *Simocephalus*. The genetic consequences of parthenogenesis in this case might be analogous to self-fertilization, increasing homozygosity from one generation to another.

Maynard Smith (1978, 1981), refering to work by Hebert (1974a and 1974b) and Young (unpublished Ph.D. dissertation, 1976), discussed the process of competitive elimination of clones. In addition to the publications mentioned above he used the results of work carried out by Suomalainen *et al.* (1973 on Insecta: Curculionidae), by Moor (1976, on gynogenetic fishes of the genus Poeciliopsis), by Neaves (1969, on lizards of the genus *Cnemidophorus*), and others. The data collected by him in general confirm previous conclusions made by Hebert and Young: reproduction without recombination definitely leads to the prevailing of a few clones, or finally even one highly heterozygous clone.

The widespread nature of polyploidy in clonal organisms

Parthenogenesis promotes polyploid individuals and their transformation into polyploid races. In common bisexual diploid species an unreduced gamete, generated as a result of accidental error, gives rise to an abnormal triploid organism (for example, a diploid chromosome set may originate from the egg and a haploid set from the sperm). Such a polyploid individual has usually no progeny due to the destruction of meiosis and an inability to produce normal gametes. In species, which in nature reproduce themselves by parthenogenesis, meiosis is known to be lost or destroyed. Therefore the difficulties of meiosis do not result in the loss of fertility. Different modifications of meiosis in unisexual all-female forms of animals are described by White, 1964, 1973; Suomalainen, Saura & Lokki, 1976, 1987; Vasetskii, 1977; Zaffagnini, 1987 and others. As a result of the absence of perfect meiosis and copulation, clonal forms reproduce themselves normally conserving the ploidy (2n, 3n, 4n, or 8n, and so on) of the clone founder. That is why a species, or a group of closely related species, which includes parthenogenetic subspecies, races, karyomorphs or clones, often have among them a number of forms of different ploidy (Suomalainen, Saura & Lokki, 1976).

Species or intraspecific forms which are in the phase of populational evolution use all the advantages of normal sexual gene recombination. They evolve by means of changes in allele frequencies, as well as by the construction and subsequent selection of new, more successful, gene combinations. Therefore they need to be diploid. In a situation where recombination is of importance any polyploid will lose the competition with a diploid. The plant *Datura stramonium* (an example from the Russian classic course of genetics – Lobashev, 1967) incudes tetraploid races with balanced, normally-operating meiosis. In a cross between two parents heterozygous at one locus (AAaa × AAaa) segregation gives 35 descendants of a phenotype corresponding to the dominant allele, and only one recessive by genotype and phenotype (proportion 35:1 instead of 3:1 as it has to be in diploid races). It seems clear that selection towards dominant as well as recessive features will be less effective in a tetraploid population than in diploid one. Therefore populational evolution proceeds much more intensively at the diploid level.

Polyploid forms of animals and plants are represented in all taxa that consist of many species if at least some of them are able to undergo parthenogenetic, gynogenetic, hybridogenetic (*sensu* Kirpichnikov, 1979, 1981; Vrijenhoek, 1979), somatic, vegetative, nucellar or apomictic reproduction. (Apomixis (*sensu str.*) – is a multiplication of plants by seeds originating from one of the somatic nuclei of the germinative sac (=ebmryo sac)).

In the Crustacea polyploids are present in Anomopoda [*Daphnia* – 2n, 4n (Beaton & Hebert, 1988)], and Anostraca [*Artemia* – 2n, 4n, 6n, 10n (Mitrofanov *et al.*, 1982)]. They are found also in wide range of other invertebrates and in various vertebrates [Pisces (Kirpichnicov, 1979, 1981), Amphibia (Borkin & Darevsky, 1980), Reptilia (Darevsky, 1992)].

Returning to *Daphnia*, one could expect the existence of polyploids within the genus. On a small island in the Canadian high Arctic (Igloolik Isl.; 69 °22′N, 81 °48′W) clonal diversity in populations of *Daphnia pulex* has been studied (Weider, Beaton & Hebert, 1987; Beaton & Hebert, 1988). From 179 tundra ponds 75 electrophoretically distinct clones were discovered (from one to 14 clones per pond). With the exception of one diploid clone, all are tetraploid. The authors believe that all populations under consideration reproduce only by parthenogenesis.

But what is the origin of such a great clonal variety? In this as well as in other cases the existence of obligate (not cyclic) parthenogenesis has been proved by absence of males and the lack of segregation at the heterozygous loci when resting eggs have been hatched. Nevertheless, for Weider and his collaborators there could be another, and perhaps much more significant, argument to prove the absence of sexual recombinations. That is the polyploidy of clones.

Geographical pattern of forms without sexual gene recombination

There are different mechanisms of recombination stoppage or limitation. Among them parthenogenesis, gynogenesis, and hybridogenesis have already been mentioned. To them some other genetical mechanisms, such as reduction of chromosome number by their fusion and suppression of crossing-over could also be added. We would like however to focus the discussion on the most easily detectable event of recombination stoppage which is certainly the best studied and illustrative.

As stated above, polyploids are either completely unable to recombine or, in any case, their life strategy is not directed to intensive recombination. Therefore polyploidy itself could be regarded as a mode of fixation of lucky gene combinations and we can consider polyploidy as a reliable indicator of reproduction without recombination.

Unlike the other mechanisms of recombination stoppage it can be discovered by counting of chromosomes or by measuring DNA-content. Thanks to that a large data set on the geographical distribution of polyploids and their comparative abundance in different taxa is available today.

Flowering plants are particularly well studied from this point of view. Examination of their geographical distribution has made some investigators suppose that polyploids are fitter in extreme environments. For example, there are some genera in which species with high chromosome number inhabit the northern parts of their areas (*Rosa* – Täckholm, 1922; *Empetrum* – Hagerup, 1927, 1932; *Agrostis* – Sokolovskaia, 1937). As confirmed by subsequent observations, the abundance of polyploids in the check lists of floras rapidly increases towards higher latitudes (Gustafsson, 1948; Löve & Löve, 1971). Gustafsson tried to explain this effect by the predominance of herblike perennials predominance in the Arctic, referring to their capacity to vegetative propagation. Cain (1944) interpreted such a correlation another and, very likely, more proper way. He supposed that the reason of polyploid plenty in Arctic and mountainous floras is connected with the history of climate. He believed that polyploid species and intraspecific forms tend to prevail in regions which have been exposed to large scale climatic fluctuations. In regions where the consequences of climatic changes have not been so destructive a good number of relict elements of the flora (palaeoendemics) have been preserved and the quota of polyploids is lower. That conclusion was drawn by Reese (1957) while investigating the flora of the northern Sahara, and by Larsen (1960) who cited as an example the flora of the Canary Islands.

The geographical distribution of weevil species *Otiorrhynchus dubius* (Coleoptera, Curculionidae) on the face of it seems to be in a good correspondence with the hypothesis of polyploid predominance in the northern parts of the species area (Fig. 1). Indeed, the triploid and tetraploid populations of the species inhabit Scandinavia, Scotland, Ireland and Iceland,

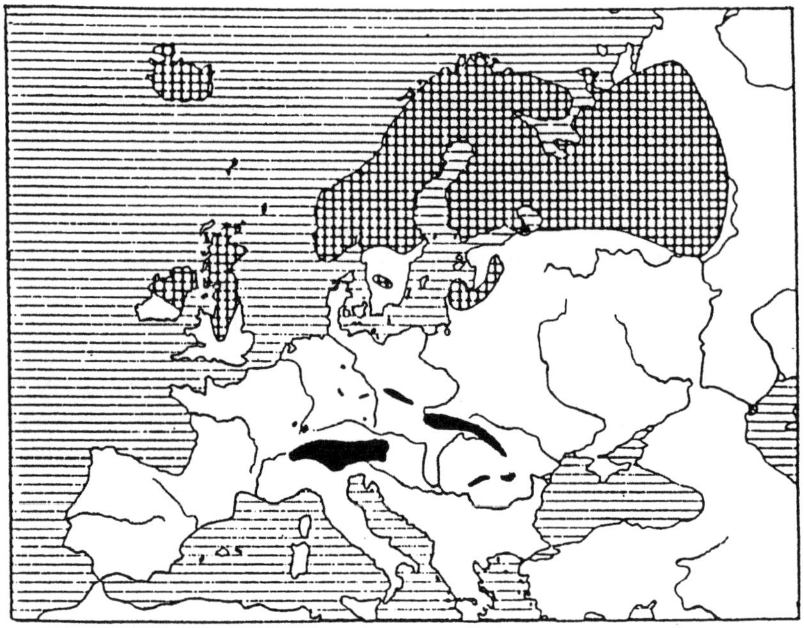

Fig. 1. The distribution of *Otiorrhynchus dubius* in Europe. The distribution area of the bisexual diploid race (Central Europe) is black and that of the parthenogenetic race (North Europe) crosshatched. (After Holdhaus & Lindroth, 1939 from Suomalainen, 1948.)

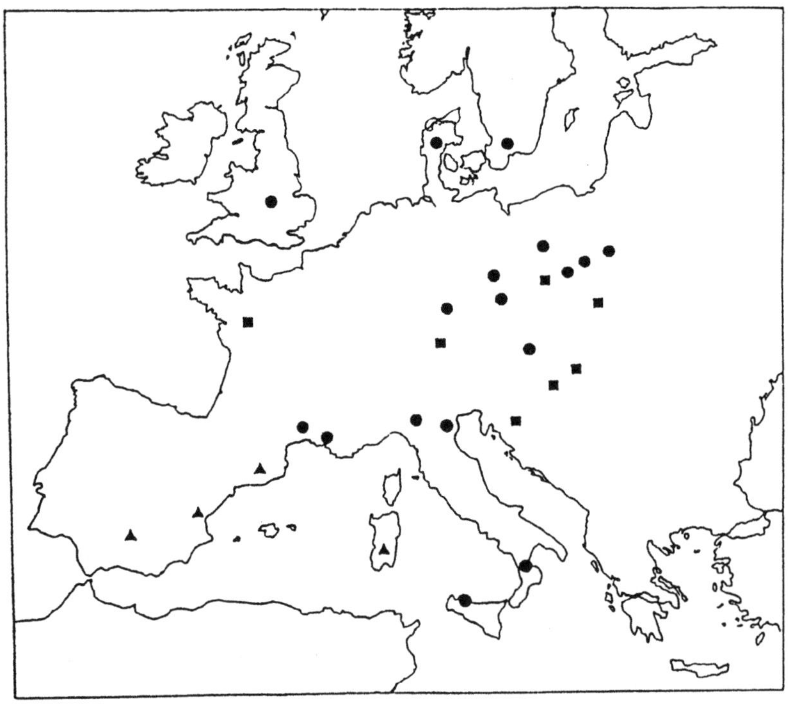

Fig. 2. European distribution and sex ratio of *Triops cancriformis*. Circle – maleless populations; triangle – normal bisexual populations; square – populations with a low percentage of males. (From Zaffagnini & Trentini, 1980.)

while the diploid bisexual race has been discovered only in the mountains of Central Europe. A similar situation found in *Triops cancriformis* (Crustacea, Notostraca – Zaffagnini & Trentini, 1980). In this case the south-western part of the range is also mainly inhabited by bisexual populations (Fig. 2). According to these authors the more northern parts of the area have been occupied by the species during postglacial periods. These northern territories are invaded by parthenogenetic all-female or female-biased populations. In *Triops* only the mode of reproduction of various geographical forms has been studied while the Curculionidae are much better explored. Northern populations of *Otiorrhynchus dubius* consist of a mixture of rather numerous, electrophoretically distinct clones, only a few of which are much more widely spread than the others. Some differences of the stations inhabited as well as different fitness to various temperature conditions have also been recorded. The detailed analysis of the distribution pattern of bisexual and parthenogenetic forms carried out by Suomalainen and collaborators (Suomalainen *et al.*, 1973, 1976, 1987) evidently proves it to be a result of the Quaternary Glaciations. The large northern part of the area undoubtedly could be inhabited by polyploid populations of *Otiorrhynchus dubius* only after glacier retreat. These are supposed to originate from diploid populations which survived to the end of the glaciation in the Central-European mountains. Some diploid populations are still capable of parthenogenetic reproduction.

Returning to *Daphnia* in the latest papers we confront the familiar questions (Weider *et al.*, 1987; Beaton & Hebert, 1988). Figure 3 demonstrates the comparative occurrence of diploid and tetraploid clones in three localities in North America which are at different latitudes. According to Beaton & Hebert's point of view these data clearly show the advantages of tetraploid *Daphnia* in high-Arctic conditions. However, I think, they indicate only that tetraploid *Daphnia* clones conserve lucky gene combinations better than their diploid ancestor. It has to be mentioned that the discovery of a single rare and depressed diploid form among some dozens of tetraploid clones in Igloolik Island is of significant importance. This diploid form was considered to be a clone incapable of recombination, like the other coexisting clones. Nevertheless it is possible that the diploid form is able to undergo normal sexual recombination, which was not distinguished because the gene polymorphism was not registered by means of enzyme analysis. The existence of this rare diploid form (2% of the pond population) makes us remember the question put by Maynard Smith (1978, 1981): Why does not the parthenogenetic non-recombinating form completely force out its bisexual ancestor? The rarity of the diploid in this case indicates that the forcing out process seems to be realized. But in the waterbodies where such a process has been

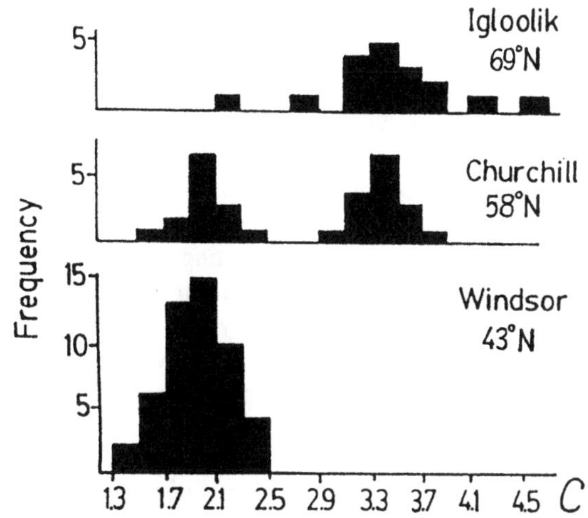

Fig. 3. Frequency distribution of clone ploidy levels in *Daphnia pulex*, from three localities, instrumentally measured after DNA-staining. Ploidy levels approximating C=2 suggest that the clone is diploid. (From Beaton & Hebert, 1988.)

completed, clonal diversity is maintained only by accidental transport of resting eggs from other sites. As it concerns the inner source of allele combination variety, it can be successfully restored by even a small number of individuals able to undergo sporadic sexual reproduction. Their presence makes a system more stable and its long-time existence more probable. So, if the diploid *Daphnia* from Igloolik Island ponds play a role similar to that of weevils from Alpine refugia the discovery of a diploid form is very probable and readily explained.

The consequences of giving up genetic recombination for the evolutionary capacity of species

Populations in which the process of recombination is stopped transform into one or a few clones. As already shown the results of competitive exclusion of clones are: (1) a low level of polymorphism, and (2) a higher level of heterozygosity in comparison with that of panmictic bisexual populations. Due to heterosis a non-recombinating form is often more viable (successful) than the recombinating one and often forces out its ancestor from the major part of its geographical or ecological area. Nevertheless it is ephemeral as it becomes extinct when the environment changes, but it is periodically replaced by another form, restored by the parental bisexual diploid panmictic form, which survives in refugia and preserves its polymorphism.

Thanks to repeated expansions and extinctions, the inhabitants of regions subjected to climatic catastrophies (subarctic, alpine regions and other places regularly devastated by glaciation, drought and so on) mainly consist of races, subspecies and species, that appear to be non-recombinating forms (polyploids, parthenogens or apomicts). Their quota in some regions and some taxa may be very significant. Thus, more than a half of the world flora of flowering plants consists of polyploid forms unable to undergo true populational evolution. So far as tundras and mountains are concerned, here the quota of polyploids in vascular plants rises to 80% or even more. The lack of morphological, biochemical or any other variability makes a population very vulnerable to considerable changes of environment. On the other hand the high uniformity of individuals together with the high heterozygosity gives them in appropriate conditions an opportunity to use natural resources more effectively. For that reason organisms whose life cycle reveals the repeated stoppage of recombination tend to undergo periodical, but not very regular, episodes of increasing abundance.

Being an elementary evolutionary event, the stoppage of recombination occurs rather often in regions of young faunas and floras where there are many empty ecological niches to be occupied. Stoppage of recombination always leads to two simple and inseparable consequences. First, it gives a certain competitive success (the so-called short term advantage); Second, it decreases the evolutionary potential of species because of the reduction in diversity.

References

Berger, E., 1976. Heterosis and the maintenance of enzyme polymorphism. Am. Nat. 110: 823–839.

Borkin, L. J. & I. S. Darevsky, 1980. Reticulate (hybridogenous) speciation in vertebrates. Zhurnal Obshchei Biologii (= Journal of General Biology). 16: 485–505 (in Russian).

Cain, S., 1944. Foundation of Plant geography. New York and London.

Darevsky, I. S., 1992. Evolution and ecology of parthenogenesis in reptiles. In K. Adler (ed.), Herpetology: Current Research on the Biology of Amphibians and Reptiles. Proc. First World Congr. of Herpetology, Oxford: 21–39.

Gustafsson, A., 1948. Polyploidy, life form and vegetative reproduction. Hereditas. 34: 1–22.

Hagerup, O., 1927. Empetrum hermaphroditum Hagerup, a new tetraploid bisexual species. Dansk. Bot. Arkiv. 5: 1–17.

Hagerup, O., 1932. Über Polyploidie in Beziehung zu Klima, Ökologie und Phylogenie und Chromosomenzahlen aus Tumbuktu. Hereditas. 16: 19–40.

Hebert, P. D. N., 1974a.Enzyme variability in natural populations of Daphnia magna II (2). Genotypic frequencies in permanent populations. Genetics 77: 323–334.

Hebert, P. D. N., 1974b.Enzyme variability in natural populations of Daphnia magna III (3). Genotypic frequencies in intermittent populations. Genetics 77: 335–341.

Hebert, P. D. N. & R. D. Ward 1972. Inheritance during parthenogenesis in Daphnia magna. Genetics 71: 639–642.

Kirpichnikov, V. S., 1979 (1987 2nd edition). Geneticheskie osnovy selectsii ryb (= Genetical foundation of fish selection). 'Nauka', Leningrad: 391 (in Russian).

Kirpichnikov, Y. S., 1981 (English edition). Genetic Bases of Fish Selection. Springer, Berlin, 410 pp.

Kühn, A., 1908. Entwicklung der Keimzellen in der parthenogenetischen Generationen der Cladoceren Daphnia pulex de Geer. Arch. Zellforsch. 1: 538–586.

Larsen, K., 1960. Cytological and experimental studies on the flowering plants of the Canary Islands. Biol Skr. Dan. Vid. Selsk. 11 (3).

Lobashev, M. E., 1967. Genetika (= Genetics). The University Press, Leningrad, 751 pp. (In Russian).

Löve, A. & D. Löve, 1971. Polyploidy et geobotanique. Nat. can. 98.

Maynard Smith, J., 1978. The Evolution of Sex. Cambridge University Press, Cambridge, London, New York, Melbourne.

Maynard Smith, J., 1981 (Russian edition). Evoliutsiia polovogo razmnozheniia. 'Mir', Moscow, 271 pp.

Mitrofanov, Yu. A., Yu. A. Ivanovskii, L. N. Lesnikova & A. M. Makarycheva, 1982. Chromosome numbers and Karyotypes of some Artemia salina populations. Tsitologia i genetika (Cytology and genetics). 16: 11–14 (in Russian).

Mortimer, C. 1936. Experimentelle und cytologische Untersuchungen über den Generationswechsel der Cladoceren. Zool. Jb., Abt. Zool. u. Physiol. 56: 323–388.

Moor, W. S., 1976. Components of fitness in the unisexual fish Poeciliopsis monacha-occidentalis. Evolution 30: 564–578.

Neaves, W. B., 1969. Adenosine deaminase phenotypes among sexual and parthenogenetic lizards in the genus Cnemidophorus (Teiidae). J. exp. Zool. 171: 175–184.

Ojima, Y., 1958. A cytological study on the development and maturation of the parthenogenetic and sexual eggs of Daphnia pulex (Crustacea, Cladocera). Kwansei Gakuin Univ. Annual Studies. 6: 123–176.

Reese, G., 1957. Über die Polyploidiespektren in der nordsaharischen Wüstenflora. Flora. 144: 598–634.

Sokolovskaia, A. P., 1937. Karyogeographical study of genus Agrostis. Botanicheskii zhurnal (=J. Bot.) 26: 457–480 (in Russian).

Suomalainen, E., 1948. Parthenogenesis and polyploidy in the weevils, Curculionidae. Ann. entomol. fenn. 14 (suppl.): 200–212.

Suomalainen, E. & A. Saura, 1973. Genetic polymorphism and evolution in parthenogenetic animals. 1. Polyploid Curculionidae. Genetics 74: 489–508.

Suomalainen, E., A. Saura & J. Lokki, 1976. Evolution of Parthenogenetic Insects. In Hecht, M. K., W. C. Steere & B. Wallace (eds), Evolutionary Biology. 9: 209–257.

Suomalainen, E., A. Saura & J. Lokki, 1987. Cytology and Evolution in parthenogenesis. CRC Press, Boca Raton, Florida, 1987, 206 pp.

Täckholm, G., 1922. Zytologische Studien über die Gattung Posa. Acta horti Bergiani. 7.

Vasetskii, S. G., 1977. Meiotic divisions (pp. 145–172). In Detlaf, T. A. Sovremennye problemy oogeneza (= Modern problems of ovogenesis). 'Nauka', Moscow, 319 pp. (in Russian)

Vrijenhoek, R. C., 1979. Factors Affecting Clonal Diversity and Coexistence. Am. Zool. 19: 787–797.

Weider, L. J., M. J. Beaton & P. D. N. Hebert, 1987. Clonal Diversity in High-Arctic Populations of Daphnia pulex, a Polyploid Apomictic Complex. Evolution. 41: 1335–1346.

White, M. J. D., 1964. Cytogenetic mechanisms in insect reproduction. In K. C. Highnam (ed.), 2nd Symposium R. Entom. Soc. London: 1–12.

White, M. J. D., 1973. Animal cytology and evolution (3nd. ed.). Cambridge University Press, Cambridge, 961 pp.

Young, J. P. W., 1979. Enzyme polymorphism and cyclic parthenogenesis in Daphnia magna. 1. Selection and clonal diversity. Genetics 92: 953–970.

Zaffagnini, F., 1987. Reproduction in Daphnia. In R. H. Peters & R. de Bernardi (eds), 'Daphnia', Mem. Ist. Ital. Idrobiol. 45: 245–284.

Zaffagnini, F. & B. Sabelli, 1972. Karyologic Observations on the Maturation of the Summer and Winter Eggs of Daphnia pulex and Daphnia middendorffiana. Chromosoma (Berl.) 36: 193–203.

Zaffagnini, F. & M. Trentini, 1980. The distribution and reproduction of Triops cancriformis (Bosc) in Europe (Crustacea Notostraca). Monit. zool. ital. 14: 1–8.

Hydrobiologia **320**: 63–70, 1996.
V. R. Alekseev & G. Fryer (eds), Diapause in the Crustacea.

Variation in torpidity of diapause in freshwater cyclopoid copepods

K. Elgmork
Department of Biology, Division of Zoology, University of Oslo, P.O. Box 1050, Blindern, N-0316, Oslo, Norway

Key words: diapause, torpidity, dormancy, impact of environment on torpidity in the sediment

Abstract

By successively filtering off animals emerging from mud cores taken to the laboratory, the degree of torpidity within the same species was found to vary considerably regarding both time and space. Variation in time from the start of dormancy follows a pattern of diapause development similar to that of insect diapause. In some localities dormancy is deep and prolonged, while in others it is less pronounced and lasts for a shorter period. The depth of torpidity during winter seems to be correlated to an induction early in the year. The intensity of diapause varies along the slope of the localities and down into the bottom mud. In some cases high temperatures seem to induce a deeper torpidity.

Introduction

The term diapause in arthropods is ambiguous. Diapause may mean an arrest in growth, development or reproduction in active individuals, or a torpid condition with inactivity. Diapause in the freshwater cyclopoids studied so far is characterized by a torpid, inactive condition in the bottom sediment in copepodid and occasionally adult stages. This type of diapause is dealt with in this study.

It is well established that diapause in arthropods varies with gross changes in latitude, climate and geographic area (Tauber *et al.*, 1986; Danks, 1987; Wyngaard, 1988; Zaslavski, 1988). Little is known about responses to proximate, small scale variations in local environments. During my studies of diapause in freshwater cyclopoids during several decades, knowledge of such variations has accumulated. Some results have been published previously, but they are scattered and not always easily available. This paper surveys earlier results together with the outcome of studies not previously published.

Localities studied

Some descriptive parametres of the localities studied are presented in Table 1. The localites are all situated in south Norway between latitudes 58–61° north. With two exceptions (C and G) they are located in the Oslo area. In the text the localities will be referred to by the capital letters used in Table 1.

Methods

Intensity of diapause is usually measured by the duration of the dormant period or by the proportion of individuals in diapause. In the present study diapause is determined by the degree of torpidity of the dormant individuals in the following way. After the mud samples were brought into the laboratory, water was added to the sample. Individuals emerging from the mud were filtered off at regular intervals, mostly one day apart, until no more individuals emerged (Elgmork, 1959). Nilssen & Elgmork (1977) have shown that the time between filterings is not very important. It is the number of physical stimuli that counts. The method has been corroborated by results from insect diapause where both high and low temperature shocks have been shown to break diapause (cited in Danks

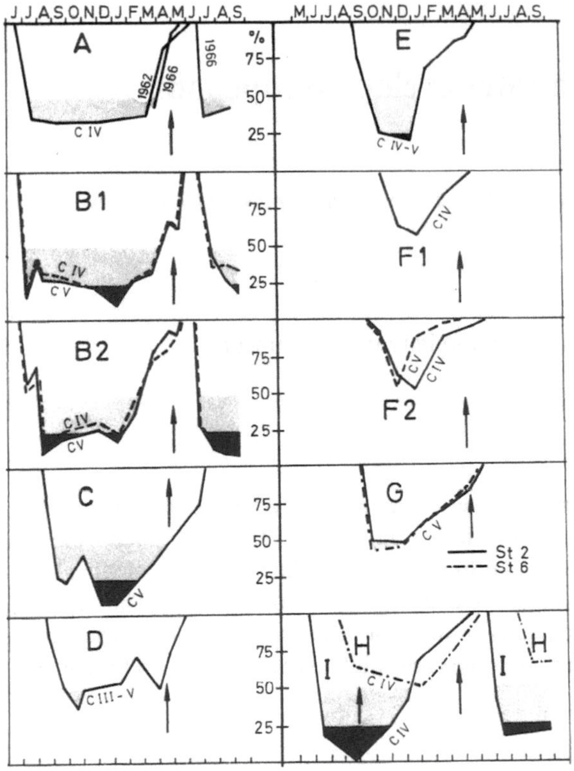

Fig. 1. Depth of torpidity and duration of diapause in resting copepodids in the sediment. Capital letters refer to localities in Table 1. *Black*: Deepest torpidity. *Shaded*: Medium torpidity. Species: A – F1: *Cyclops scutifer*, F2: *Mesocyclops leuckarti*, G: *Cyclops abyssorum*, H-I: *Cyclops strenuus*. B1: Deepest meromictic basin, B2: shallow dimictic basin. *Arrow*: first postdiapause individuals recorded in the open water in spring. (From Elgmork & Nilssen 1978).

(1987): 150,155). The rate of emergence defined by the percentage of emerging individuals each day gives an indication of the degree of torpidity in the population. A rapid emergence with high percentages in the first filterings indicates a less torpid state than when most individuals emerge after several filterings. Results using this method have been published previously (Elgmork, 1959; Nilssen & Elgmork, 1977; Elgmork & Langeland, 1980).

In order to simplify presentation the percentages emerging in the first three filterings were added together and the sum defined as the *torpidity index*. This ranges from zero, with a very deep torpidity, to nearly 100 when torpidity is weak. As this index is based on ratios, the statisical method used was a $2 \times n$ chi-squared test for independence.

Variation with time in diapause

A general feature in the variation of diapause with time is high torpidity during the early phase of the diapause period and a gradual decrease in diapause intensity with passing time (Tauber *et al.*, 1986:56). A weak torpidity is found preceeding termination. This gradual reduction in torpidity with time is interpreted as the result of a diapause development (Andrewartha, 1952). Physiological changes take place during dormancy leading to a less torpid condition as diapause proceeds.

Phases defined for diapause in insects (Mansingh, 1971; Tauber *et al.*, 1986) are analogous in cyclopoid diapause (Elgmork & Nilssen, 1978). After a preparatory phase with accumulation of fat reserves, a short induction phase soon passes into the maintenance phase. The length and depth of torpidity in this phase vary considerably (Fig. 1). Deep torpidity is usually correlated with a longer time in diapause ($r = 0.82$; $p < 0.01$). The maintenance phase is followed by the transitional phase that also lasts for variable time spans. After a long and deep maintenance phase the transitional phase can be short (Fig. 1 A and B). With reduced maintenance phases, the transitional phase extends over relatively longer periods (Fig. 1 D, F and G). Diapause finally ends in the termination phase when the great majority of individuals have left the sediment and resumed active life.

Variation with distribution on the slope

Studies of variation in torpidity along the slopes of the localities were performed in Loc. A and H. In Loc. A samples were taken in the deepest part at 16 m, and on the slope at 15m and 13 m depth (map in Elgmork, 1981). Except in the April sample, when the difference was not significant, torpidity was significantly deeper ($p < 0.0001$) at the deepest station than on the slope (Fig. 2). The difference was greatest at the start of diapause in summer and diminished gradually to being similar at the termination of diapause in spring.

A similar trend was demonstrated in the shallow Loc. H during the very hot summer of 1955 (Fig. 3). In the deepest part, at about 65 cm depth, the torpidity index was zero, whereas on the slope about one m from the shore at a depth of 30 cm, the index ranged from 10–14. A couple of weeks later samples taken on dry land previously flooded, showed copepodids with a very low torpidity (Fig. 3). The torpidity indices were significantly higher ($p < 0.01$) than those from sam-

Table 1. Species studied and localities.

Localities	A Store Tryvann	B Store Åklungen	C Kjosetjern	D Svinsjøen
Species studied	*Cyclops scutifer*	*C. scutifer*	*C. scutifer*	*C. scutifer*
Elevation m a.s.l.	395	293	18	180
Surroundings	Spruce forest	Spruce forest	Coniferous-deciduous	Coniferous-deciduous-cultural
Surface data	44	132	27	110
Max. depth m	16	32.5	20.5	33
Trophic level	Oligotrophic	Oligotrophic	Oligotrophic	Mesotrophic
Humic contents	Mesohumic	Mesohumic	Oligo-mesohumic	Oligohumic
Type of circulation	Dimictic	One basin dimictic, other part meromictic	Dimictic	Meromictic
Ref. to environments	Kjensmo (1967)	Kjensmo (1967)		Kjensmo (1968)
Ref. to species	Elgmork (1981)	Elgmork (1973)		Elgmork (1973)
Collection	Elgmork	Sirnes	Øvrevik	Lie

Table 1. Continued.

E Blankvann	F Cabin Pond A	G Lønavatn	H Pond A	I Bergstjern
C. scutifer	*C. scutifer* *Mesocyclops leuckarti*	*C. abyssorum*	*C. strenuus*	*C. strenuus*
350	810	78	70	280
Spruce forest	Boreal forest	Deciduous	Deciduous	Cultural
84	1	3020	0.6	6.5
55	2.7	27	0.7	5.5
Oligo-mesotrophic	Oligotrophic	Oligotrophic	Mesotrophic	Eutrophic
Oligohumic	Polyhumic	Oligohumic	Oligohumic	Oligohumic
Meromictic	Dimictic	Dimictic	Polymictic	Spring-meromictic
Strøm & Østtvedt (1948)	Elgmork *et al.* 1990	Matzow (1967)	Elgmork (1959, 1964)	Elgmork (1959)
Larsson (1971)	Elgmork *et al.* 1990	Nilssen (1977)	Elgmork (1959, 1964)	Elgmork (1959)
Elgmork (1973)		Nilssen & Elgmork (1977)	Elgmork & Halvorsen (1976)	
Langeland	Elgmork	Nilssen	Elgmork	Elgmork

ples taken in the pond later in the season, excluding the possibility that the higher values for samples on dry land were part of a seasonal change in torpidity. The considerably lower torpidity in animals on dry land may have been induced by the more fluctuating temperatures on dry land in contrast to the more uniform conditions in the pond. Individuals from the three locations in Loc. H with so different torpidity originated from the same cohort.

There thus seems to be a general tendency for diapausing copepodids of *C. scutifer* and *C. strenuus* to acquire a deeper torpidity in the deepest part of the two localities studied. The same trend was also found between the two basins 32.5 m and 16 m deep in Loc. B, with a longer maintenance period in the deepest basin (Fig. 1). In the two basins with about the same depth in Loc.G, however, there was no difference in torpidity in *C. abyssorum* (Fig. 1).

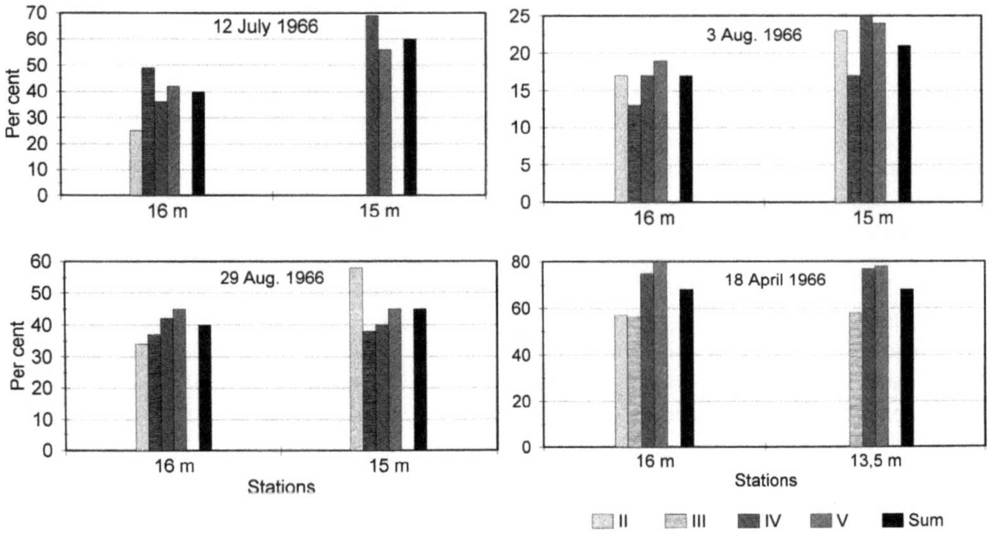

Fig. 2. Degree of torpidity (per cent emerging in the three first filterings) in copepodids II – V in *C. scutifer* in the deepest part and on two stations on the slope in Loc. A.

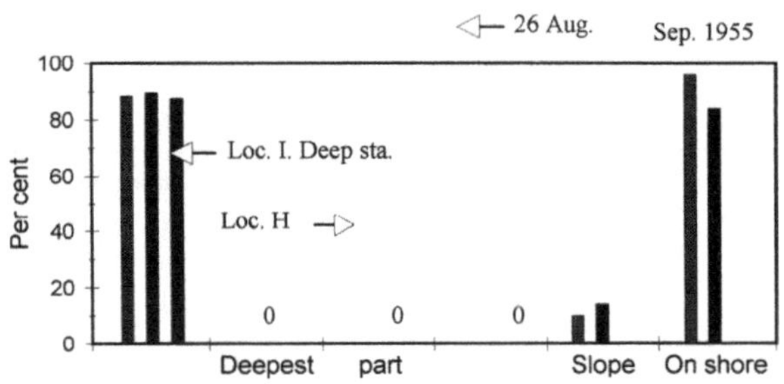

Fig. 3. Comparison of the degree of torpidity (per cent emerging in the three first filterings) in *C. strenuus* in the shallow pond Loc. H and in the lakelet Loc. I in the hot summer 1955.

In Loc. A the reducing power of the lake water is greater in the deepest part (Elgmork, 1981). The same applies to Loc. B that is of the same lake type (Table 1) and situated only three km away. It is doubtful that this is the case in the shallow Loc.H as an oxygen sample in the water taken two weeks before the mud samples showed an oxygen saturation of 108%. Temperature differences between the two stations in all three localities A, B and H were insignificant.

Variation in torpidity along the slope was also found in Mt. Tom Pond in Connecticut, USA, a 14 m deep eutrophic lake (Elgmork, 1967b, 1991). Relatively small samples did not give significant results. However, the same tendencies were repeated in several samples and were consistent between two species, *Cyclops vernalis* and *Mesocyclops edax*. A slightly lower torpidity was found in diapausing females of both species in the deeper part than on the slope at 9 m depth. This is contrary to the results from the Norwegian localties presented above. Temperature differences between the two stations were also insignificant, but oxygen tension in the water outside the station on the slope was greater (Elgmork, 1967b).

Variation in relation to depth in the sediment

Changes in torpidity at various depths in the bottom mud was studied in *C. scutifer* in Loc.A and in *C. strenuus* in Loc.H. In Loc.A mud samples with diapausing copepodids could be sectioned down to a depth of 30 cm at the deepest point of the lake (Fig. 4). In the December sample there was a clear and highly significant ($p < 0.0001$) variation with a gradual increase in torpidity from the surface of the mud down to to about 5–10 cm. Then follows a decrease down to the lowest torpidity at 20–30 cm. The same tendency was repeated in the April sample with the maximum in torpidity again in the middle section. In December torpidity is higher in the upper than in the lower strata while in April this pattern is reversed. In the April sample there was a considerably lower torpidity at all depths in the mud in accordance with the approach of the termination of diapause.

The maximum torpidity occurring in the 7.5–10 cm zone is extraordinary and difficult to explain. Mortimer

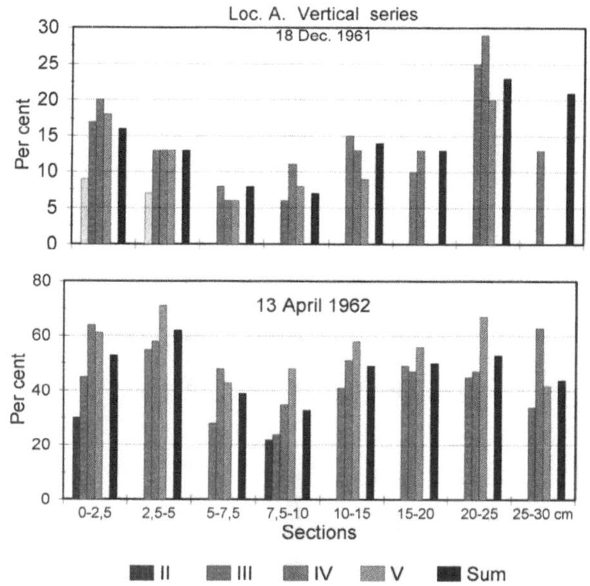

Fig. 4. Variation in the degree of torpidity (per cent emerging in the three first filterings) in the copepodid instars II – V of *C. scutifer* in Loc. A down to 30 cm depth in the mud.

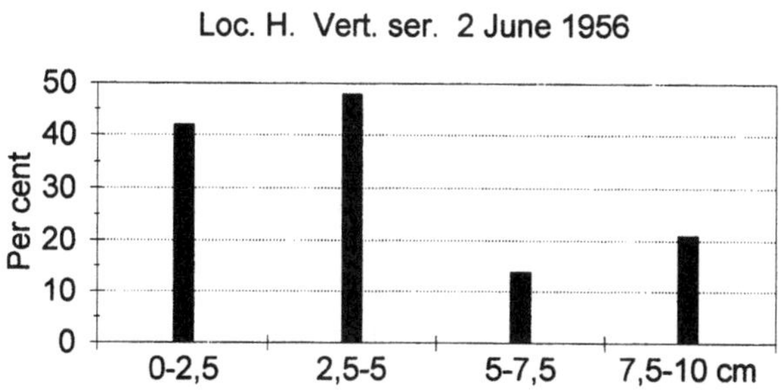

Fig. 5. Variation in the degree of torpidity (per cent emerging in the three first filterings) in copepodid IV of *C. strenuus* in the shallow pond Loc. H down to 10 cm in the mud.

(1942) found in Windermere and Esthwaite Water that the redox potential in the mud was at a minimum about 4–6 cm in the mud and increased again in deeper strata. That a similar distribution should occur in the sediment of Loc. A is, however, less probable. In a lake of the same type situated less than 2 km away (Skjennungen), redox potentials were found to decrease gradually to at least 30 cm down into the mud (J. Kjensmo pers. comm.)

The vertical section in Loc.H in June (Fig. 5) revealed greater torpidity ($p < 0.05$) in the 5–10 cm zone than in the 0–5 cm zone.

Variation among instars

Diapause occurs in 4 copepod instars, II – V in *C. scutifer*, which is exceptional in cyclopoids (Elgmork, 1962, 1967, 1981). The intensity of diapause among instars has been examined at the start of, in the middle of, and near termination of diapause in Loc.A (Fig. 6). The differences found are highly significant ($p < 0.0001$). When considering the results it should be kept in mind that cop. II and III mainly belong to the 2-yr-old cohort, whereas cop. IV and V belong to the 3-yr-old cohort (Elgmork, 1981).

Near induction time in July the torpidity is slightly less in cop. II and III. In the middle of the maintenance phase in December the situation changed in both years; torpidity became deeper in the two younger instars. This situation prevailed into April, when the difference between the young and old copepodid instars was especially clear. Except for the early period of diapause, torpidity thus seems to be deeper in the 2-yr-old generation than in the 3-yr-old. The tendencies of variation in torpidity among the instars are the same in different years, supporting the reliability of the results.

The relatively small differences observed between cop. IV and V in Loc. A was corroborated in the nearby very similar Loc. B. There was only a slight difference in torpidity between these instars both in the meromictic and dimictic basin (Fig. 1).

In Loc.F, a small humic lake, cop.IV of *Mesocyclops leuckarti* showed a deeper torpidity ($p < 0.05$) than cop. V (Fig. 1).

Correlation with environmental factors

The degree of torpidity in the localities presented in Fig. 1 has been correlated with various environmental factors. No significant correlations were found with temperature at the diapause site, average oxygen at site, oxygen content at the start of diapause, trophic level or humic content. A significant negative correlation ($r = -0.83$; $p < 0.01$) was found with the induction time of diapause (Fig. 7). Diapause initiated from May to August showed a higher degree of torpidity (low torpidity index) than diapause initiated during September to November. This is in agreement with results from insect diapause. The later in the season that diapause is initiated in some species, the shorter time it lasts (cited in Danks (1987: 135).

The correlation with induction time may be related both to photoperiodicity and to temperature. In certain

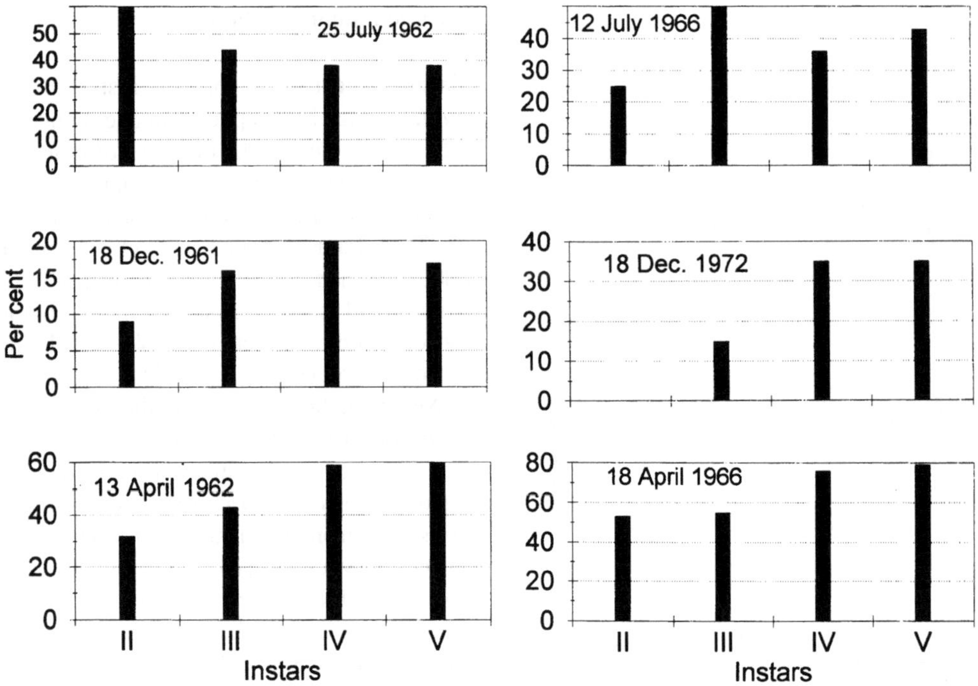

Fig. 6. Variation in the degree of torpidity (per cent emerging in the three first filterings) among copepodid II – V in *C. scutifer* in Loc. A from time of induction to termination.

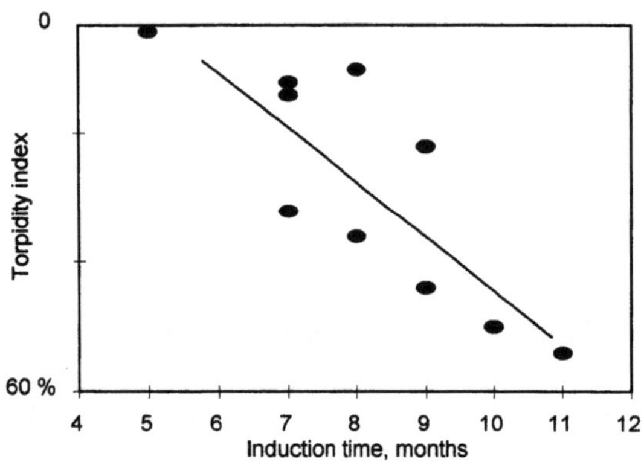

Fig. 7. Degree of torpidity (per cent emerging in the three first filterings) correlated with the time of induction of diapause in the localities presented in Table 1 and in Fig. 1.

insects it has been shown that winter diapause induced in long days is more intense, and less intense if exposed to short days (cited in Danks (1987: 136). This is in agreement with the present results on cyclopoids. The surface temperature in the localities was significantly correlated with month ($r = -0.70$; $p < 0.05$). An influence of temperature may thus also be predicted. Danks (cited in 1987:137) reports that diapause in certain species of insects becomes more intense at higher temperatures during induction.

Contemporaneous variations in different localities may give some indication of influencing factors. An example is a comparison of torpidity in *C. strenuus* in the deepest part of the shallow pond, Loc. H and in the stratified lakelet, Loc.I. Samples were taken on the same day, 26 August 1955, in both localities and treated exactly alike in the laboratory during the filtering period. In the pond the torpidity index in the deepest part was zero which was the deepest torpidity observed in the study (Fig. 3). In the lakelet on the other hand, the index was about 88%, indicating a radical difference in torpidity between the two populations. This was an exceptionally warm summer with temperatures in the shallow Loc.H up to 27° C at the mud-water interface. In Loc.I the bottom temperature was 7–8°. The temperature difference may thus have influenced the marked contrast in torpidity. Also in insect diapause increased dormancy has been recorded at high temperatures (Danks, 1987:137). Another difference was an anoxic environment in the lakelet and supersaturation in the pond up to 108% oxygen. As temperature decreased during the following autumn, some copepodids emerged also in Loc.H. After being exposed to low temperatures during the winter, the torpidities were alike (Elgmork, 1959).

Discussion

This study shows that there is considerable variation in the torpidity of diapause within species of freshwater cylopoids. There are differences between instars, between localities, and along bottom gradients and depths in the sediment. The environment obviously exercises a profound influence on the degree of torpidity, even if single factors are difficult to isolate. This is evident from comparing the torpidity in various populations of *C. scutifer* (Fig. 1). Torpidity was deep and lasted for a long time in the larger lakes Loc. A, B and C, in contrast to the small humic pond Loc. F, where torpidity was very restricted and similar in the two species *C. scutifer* and *M. leuckarti*. As these species have quite different seasonal cycles elsewhere, this correspondence indicates a strong impact of the environment. Both species are forced into the same pattern of diapause and seasonal cycle (Elgmork *et al.*, 1990). Furthermore, the torpidity of the different instars in *C. scutifer* in the vertical series in the sediments (Fig. 4) are in agreement at each depth, whereas there is considerable variation for each instar in the vertical series. This also indicates a strong influence of the environment by proximate and local factors.

The torpid condition in resting copepodids of cyclopoids seems to be sensitive even to minute variations in environmental factors. The very small differences at 16 and 15 m depth in Loc. A, and the few centimetres more shallow station in Loc. H, gave measurable and significant differences in torpidity. Also a few centimetres' variation in depth in the sediment gave significant differences.

The adaptive significance of the variations in torpidity is not easy to elucidate. The lower intensity of diapause when it is initiated later in the season (Figs 1 and 7) may be interpreted as an adaptation to a shorter time in diapause. In order to fulfil the diapause development in a shorter period, the diapause ought not to be too intense. In the variation along the slope and down into the sediment, torpidity seems to be deeper under suboptimal conditions, as in a more reduced environment (Figs 2 and 3) and with very high temperatures (Fig. 3). However, the decrease in torpidity from about 10 to 25 cm in the mud in Loc. A (Fig. 4) does not fit this hypothesis as discussed above. Also under very reduced conditions in the deepest part of Loc. I the torpidity was relatively low (Fig. 3). Further results are obviously needed to evaluate more thoroughly the adaptive significance of the variations in torpidity in freshwater cyclopoids.

Acknowledgement

I am thankful to Dr Jon Swenson, Norwegian Institute for Nature Research, Trondheim, for refining the English style.

References

Andrewartha, H. G., 1952. Diapause in relation to the ecology of insects. Biol. Rev. 27: 50–107.

Danks, H. V., 1987. Insect dormancy: an ecologicl perspective. Biol. Surv. Canada Monogr. Ser. 1: 1–439.

Elgmork, K., 1959. Seasonal occurrence of Cyclops strenuus strenuus in relation to environment in small water bodies in southern Norway. Folia limnol. scand. 11: 1–196.

Elgmork, K., 1962. A bottom resting stage in the planktonic freshwater copepod Cyclops scutifer Sars. Oikos 13: 306–310.

Elgmork, K., 1964. Dynamics of zooplankton communities in some small inundated ponds. Folia Limnol. Scand. 12: 1–83.

Elgmork, K., 1967a. Ecological aspects of diapause in copepods. Proc. Symp. Crust. 3: 947–954.

Elgmork, K., 1967b. On the distribution and ecology of Cyclops scutifer Sars in New England (Copepoda, Crustacea). Ecology 48: 967–971.

Elgmork, K., 1973. Bottom resting stages of planktonic cyclopoid copepods in meromictic lakes. Verh. int. Ver. Limnol. 18: 1474–1478.

Elgmork. K., 1980. Evolutionary aspects of diapause in freshwater copepods. In Kerfoot, W. C. (ed.), Evol. Ecol. Zooplankton Comm., Univ. Press New England. pp. 411–417.

Elgmork, K., 1981. Extraordinary prolongation of the life cycle in a freshwater planktonic copepod. Holarct. Ecol. 4: 278–290.

Elgmork, K., 1991. Winter reproduction strategies in freshwater cyclopoids. Verh. int. Ver. Limnol. 24: 2844–2846.

Elgmork, K. & G. Halvorsen, 1976. Body size of free-living copepods. Oikos 27: 27–33.

Elgmork, K. & J. P. Nilssen, 1978. Equivalence of copepod and insect diapause. Verh. int. Ver. Limnol. 20: 2511–2517.

Elgmork, K. & A. Langeland, 1980. Cyclops scutifer Sars – one and two-year life cycles with diapause in the meromictic lake Blankvatn. Arch. Hydrobiol. 88: 178–201.

Elgmork, K., G. Halvorsen, J.A. Eie & A. Langeland, 1990. Coexistence with similar life cycles in two species of freshwater copepods (Crustacea). Hydrobiologia 208: 187–199.

Kjensmo, J., 1967. The development and some main features of "iron-meromictic" soft water lakes. Arch. Hydrobiol. Suppl. 32: 137–312.

Kjensmo, J., 1968. The primary production and its influence on the meromictic stability in Lake Svinsjøen. Schweiz. Z. Hydrol. 30: 297–317.

Larsson, P., 1971. Vertical distribution of planktonic rotifers in a meromictic lake, Blankvatn near Oslo, Norway. Norw. J. Zool. 19: 47–75.

Mansingh, A., 1971. Physiological classification of dormancies in insects. Can. Ent. 103: 983–1009.
Matzow, D. (ed.), 1976. Investigations of freshwater biology in Lønavatn lake and Strandaelva river. Voss Project 1: 1–235. Zool. Inst. Univ. Oslo. (Mimeogr. In Norwegian with English summary.)
Mortimer, C. H., 1942. The exchange of dissolved substances between mud and water in lakes. J. Ecol. 30: 147–201.
Nilssen, J.P., 1977. Cryptic predation and the demographic strategy of two limnetic cyclopoid copepods. Mem. Ist. Ital. Idrobiol. 34: 187–196.
Nilssen, J. P. & K. Elgmork, 1977. Cyclops abyssorum – Life cycle dynamics and habitat selection. Mem. Ist. ital. Idrobiol. 34: 197–238.
Strøm, K. & H. Østtveit, 1948. Blankvatn. A meromictic lake near Oslo. Skr. norske Vidensk.-Akad. mat.-nat. kl. 1: 1–41.
Tauber, M.J., C. Tauber & S. Masaki, 1986. Seasonal adaptations of insects. Oxford University Press, New York, Oxford 411 pp.
Wyngaard, G. A., 1988. Geographical variation in dormancy in a copepod: evidence from population crosses. Hydrobiologia 167/168 (Dev. Hydrobiol. 47): 367–374.
Zaslavski, V.A., 1988. Insect development. Photoperiodic and temperature control. Springer-Verlag, Berlin, Heidelberg, New York, 187 pp.

Hydrobiologia **320**: 71–79, 1996.
V. R. Alekseev & G. Fryer (eds), Diapause in the Crustacea.

Feeding and digestive activity of cyclopoid copepods in active diapause

Piotr I. Krylov[1], Victor R. Alekseev[1] & Oleg A. Frenkel[2]
[1]*Zoological Institute, Russian Academy of Sciences, St. Petersburg 199 034, Russia*
[2]*Biological Faculty, Moscow State University, Moscow, Russia*

Key words: diapause, feeding, functional response, digestive enzymes, cyclopoids, zooplankton

Abstract

Three different methods were used to compare the feeding patterns of active cyclopoids and cyclopoids in active diapause from different habitats. First, we offered ^{14}C- labeled algal detritus to stage III, stage IV (two different physiological groups), and adult females of *Acanthocyclops vernalis* from temporary water-bodies of the Volga Delta. Second, we compared the functional responses of active and actively diapausing cyclopoids preying on copepod nauplii. Stage III, stage IV (various physiological groups), stage V, adult males and females of *Acanthocyclops viridis, Cyclops kolensis, C. abyssorum sevani* from the Volga Delta, Lake Sevan (Armenia), and Mozhaisk Reservoir (Moscow District) were tested. Finally, we examined the activity of digestive enzymes (proteases) of CIV of *C. kolensis* from Mozhaisk Reservoir at different temperatures and temperature acclimation regimes. All three approaches used have led to similar conclusions. Actively diapausing cyclopoids were capable of feeding on both plant and animal prey; however, their feeding rates and digestive enzyme activity were 2–8 times lower than those of active cyclopoids of similar size. At the initial stage of active diapause the feeding rates were comparable with those of active cyclopoids then declined to a low level during approximately two weeks. Transfer of cyclopoids undergoing winter diapause from ambient temperature (4 °C) to warmer water (20 °C) resulted in dramatic increase of proteases activity only if the acclimation period was relatively long (two weeks).

Introduction

Diapause of late copepodid instars is a widespread phenomenon among freshwater cyclopoid copepods. Dormant stages of many species of cyclopoids have frequently been found in the profundal zone and bottom sediments of lakes and ponds (see Elgmork, 1967, 1980; Nilssen, 1978; Alekseev, 1990 for review). Several observations show that the dormant stages of cyclopoids in the sediments remain completely inert and that their guts do not contain food (Elgmork, 1959; Wierzbicka, 1962).

However, besides typical dormancy, many species of cyclopoids exhibit so-called *active diapause*. According to Elgmork (1980), the term *active diapause* indicates an arrested growth and development without dormancy in the sediments. Cyclopoids in active diapause occur in the plankton or in the profundal zone in the near-bottom layers of water.

In experiments, cyclopoids in active diapause consumed oxygen, although at a lower rate than non-diapausing individuals of the same stage and species (Alekseev, 1990). Thus, it seems logical to expect that the diapausing cyclopoids should be capable of feeding on any available food to cover their energy losses for respiration and activity.

Our aim of the investigation was to test this hypothesis with several species of cyclopoids from different water-bodies. We have examined the feeding of diapausing stages on both plant material (dead *Chlorella* cells) and live animal prey (copepod nauplii). We have also carried out enzyme assays to determine the levels of non-specific proteases at different temperatures and temperature acclimation regimes.

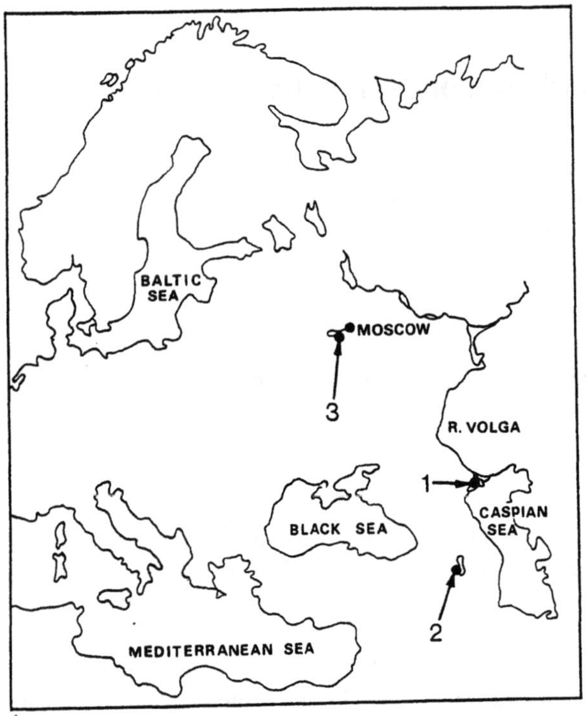

Fig. 1. The location of study areas: (1) the temporary water- bodies of the Volga Delta, (2) Lake Sevan, Armenia, and (3) the Mozhaisk Reservoir west of Moscow.

Study sites and material

Radiocarbon and predation experiments were conducted with several species of cyclopoids from temporary water-bodies of the Volga Delta (Fig. 1). These water-bodies are shallow (often less than 1 m deep) and well warmed during the summer season (water temperatures usually range from 20 °C to 25 °C, sometimes up to 30 °C). The filling with water occurs in spring (April–May) as a result of River Volga flood. Later in the season these water-bodies separate from the river. The level of water decreases due to evaporation and complete drying is usually observed at the end of July. The general biology of zooplankton of these water-bodies is described by Kosova (1960, 1965).

Radiocarbon experiments with *Acanthocyclops vernalis* were conducted in May and June 1980. Preliminary observations have shown that this species in the Volga Delta has a summer diapause at copepodid stage IV, and that the specimens at the very beginning of diapause differ from those at more advanced stages of diapause by colour and amount of oil droplets (Alekseev, 1980, 1981). Thus, we compared the feeding rates of two physiologically different groups of stage IV copepodids. For comparison, we also examined the feeding rates of stage III copepodids and adult females of *A. vernalis*.

Predation experiments with *Acanthocyclops viridis*, *Cyclops kolensis*, and *Cyclops vicinus* were conducted in the same location in May 1989. In these species, both diapausing and non-diapausing stage IV copepodids were available. For purposes of comparison we also examined the functional responses of the following, stage V, copepodids.

The next series of predation experiments was conducted with *Cyclops vicinus* and *C. abyssorum sevani* from the Lake Sevan (Fig. 1, 2). Lake Sevan is a large (surface area 1248 km^2 in 1983) and deep (maximum depth 80.2 m) mountain lake in Armenia. Meshkova (1952) and Simonijan (1991) provide detailed information on the zooplankton ecology of this lake. As show by Alekseev (1990) and Alekseev *et al.* (1991) *Cyclops vicinus* in Lake Sevan has a summer diapause. Diapausing stage IV copepodids remain in the meiobenthos for approximately two months. Several characteristics of stage IV copepodids of *C. abyssorum sevani* (accumulated oil droplets, decreased oxygen consumption, darkened integument) indicate the presence of summer diapause in this species. However, they were never found in the meiobenthos in large numbers. Thus, the physiological status of these animals remained uncertain.

Predation experiments were conducted in July 1990 when the surface water temperature in the lake ranged from 13.5 to 15.6 °C decreasing to 4.5 °C near the bottom. Copepodids of both species were collected at the station in the centre of Maliy Sevan where the water was 45 m deep. Three groups of stage IV copepodids were tested in our experiments: (i) *C. vicinus* from the meiobenthos and near-bottom layers of water; (ii) *C. vicinus* from the plankton; (iii) *C. abyssorum sevani* from the plankton.

Predation experiments and enzyme assays were also carried out with *Cyclops kolensis* from Mozhaisk Reservoir (surface area 31 km^2, maximum depth 22 m), the source for the Moscow water supply (Fig. 1, 3). The extensive data on the ecology of the aquatic communities of this reservoir, including zooplankton, are summarized by Bykov & Edelstein (1979) and Sakharova *et al.* (1993).

Cyclops kolensis in Mozhaisk Reservoir has a prolonged winter diapause, disappearing from the plankton for 8–10 months. Diapausing stage IV copepodids migrate to the bottom where their density may reach 7×10^6 ind m^{-2}. They, however, remain active, probably because of favourable oxygen conditions (Sakharova, 1973, 1979).

Functional response experiments and enzyme assays were conducted in February (1991). Diapausing stage IV copepodids of *C. kolensis* collected from the near-bottom strata and stage V copepodids recovered from diapause and collected from the plankton of the reservoir were tested. For purposes of comparison, we also determined digestive enzyme activity in the non-diapausing winter species, *Megacyclops gigas*.

Methods

Radiocarbon experiments

A laboratory culture of *Chlorella* sp. was radioactively labelled with $NaH^{14}CO_3$ using procedures described in Krylov (1979) and Alekseev & Krylov (1981). Labelled algal suspensions were centrifuged at least twice, resuspended in non-radioactive medium, concentrated, sterilized, and sealed in glass ampoules. The ampoules were stored in darkness until the beginning of the experiment.

Each ampoule contained 1.04 mg (dry weight) of algal material. Before the experiment, the contents of one ampoule were added to 1 litre of 0.6 μm filtered lake water and stirred carefully. The aliquots (50 ml each) of this labelled food suspension were placed into 7 experimental flasks. The contents of two flasks were filtered onto 0.6 μm SYNPOR membrane filters and then used for determination of the initial radioactivity of the food suspension and its specific activity.

Copepods freshly collected from the plankton were pipetted into three of the remaining flasks (30 to 45 copepodids or 8 females per flask). Flasks with and without animals ('final controls') were incubated for 6 or 24 h (in experiments with females and copepodids, respectively) at ambient temperature (19.5 °C and 24 °C in experiments with females and copepodids, respectively).

At the end of the incubation period the contents of the flasks were poured through a mesh to remove the copepods and then filtered onto 0.6 μm SYNPOR filters. Filters were placed into glass scintillation vials. Methods of scintillation counting and calculation of ingestion rates were those of Krylov (1979, 1980) and Alekseev & Krylov (1981).

Predation experiments

The experimental procedures in predation trials with cyclopoids from all three locations were similar. A known number of prey (cyclopoid nauplii) were pipetted into glass flasks filled with 10, 15 or 30 ml of filtered lake water. One individual predator of a known age and species was placed into the experimental flask and its feeding behaviour was monitored over a period of 3 to 8 days. At the end of each day the remaining nauplii were counted; the prey consumed by the predator were replaced by new individuals. The difference between the initial and final number of nauplii in the experimental flasks after correcting for the percentage recovery in controls was used to calculate the predation rates.

The results of experiments in which the predator either died or moulted to the next stage during the period of observations were omitted from the calculations. In some experiments the time-series feeding rate estimates of an individual predator permitted the changes in predation rates in the course of the transition to diapause to be recorded (see Results).

Determination of proteolytic enzyme activity

Studies on the proteolytic activity of a complex of enzymes obtained from cyclopoids were carried out according to the method described by Savitskiy (1990). This technique and its application for estimating proteolytic activity in bacterio-, phyto-, and zooplankton samples was described in detail in Sadchikov *et al.* (1994) and Frenkel (1994); only a brief account is given here.

The method is based on the measurement of changes in the polarization of fluorescence of the fluorescent labelled protein, immunoglobuline G (IgG), in the presence of proteases. Fluoresceinisothiacyanate strain (Fitc) was used as a fluorescent tracer. The concentration of the working solution of IgG-Fitc was 1.25 μg ml^{-1}. Initial and final (after 3 h incubation) values of polarization of fluorescence were measured with TDX Analyzer (ABBOT Laboratories, USA). They were used to calculate changes in the polarization of fluorescence of a given sample according to the equation

$$Y = \frac{\text{pol}_{(o)} - \text{pol}_{(t)}}{\text{pol}_{(o)}t}, \tag{1}$$

where Y is the rate of proteolytic reaction, $pol_{(o)}$ and $pol_{(t)}$ are initial and final values of polarization of fluorescence, respectively, and t is the incubation time.

Since only comparative estimates of proteolytic activity were used in this study, the results were expressed in polarization units or in polarization units per unit of mass for mass- specific estimates.

Calculations

To calculate ingestion rates in terms of wet mass and mass- specific ingestion rates, body length-wet mass regressions (Balushkina & Winberg, 1979) were used.

In some series of experiments the observed ingestion rates were compared with predicted ingestion rates. Estimates of the predicted rates were made using the bioenergetic approach according to the equation

$$I = (P + R)/a, \qquad (2)$$

where I is ingestion, P is growth, R is respiration and a is assimilation efficiency. The assessment of R and P was based on the experimental data on respiration and growth (without diapause) for particular species of cyclopoids (Alekseev, 1990 and unpublished data). Assimilation efficiency was assumed to be equal to 0.6 and 0.8 for plant and animal food, respectively.

Results and discussion

Radiocarbon experiments

In ^{14}C uptake experiments with *Acanthocyclops vernalis* from temporary water-bodies of the Volga Delta, the ingestion rates differed considerably between four groups studied (Table 1). The highest rate was in females fed on detritus, while the lowest ingestion rate was observed in *Chlorella* typical diapause stage IV' group.

Figure 2 shows the mass-specific ingestion rates observed in these experiments compared with daily food requirements calculated on the basis of energy losses for respiration and normal growth. The ingestion rates of stage IV 'typical diapause' copepodids were 2.6 times lower than the ingestion rates of CIV at the initial stages of diapause, and 1.9 times lower than the ingestion rates of adult females. The observed mass-specific ingestion rates of 'typical diapause CIV' were 1.8 times lower than those required to cover daily energy losses for normal growth and respiration while the ingestion rates of adult females and CIV at the beginning of diapause were very close to calculated values.

In these experiments we did not attempt to prevent sinking of *Chlorella* cells. Observations have shown that at the end of the experiment cyclopoids were concentrated on the bottom of the vessels covered with a thin layer of algal cells. Although late copepodid instars and adult cyclopoids are usually considered as predators there are many indications on the importance of algae in their diet (Krylov, 1989). One may speculate that pre-diapausing cyclopoids and those in active diapause are even more adapted to feed on plant material than planktonic forms.

Table 1. Ingestion rates (IR) of various developmental stages of *Acanthocyclops vernalis* from temporary water-bodies of the Volga Delta feeding on ^{14}C-labelled detritus. CIII - stage III copepodids, CIV_0 - stage IV copepodids at the very beginning of the diapause, CIV_t - typical diapausing stage IV copepodids.

Stage	Temperature (°C)	IR (μg wet weight ind^{-1} d^{-1})	Replicates
CIII	24.0	14.1	3
CIV_0	24.0	21.7	3
CIV_t	24.0	8.4	3
Adult females	19.5	47.2	1

Table 2. Activity of proteases (10^3 polarization units mg^{-1}) of *Cyclops kolensis* and *Megacyclops gigas* from Mozhaisk Reservoir at different temperatures.

Stage and species	Temp. (°C)	Activity of proteases (mean ± SE)
Cyclops kolensis		
CIV	4	67.5±7.1
CIV	20	166.3±13.4
CV	4	51.6±2.9
CV	20	70.3±8.6
Adult females	20	63.6±7.4
Megacyclops gigas		
CIII	20	71.6±18.3
CIV	20	31.1±8.4

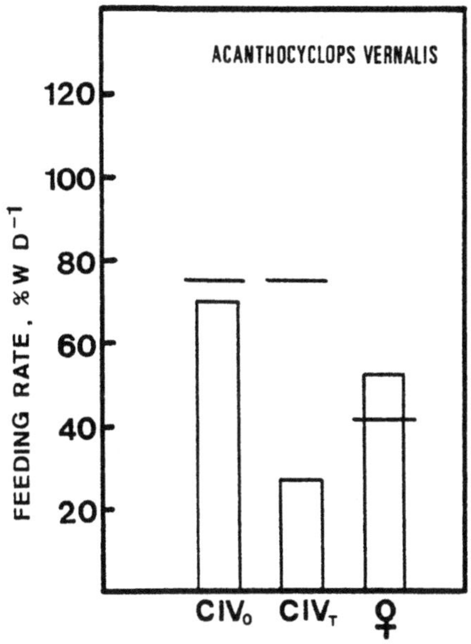

Fig. 2. Specific feeding rates of various developmental stages of *Acanthocyclops vernalis* from temporary water- bodies of the Volga Delta on ^{14}C-labelled algal detritus. CIV_o — CIV at the beginning of the diapause; CIV_T — typical diapause CIV; ♀ — adult females. Horizontal bars show the calculated specific feeding rates required for respiration and normal growth.

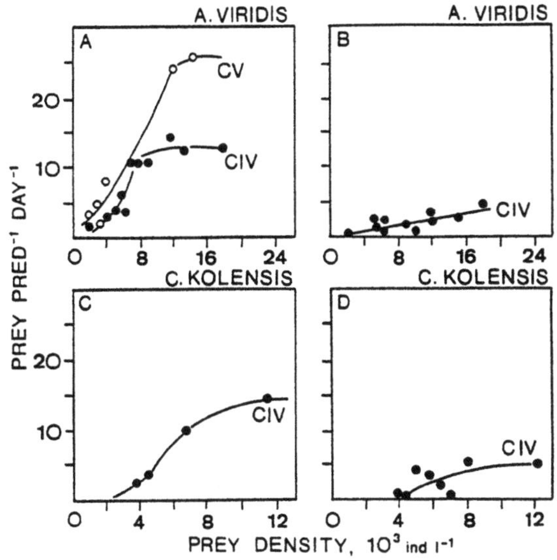

Fig. 3. Functional responses of *Acanthocyclops viridis* and *Cyclops kolensis* from temporary water-bodies of the Volga Delta preying on nauplii. A: CV (open circles) and non- diapausing CIV (black circles) of *A. viridis*. B: diapausing CIV of *A. viridis*. C: non-diapausing CIV of *C. kolensis*. D: diapausing CIV of *C. kolensis*. Curves are fitted by eye.

Predation experiments

Predation rates of *Acanthocyclops viridis*, *Cyclops kolensis* and *C. vicinus* from temporary water-bodies of the Volga Delta increased with increasing prey concentration (Figs. 3 and 4). In copepodids V and non-diapausing copepodids IV maximum daily predation rates (10–25 prey per predator) were attained at a concentration of approximately 10^4 nauplii per litre. Diapausing stage IV copepodids of all three species preyed upon nauplii at a lower rate. Daily predation rates of diapausing cyclopoids were always below 10 prey per predator. There was less evidence that a saturation was attained in the range of prey concentrations examined (up to 2×10^4 nauplii per litre) (Figs 3 and 4).

Predation rates of *C. vicinus* CIV at the initial stages of diapause were higher than those of CIV at more advanced stages of diapause (Fig. 4A). Figure 4B illustrates a day-by- day decrease in predation rates of diapausing CIV. In the last case the time factor seemed to be more important than prey concentration.

A similar pattern was demonstrated by stage IV copepodids of *Cyclops vicinus* from the plankton of Lake Sevan (Fig. 5). Their predation rates declined sharply with time and were unaffected by prey concen-

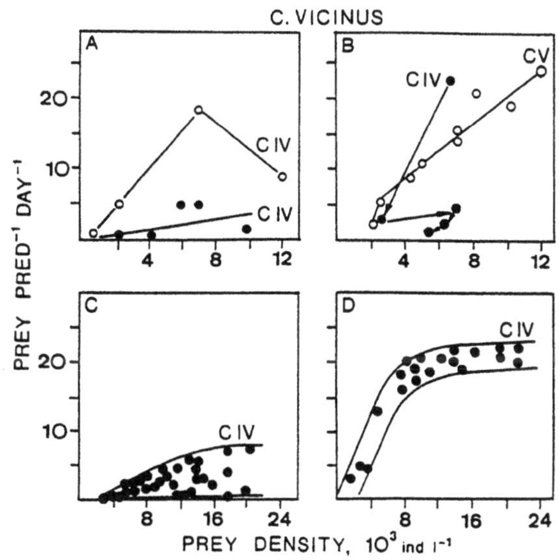

Fig. 4. Functional responses of copepodids of *Cyclops vicinus* from temporary water-bodies of the Volga Delta preying on nauplii. A: CIV at the very beginning of diapause (open circles) and 5–7 days later (black circles). B: CV (open circles) and diapausing CIV copepodids (black circles); arrows show the time sequence of experiments. C: the pooled data for diapausing CIV. D: the pooled data for non-diapausing CIV. Curves are fitted by eye.

tration. The predation rates of CIV *C. vicinus* collected from the meiobenthos of the lake were also low at all prey concentrations studied.

Stage IV copepodids of *Cyclops abyssorum sevani* demonstrated a typical functional response with saturation of predation rates at approximately 8×10^3 prey l^{-1}. However, the maximum daily predation rates of CIV *C. abyssorum sevani* were 1.3–2.3 times lower than those required to cover the energy losses for growth and respiration (Fig. 5).

Predation rates of CIV and CV of *Cyclops kolensis* collected in Mozhaisk Reservoir and acclimated to 20 °C over two weeks increased with increasing prey concentration (Fig. 6). As in the other experiments, the predation rates of diapausing stage IV copepodids were always lower than the predation rates of post-diapausing CV. At the highest prey concentration, 8×10^3 per litre, the ratio of CV and CIV predation rates ranged from 2.1 to 5.5.

The maximum feeding rates of non-diapausing cyclopoids (15–30 prey per day) were comparable with those observed in other functional response experiments with similar sized prey (Krylov, 1989). Saturation of feeding rates occurred at a concentration of approximately 8–10 10^3 prey per litre. Assuming the body mass of one nauplius to range between 0.5 and 1.0 μg (wet weight), this corresponds to a concentration of 4–10 mg per litre. These values of saturation prey density fell within the range reported for cyclopoids (0.4–20 × 3 mg l^{-1}; Krylov, 1989).

These comparisons suggest that the functional responses exhibited by non-diapausing cyclopoids in our experiments may be considered as 'typical' for cyclopoids preying on relatively small and mobile prey. The results of predation experiments show that cyclopoid copepodids in active diapause are capable of preying on planktonic animals, though at a much lower rate than non-diapausing stages.

Activity of proteolytic enzymes

At ambient temperature, 4 °C, enzyme activity in diapausing CIV of *Cyclops kolensis* from the near-bottom layers of Mozhaisk Reservoir was 1.96 times lower than in active CV from the pelagial. A physiological difference between these two groups of copepods was also evident when the copepods were transferred to the warmer water. After 15 days acclimation to 20 °C an almost 3-fold increase in proteolytic enzyme activity was observed in CIV while that of CV did not change significantly (Fig. 7).

In a further series of experiments the effect of temperature acclimation regimes on enzyme activity induction in stage IV copepodids of *C. kolensis* was tested (Fig. 8). Three-day acclimation to 20 °C did not change the level of proteases activity. However, as in the previous assays, after 15 days of exposure at 20 °C a considerable increase in proteolytic activity of enzymes was observed. Transfer from 20 ° to 4 ° resulted in dramatic decline of enzyme activity. The values observed in stage IV copepodids re-acclimated to 4 °C did not differ from the background enzyme activity of the substrate in the control series (Fig. 8). In relation to the body mass of the copepods, the activity of proteases in diapausing copepodids IV of *C. kolensis* acclimated over two weeks to 20 °C was higher than in any other stage of this species or in the larger *Megacyclops gigas* (Table 2).

Relatively little is known about variations in digestive enzymes in freshwater cyclopoids. It has been demonstrated that digestive enzyme activity in predatory cyclopoids is relatively high in comparison with that in herbivorous planktonic crustaceans (Dabrovski & Glogowski, 1977; Wynne & Gophen, 1981) and that both diversity and activity of enzymes increase with age (Defaye & Dussart, 1981). Most of the information concerning digestive enzyme activity in copepods comes from investigations on marine calanoids. The results of Hallberg & Hirche (1980), Hirche (1981, 1983), Tande & Slagstad (1982) and Head & Conover (1983) documented low enzyme activity during copepod diapause. Enzyme induction in copepods recovering from diapause requires 1 to 3 weeks, is a function of both presence of food and temperature, and is probably related to the development of gut epithelial cells (Hallberg & Hirche, 1980; Head & Conover, 1983). Hasset & Landry (1990) have demonstrated that in post-diapause period enzyme induction and feeding capabilities develop concur- rently. However, they become less closely coupled later in the season.

The results of this study are in part consistent with these findings. At ambient temperature, the activity of non-specific proteases in diapausing stage IV copepodids of *Cyclops kolensis* was half that of post-diapausing stage V copepodids (Fig. 7). Increased temperature in the presence of food had a positive effect on enzyme activity. Approximately two weeks acclimation to 20 ° was required to induce changes in proteases activity.

Similar effect of temperature acclimation regime on respiratory enzyme activity was observed in experiments with diapausing crayfishes, *Astacus astacus*.

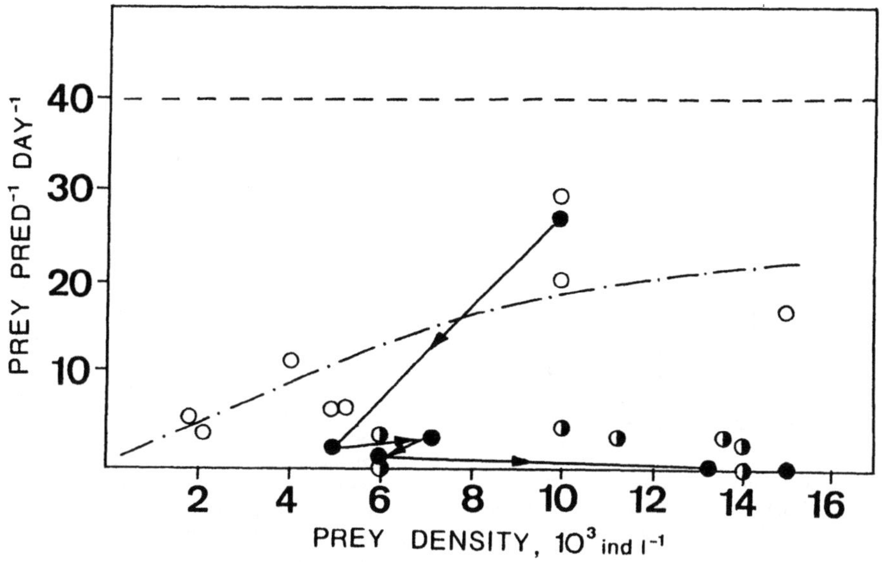

Fig. 5. Functional responses of stage IV copepodids of *Cyclops abyssorum sevani* and *Cyclops vicinus* from Lake Sevan, Armenia, preying on nauplii. Open circles: *C. abyssorum sevani* from the plankton. Black circles: *C. vicinus* from the plankton. Black and white circles: *C. vicinus* from the meiobenthos. Arrows show the time sequence of experiments. The broken line at top shows the predation rate required to cover the losses of energy for respiration and normal growth. Curves are fitted by eye.

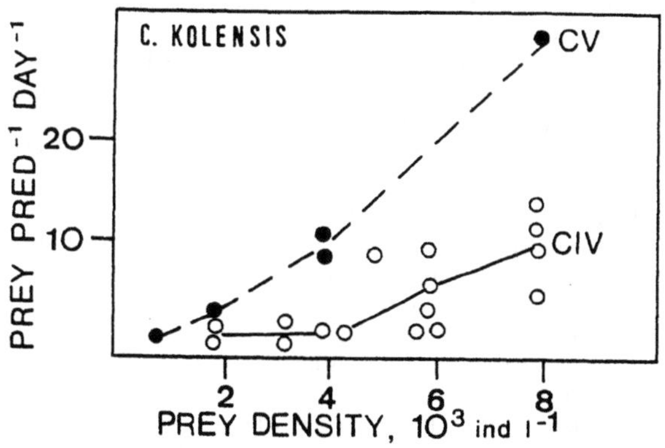

Fig. 6. Functional responses of CIV (open circles) and CV (black circles) of *Cyclops kolensis* collected in the meiobenthos of Mozhaisk Reservoir and preying on nauplii. Curves are fitted by eye.

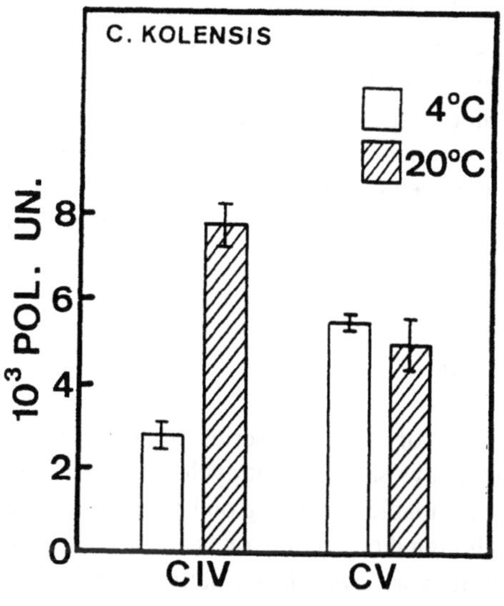

Fig. 7. Activity of proteolytic enzymes of stage IV and stage V copepodids of *Cyclops kolensis* from Mozhaisk Reservoir at 4 °C (open bars) and after 15 days acclimation to 20 °C (hatched bars). Values represent means ± 1SE.

The short 5 days acclimation to increased temperature did not change significantly the level of respiratory enzyme, succinat dehydrogenase (SDH). However, 10 days acclimation resulted in dramatic changes in SDH (Alekseev, 1989).

On the other hand, the observed increase in enzyme activity in diapausing stage IV copepodids of *Cyclops kolensis* was not accompanied by a similar rise in predation rates (Fig. 6). Although the factors that control the breaking of diapause in copepods are not well understood, it has been suggested that some internal rhythms are involved. Sudden increase in temperature, as in our experiments, may thus not be essential for the induction of a complex of physiological reactions associated with the end of diapause. Published data on crustaceans and insects provide evidence that sudden changes in conditions of overwintering may

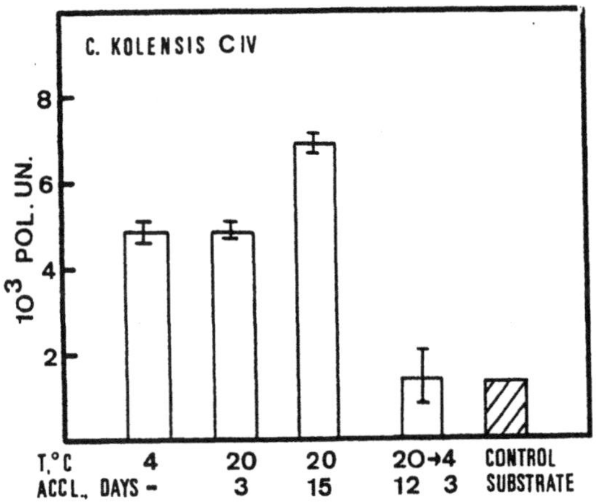

Fig. 8. Activity of proteolytic enzymes of stage IV copepodids of *Cyclops kolensis* from Mozhaisk Reservoir at different temperatures and acclimation times (see text for details). Values represent means ± 1 SE.

result in abnormal behaviour, including cannibalism, and increased level of mortality (see Danilevski, 1961; Alekseev, 1990 for review). One may suggest that temperature-dependent induction of enzyme activity without simultaneous increase in predation rates in *Cyclops kolensis* from Mozhaisk Reservoir was the consequence of unusual timing of emergence from diapause.

Conclusion

The results of all these series of experiments and enzyme assays showed that cyclopoids in active diapause are capable of feeding on both plant and animal food although their feeding rates and activity of digestive enzymes are lower then in active individuals. They also illustrate the importance of the time factor in determining the levels of feeding and digestive activity of cyclopoid copepods at both the beginning and the end of the diapause.

Acknowledgments

We are grateful to G. Fryer, K. Elgmork & M. B. Ivanova for their comments, constructive criticism and for improving the English text. T. L. Lazareva helped to draw the figures and to prepare the manuscript.

References

Alekseev, V. R., 1980. Osobennosti razvitiya veslonogikh rakov vo vremennykh vodoyomakh delty Volgi [Pecularities of the development of copepods from the temporary water-bodies of the Volga Delta]. Biol. vnutr. vod. Inform. byul. 47: 30–32 (in Russian).

Alekseev, V. R., 1981. Vliyanie diapauzy ciklopov na ikh biologiyu i produkciyu vo vremennykh vodoyemakh delty Volgi [Effect of diapause in cyclopoids on their biology and productivity in temporary water-bodies of Volga Delta]. In G. G. Winberg (ed.), Osnovy izucheniya presnovodnykh ekosistem. Zoological Institute, Leningrad: 74–80 (in Russian).

Alekseev, V. R., 1989. Vliyanie diapauzy na uroven' dykhatel' noy aktivnosti rechnykh rakov [Influence of diapause on the level of respiratory activity in crayfishes]. Sbornik nauchnykh trudov GosNIORKH, 300: 80–94 (in Russian).

Alekseev, V. R., 1990. Diapauza u rakoobraznykh: Ecologo-fiziologicheskie aspekty [Diapause in crustaceans: Ecological and physiological aspects]. Nauka, Moscow, 144 pp. (in Russian, English summary).

Alekseev, V. R. & P. I. Krylov, 1981. Vliyanie fiziologicheskogo sostoyaniya ciklopov na skorost' ikh pitaniya [Influence of the physiological state of cyclopoid copepods on their feeding rate]. Sbornik nauchnykh trudov GosNIORKH, 161: 130–134 (in Russian).

Alekseev, V. R., A. A. Simonyan & K. Djenderedjian, 1991. The role of diapause in the life cycle of a planktonic cyclopoid in Lake Sevan. Mat. Int. Symp. Limnol. of Mountain Lakes, Stara Lesna: 6.

Balushkina, E. V. & G. G. Winberg, 1979. Zavisimost' mezhdu massoy i dlinoy tela u planktonnykh zhivotnykh (Relationship between mass and body length in planktonic animals). In G. G. Winberg (ed.), Obschie osnovy izucheniya vodnykh ekosistem. 'Nauka', Leningrad: 169–171 (in Russian).

Bykov, V. D. & K. K. Edelstein (eds), 1979. Mozhaiskoye vodokhranilishche [Mozhaisk Reservoir] (Complex investigations of water reservoirs, vol. 3). Moscow University Press, Moscow, 399 pp (in Russian).

Dabrowski, K. & J. Glogowski, 1977. Studies on the proteolytic enzymes of invertebrates constituting fish food. Hydrobiologia 52: 171–174.

Danilevsky, A. S., 1961. Fotoperiodism i sezonnoe razvitie nasekomykh [Photoperiodism and seasonal development in insects]. Izdat. Len. Gos. Univ., Leningrad, 244 pp. (in Russian).

Defaye, D. & B. Dussart, 1981. Variations de quelques activités enzymatiques au cours du dévelopement de *Macrocyclops albidus* (Crustacé, Copépode). Vie Milieu 31: 41–47.

Elgmork, K., 1959. Seasonal occurrence of *Cyclops strenuus strenuus*. Folia limnol. scand. 11: 1–196.

Elgmork, K., 1967. Ecological aspects of diapause in copepods. Proc. Symp. Crustacea 3: 947–954.

Elgmork, K., 1980. Evolutionary aspects of diapause in freshwater copepods. In W. C. Kerfoot (ed.), Evolution and Ecology of Zooplankton Communities. The University Press of New England, Hanover (N.H.): 411–417.

Frenkel, O. A., 1994. Bakterial'naya transformaciya organicheskikh veschestv pri geterotrofnom pitanii fitoplanktona. [Bacterial transformation of organic substances in heterotrofic feeding of phytoplankton]. Diss. kand. biol. nauk. Moscow State University, Moscow, 199 p. (in Russian).

Hallberg, E. & H.-J. Hirche, 1980. Differentiation of the mid-gut in adults and over-wintering copepodids of *Calanus finmarchicus*

(Gunnerus) and *C. helgolandicus* Claus. J. exp. mar. Biol. Ecol. 48: 283–295.

Hassett, R. P. & M. R. Landry, 1990. Seasonal changes in feeding rate, digestive enzyme activity, and assimilation efficiency of *Calanus pacificus*. Mar. Ecol. Progr. Ser. 62: 203–210.

Head, E. J. H. & R. J. Conover, 1983. Induction of digestive enzymes in *Calanus hyperboreus*. Mar. Biol. Lett. 4: 219–231.

Hirche, H.-J., 1981. Digestive enzymes of copepodids and adults of *Calanus finmarchicus* and *C. helgolandicus* in relation to particulate matter. Kieler Meerresforsch., Sonderh. 5: 174–185.

Hirche, H.-J., 1983. Overwintering of *Calanus finmarchicus* and *C. helgolandicus*. Mar. Ecol. Progr. Ser. 11: 281–290.

Kosova, A. A., 1960. Sezonnye izmeniya planktona i bentosa na poloyakh nizhney zony del'ty Volgi [Seasonal changes in plankton and benthos dynamics in the bayou lakes of the lower Volga Delta]. Trudy VGBO 10: 102–135 (in Russian).

Kosova, A. A., 1965. Zooplankton zapadnoy chasti del'ty Volgi v period zaregulirovaniya stoka [Zooplankton of the western part of Volga Delta during the period of flow regulation] In: Izmeneniye biologicheskikh kompleksov Kaspiyskogo morya za posledniye desyatiletiya [Changes in biological complexes of Caspian Sea during last decades]. Nauka, Moscow: 98–138 (in Russian).

Krylov, P. I., 1979. Intensivnost' pitaniya rastitel'nym detritom planktonnykh rakoobraznykh v Koporskoy gube [Intensity of feeding of planktonic crustaceans on plant detritus in Koporskaya Bay]. In G. G. Winberg (ed.), Experimental and field investigations of the logical bases of lake productivity. Zoological Institute, Leningrad: 41–49 (in Russian).

Krylov, P. I., 1980. Radioisotopic methods for the study of the nutrition of zooplankton and its transformation of food energy. Hydrobiol. J. 16: 52–68.

Krylov, P. I., 1989. Pitanie presnovodnogo khischnogo zooplanktona [Feeding of freshwater predaceous zooplankton]. VINITI Inf. Publ. Ser. General Ecology. Biocenology. Hydrobiology. Vol. 7. VINITI Publishers, Moscow, 146 pp. (in Russian).

Meshkova, T. M., 1953. Zooplankton ozera Sevan: Biologiya i produktivnost' [Zooplankton of Lake Sevan: Biology and productivity]. Tr. Sevansk. Ghidrobiol. Stancii 13: 5–170 (in Russian).

Nilssen, J. P., 1978. On the evolution of life histories of limnetic cyclopoid copepods. Mem. Ist. Ital. Idrobiol. 36: 193–214.

Sadchikov, A. P., O. A. Frenkel & S. A. Eriomin, 1994. Opredelenie fermentativnoy aktivnosti v vodnykh ecosistemakh metodom poliarizacii fluorescencii. [Determination of enzymatic activity in aquatic ecosystems by the method of polarized fluorescence]. Hidrobiol. J., *in press* (in Russian, English summary).

Sakharova, M. I., 1973. Donnaya faza v zhiznennykh ciklakh massovykh vidov pelagicheskikh ciclopov Uchinskogo vodokhranilishcha [Benthic stage in the life cycle of mass species of pelagic cyclopoids from Ucha Reservoir]. In: V. D. Bykov *et al.* (eds.). Complex investigations of water reservoirs 2: 122–129 (in Russian).

Sakharova, M. I., 1979. Microbenthos. In: V. D. Bykov & K. K. Edelstein (eds.). Mozhaisk Reservoir. Complex investigations of water reservoirs 3: 301–305 (in Russian).

Sakharova, M. I., N. Y. Sokolova & K. K. Edelstein, 1993. Monogoletnie izmeneniya struktury soobshchestv i nekotorye meropiyatiya po snizheniyu tempov evtrofirovaniya [Long-term changes in community structure and some measures for decreasing the rate of eutrophication]. Biol. Nauki 1: 66- -73 (in Russian), English summary).

Savitskiy, A. P., 1990. Fluorescentniy analiz: immunoanaliz, gibridniye DNK, biosensory [Fluorescent analysis: immun-analysis, DNA-hybrids, biosensors]. Uspekhi Biol. Khimii 31: 209–240 (in Russian).

Simonyan, A. A., 1991. Zooplankton ozera Sevan [Zooplankton of Lake Sevan]. Izdat. AN Armenii, Erevan, 299 pp. (in Russian).

Tande, K. S. & D. Slagstad, 1982. Ecological investigation on the zooplankton community of Balsfjorden, northern Norway. Seasonal and short-time variations in enzyme activity in copepodite stage V and VI males and females of *Calanus finmarchicus* (Gunnerus). Sarsia 67: 63–68.

Wierzbicka, M., 1962. On the resting stage and mode of life of some species of Cyclopoida. Pol. Arch. Hydrobiol. 10: 215–229.

Wynne, D. & M. Gophen, 1981. Phosphatase activity in freshwater zooplankton. Oikos 37: 369–376.

Hydrobiologia **320**: 81–82, 1996.
V. R. Alekseev & G. Fryer (eds), Diapause in the Crustacea.

On various types of antifreezes in active and resting stages of organisms

V. V. Khlebovich
Zoological Institute, Russian Academy of Sciences, St.-Petersburg 199034, Russia

Key words: ice-nucleating agents, biological antifreezes, resting stages

Abstract

A group of monomolecular antifreezes, glycerol, amino acids and glucose may be functioning in this role only at the stages of diapause. Use of high-molecular specific glycoproteins or proteins as cryoprotectors is characteristic for the active stages of hydrobionts.

At high latitudes, low temperatures threatening organismic freeze-up, represent rigid ecological conditions controlling a set of special adaptations of animals and plants. Since a most important threat is the formation of ice crystals inside cells, such a type of adaptation is represented by the recently described so called ice nucleating agents, which, by provoking intercellular ice formation, inhibit the formation of intracellular ice (Aunaas & Denstad, 1988; Durñan, 1988; Storey, 1988). The commonest ice nucleating agents are polyoles, particularly sorbitol. It is perfectly evident that the substances provoking ice nucleating have no right to be termed antifreezes: they are only organizing the process in such a way that it brings the least peril to cells and hence to the organism as a whole.

Actual antifreezes may be classed into two large groups. The first incorporates widely distributed low molecular substances: amino acids, glycerol, glucose. They are formed in tissues by the winter period from the reserve of proteins, fats and glycogen previously accumulated in autumn. Since, according to the laws of osmosis, reduction of the freezing point of a solution is a function of particle concentration irrespective of their size, the decomposition of polymer molecules into hundreds and thousands of monomer molecules results in a considerable reduction of ice formation temperature. Besides, glycerol itself has a very low freezing point, and the same characteristic, though to a lesser degree, is shared by its aqueous solutions. Therefore glycerol is the most widespread biological antifreeze of a monomolecular nature forming in autumn from the accumulated stock of macromolecular substances. Antifreezes of first group are present in practically all classes of terrestrial and freshwater invertebrates as well as in a number of fishes, amphibians and reptiles wintering at negative temperatures (see, for example, Prosser, 1973; Precht *et al.*, 1973; Kuzmin, 1990).

The second group has been known only since the middle of this century. These antifreezes are high-molecular glycoproteins or, less commonly, proteins capable of forming a kind of meshwork on the surface of immediately-formed ice crystals, thus preventing further organismic damage. So, isolated, an ice crystal neither proceeds to grow nor becomes a centre of crystallization. In this way organismic freezup is blocked. Antifreezes of the second group are mostly typical for fishes of polar zones inhabiting the region year-round or only in winter at temperatures below blood plasma freezing point (DeVries, 1976; MacDonald *et al.*, 1987; Johnston, 1990). Antifreeze-glycoproteins have also been described in an invertebrates, the bivalve *Mytilus edulis* in winter (Theede *et al.*, 1976).

In our opinion, the existence of two types of antifreezes may be explained by differences in biochemical activity of the corresponding substances, and differences in the mode of life of animals.

Antifreezes of the first group are always formed in inactive stages: cysts, resting eggs, or during hibernation. They are antifreezes of winter diapause connected with the maximum depression of metabolism. Glucose, amino acids and glycerides are the primary elements of anabolism and the final elements of catabolism. It is in this form that the substances get into the organism directly through the alimentary system and are distributed within it. Control of the flow of glucose, amino acids and glycerides (substances of tremendous biological activity) is the most important task of the organisms (for example, one should recall the problem of diabetes). Oversaturation of an organism with these substances is more dangerous than, say, the cooling of an automobile engine with gasoline instead of water. Antifreezes of this group cannot be acting as such in other instances but only at the resting stage with depressed metabolism.

Conversely, antifreezes of the second group are metabolically relatively immobile and are characteristic for active life stages which are represented by fishes constantly living under the threat of freezing up, as well as molluscs — filtrators continuously feeding by the activity of ciliary epithelium of the gills in conditions of a freezing intertidal zone. In these conditions, use of glycerol, glucose and amino acids is evidently unsatisfactory.

For practical purposes, the concept of two types of antifreezes may have predictive value as, judging from the mode of life, we may assume the presence of antifreezes of one or another type.

References

Aunaas, T. & J.-E. Denstad, 1988. Triggering of ice nucleator activity in the blue mussel (*Mytilus edulis*). CRYO-Lett. 9, 427 p.

DeVries, A. L., 1976. Antifreezes in cold-water fishes. Oceanus 19: 23–31.

Duman, J. G., 1988. Insect and centipede ice nucleating agents. CRYO-Lett. 9, 430 pp.

Johnston, I. A., 1990. Cold adaptation in marine organisms. Phil. Trans. Roy. Soc. London 326 B: 655-666.

Kuzmin, S. L., 1990. Cold resistance in animals. Priroda (Moscow) 10: 108–109. (in Russian).

MacDonald, J. A., J. C. Montgomery & R. M. G. Wells, 1987. Comparative physiology in antarctic fishes. Adv. Marine Biol. 24: 321–388.

Precht, H., J. Christophersen, H. Hensel & W. Larcher, 1973. Temperature and life. Springer Verlag, Berlin, 779 pp.

Prosser, C. L., 1973. Temperature. In C. L. Prosser (ed.), Comparative Animal Physiology. W. B. Saunders Company, Philadelphia *et al.*: 356–432.

Storey, K. B., 1988. Biochemistry of cold hardiness in invertebrates. CRYO-Lett. 9: 438–439.

Theede, H., R. Schneppenheim & L. Béress, 1976. Frostschutz-Glukoproteine bei *Mytilus edulis*? Marine Biol. 36: 183–189.

Hydrobiologia **320**: 83–84, 1996.
V. R. Alekseev & G. Fryer (eds), Diapause in the Crustacea.

The susceptibility to loss of diapause capacity in hydrobionts of ephemeral waterbodies

V. V. Khlebovich
Zoological Institute, Russian Academy of Sciences, St. Petersburg 199 034, Russia

Key words: ephemeral water bodies, life cycle, resting stages

Abstract

The life cycle strategy of inhabitants of intermittent water bodies includes a short period of intensive growth and reproduction and an indispensable long term pause. The frequent development of so called laboratory cultures of species originating from such habitats which are able to reproduce persistently without undergoing such diapause indicates that this reproductive ban can be lost relatively easily.

Researchers who have had the good fortune to work with fauna of small tundra lakes and intermittent spring pools or ponds of the arid zone, have always been amazed by a surprising similarity between those complexes so far apart from each other. This similarity is explained by the predominance of taxa capable of activity for no more than two-three months a year. Primarily these are species of Anostraca, Notostraca and Cladocera. According to environmental conditions, the life cycle strategy of inhabitants of the two types of ephemeral water bodies, arctic and arid, lies in strict observation of two principles.

The first consists of the maximal and intensive utilization of short term conditions favourable for life activity and the awift accumulation of biomass and production of offspring.This principle is best satisfied by continuous asexual or parthenogenetic reproduction characteristic of the majority of the inhabitants of open water area of intermittent water bodies.

The second principle is the requirement for cessation of reproduction and population growth, and transition into resting stages prior to the time of complete freezing up of the water body in autumn in the North or its complete drying in summer in the South. A ban on year-round reproduction is absolute in these cases, and violators are inevitably eliminated the same year through freezing in one case and drying in another. It is common knowledge that, in the drying conditions, the resting stages of Crustacea are equivalent to embryonic diapause (Makrushin, 1981; Alekseev, 1990); evidently, the same phenomenon occurs upon adaptation to complete freezing up of water basins.

The ecological group of ephemeral water basin organisms present much interest for purposes of practical hydrobiology, i.e. aquaculture, if a possibility is found of retaining a high productive population potential and of removing the reproduction ban. In this case the activity should be aimed at eliminating the resting stage in the life cycle.

This may be assumed to be theoretically feasible, though in the rigid conditions of persitent freezing or drying of a water body, the genetic diversity with respect to the presence or absence of resting stage may prove inadaptive. Diapause is based on a multichain cascade of events which may be determined polygenically. Total loss of the ability to undergo diapause developed through high evolutionary endeavour, may result from of a single disturbance in any chain of biosyntheses. Evidently, this occur frequently in water bodies, but, lacking sufficient capabilities to form resting stages, the individuals are eliminated as totally inadaptive to the particular conditions of ephemeral water basins. The relative facility of formation of so called laboratory cultures from conventional diapausing species may serve an implicit proof of the above assumption. Evidently, it is not without purpose that a large number of promising laboratory cultures has been originated from species of inhabitants of ephemeral water bodies or those ecologically close thereto (intermittent ponds, high salinity lakes and lagoons). A most amazing feature in this case is that such types of laboratory cultures have originated as if by themselves; without the special efforts of researchers. When their owners are asked whether these cultures are capable of producing resting stages, cysts, eggs or ephippia, they remain without any definite answer or description of the procedure. Some information on laboratory culture developed on the basis of loss of capability for formation of resting stages in given below:

Fabrea salina *(Infusoria, Heterotricha).* A typical inhabitant of saline lakes and marine lagoons commonly intermittent. In natural conditions, it variably forms cysts as an adaptation to probable drying out. From the École Normale Supérieure (Paris), where investigations of genetic variability in terms of encystment conditions were made (Génermont & Wright, 1965), I received a laboratory culture in 1972. From this I failed to obtain any cysts.

Daphnia magna *(Crustacea, Anomopoda).* A widely distributed inhabitor of freshwater ponds and lakes, with seasonal diapause in a winter foodless period. A laboratory culture of Daphnia which has lost the capability for the formation of ephippia has been kept in the Laboratory of Toxocology of State Institute of Lake and River Fisheries in St.-Petersburg since the mid-60 (Lesnikov, 1967).

Moina macrocopa *(Crustacea, Anomopoda).* An inhabitant of continental, often highly mineralized, temporary water bodies of the arid zone, where the formation of ephippia is indispensable. A laboratory culture kept for a long time in the Institute of Biology of Inland waters of the Russian Academy of Sciences (Borok) as well as at the Institute of Zoology, Byelorussian Academy of Sciences (Minsk) has evidently lost the ability to form resting stages.

Artemia parthenogenetica *(Crustacea, Anostraca).* The species has been separated from the widely known *A. salina* from which, as is clear from its name, it is distinguished by dominance of parthenogenetic over sexual reproduction. Initially observed from saline lakes of China (Cai, 1985), the species has now been reported from other sites, paricularly from the Mediterranean coast (MacDonald & Browne, 1990). Evidently, this species is more promising in the capacity of a laboratory species than *A. salina.*

Brachionus plicatilis *(Rotatoria).* An inhabitant of intermittent saline lakes and lagoons of the arid zone. In natural conditions, intensive parthenogenetic reproductions, prior to water basin drying out, and before winter, alternates with bisexual reproduction with the formation of resting eggs. Mutation (N.R.E.) of this rotifer is incapable of sexual reproduction and hence of diapause, and is used in laboratory culture in Israel for its high-productivity and stability (Minkoff *et al*, 1984–85).

The cited examples, far from being complete, generally validate the concept of relative facility of dropout of dormant stages from the life cycle of inhabitants of ephemeral and similar water bodies. The high efficiency of artificial selection of the principle of diapause was interpreted by Danilevsky (1961) as a proof of the genetic nature of variations from the normal type of seasonal development. The probable inadaptability of such genetic variability in populations of ephemeral water bodies deserves emphasis. The relatively easy loss of dorminant stages may be explained by either direct mutation leading to 'breakage' of some sections in a chain of biosynthesis controlling the diapause, or loss of phenotypically controlled genes in their dormant condition when they are not subject to selection (Labas & Khlebovich, 1976).

References

Alekseev, V. R., 1990. Diapause of Crustacea. Ecological and physiological issues. Nauka, Moscow, 144 pp. (in Russian).

Cai, Y., 1985. Observations o, parthenogenetic and bisexual brine shrimp from the People's Republic of China. Abstr. Second Internat. Symposium Brine Shrimp *Artemia*, 85 pp.

Danilevsky, A. S., 1961. Photoperiodicity and seasonal development of insects. Ed. by Leningrad University, 243 pp. (in Russian).

Génermont, J. & M. Wright, 1965. Aspects génétiques de la multiplication et de l'enkystment chez *Fabrea salina* Henneguy (Ciliata Heterotricha). Protistologica 1: 105–119.

Labas, Ju. A. & V. V. Khlebovich, 1976. 'Phenotypical window' of gnome and progressive evolution. Exploration of the fauna of the seas. Zoological inst. Acad. Sci. USSR 17 (25): 4–25. (in Russian).

Lesnikov, L. A., 1967. Daphnia as test-organism when establishing the degree and character of impact of waste water on fishery basins. Author's abstract of Doctor degree thesis, State Institute of Lake and River Fisheries, Leningrad, 21 pp. (in Russian).

MacDonald, G. H. & R. A. Browne, 1990. Population dynamics of an asexual brine shrimp *Artemia* population. J. Exp. Mar. Biol. Ecol. 133: 169–188.

Makrushin, A. V., 1981. Dormant embryos and buds of protoaqueus metazoan invertebrates. J. General Biol. (Moscow) 42: 834–843. (in Russian).

Minkoff, G. E., E. Lubzens & E. Meragelman, 1984–85. Improving asexual reproduction rates in a rotifer (*Brachionus plicatilis*) by salinity manipulations. Isr. J. Zool. 33: 195–203.

Hydrobiologia **320**: 85–97, 1996.
V. R. Alekseev & G. Fryer (eds), Diapause in the Crustacea.

Diapause, quiescence, hatching requirements: what we can learn from large freshwater branchiopods (Crustacea: Branchiopoda: Anostraca, Notostraca, Conchostraca)

Luc Brendonck
Royal Belgian Institute of Natural Sciences, Freshwater Biology, Vautierstraat 29, B-1040 Brussels, Belgium

Key words: Branchiopoda, diapause, quiescence, hatching requirements, adaptations

Abstract

The extent to which dormancy in large freshwater branchiopods is controlled endogenously (diapause) or exogenously (quiescence) is not always clear. It is assumed that both processes occur even within the same brood. Based on the effectiveness of common diapause-deactivating processes such as desiccation, hibernation, and resting, it can be stated that diapause is not a general process controlling responsiveness of large freshwater branchiopod eggs. Only in limited cases unequivocal evidence for the positive influence of these treatments is found.

With few exceptions, hatching of activated cysts is effected by specific conditions that may even differ among conspecific populations. Generally, each species (or even population) has a specific temperature range or regime for optimal hatching performance. In a suitable thermal environment with sufficient light and oxygen, hatching is invariably invoked by a low osmotic medium.

The erratic hatching pattern in most species is thought to be an adaptation to the variable temporary habitat. Hatching is generally spread over several days or even weeks, but the highest peak usually occurs on the first or second day of hatching. Low hatching percentages were found only in subtropical/desert species and may be a reflection of the low chances for successful reproduction. Generation carry-over of propagules in the egg bank by dormancy, and hatching at low conductivity, are, together with obligate oviparity and absence of an asexual life cycle phase, thought to be highly adaptive to the temporary environment. This enabled large branchiopods to survive since the Upper Cambrian.

Introduction

Large freshwater branchiopods, descriptively termed phyllopods, typically inhabit aquatic environments which regularly or irregularly dry out, freeze, or show striking changes in water level. Such astatic waters occur in all climatic regions except the tropics, but are most abundant in arid and semi-arid regions.

The hazardous, often unpredictable nature of the temporary habitat is the most important factor determining the life-history of these species. Some basins may be dry for several years; others are inundated for a few weeks or days only. Some ponds may be flooded several times a year; others only have a single wet phase. Freshwater phyllopods survive the adverse period by the production of resistant shelled dormant embryos, so-called resting eggs or cysts.

After deposition, resting eggs generally sink to the bottom or are attached to plant leaves or sediment as in notostracans (Longhurst, 1955). These propagules are usually produced in large quantities from an early stage onwards to build up a cyst bank. Part of the cyst bank hatches in response to suitable environmental conditions and synchronizes the life cycle with a favourable environment for growth and reproduction.

A major threat for many temporary pond species is early drying of the pond. Several papers on field studies, indeed, report that drying occurred before the organisms attained sexual maturity (Moore, 1955; Bishop, 1967b, 1968; Hildrew, 1985). Such so-called abortive hatchings may considerably reduce the available cyst bank or eventually lead to extinction of the population. This is especially the case in small subtropical pools where egg banks are often very small (pers. observation), or at the onset of colonization of a new habitat. Large branchiopods, however, have successfully survived since upper Cambrian, in the extreme, but at the same time almost enemy-free space (Fryer, 1986) provided by temporary pools by a combination of peculiar quantitative and qualitative reproductive traits (for a review see Brendonck & Persoone (1993)). Here, only some qualitative characteristics of the resting eggs will be considered. Millions of years of selection on the dormant stages have resulted in complex strategies that help to ensure the perpetuation of the group.

Since the phenology of ponds (duration of wet phase, number of floodings, etc.) is largely determined by pond morphometry and climate, we expect different hatching patterns (rate, percentage, variability) and hatching requirements in different types of ponds (e.g. small rock pools, large pans, etc.) at different altitudes and climatic regions.

As a first step in a study of how dormancy and hatching may be involved in the adaptation to the habitat, I present here a summary of most available literature on hatching requirements and phenology of hatching in large freshwater branchiopods. About 150 papers were considered. This data base may help to organize the plethora of available information as a basis for future ecological studies.

Definitions, terminology

The terms resting or dormant as used above are very general and need further specification. Many definitions and distinctions of dormant stages exist that differ according to author and to group of organisms. Here the definitions by Lavens & Sorgeloos (1987), and Drinkwater & Clegg (1991) as proposed for the brine shrimp *Artemia* are adopted. Diapause is taken to mean 'the state of dormancy where an arrest of development is initiated by internal factors'. Quiescence refers to 'a hypo-metabolic state imposed by unfavourable external conditions'. In the latter case, normal metabolism and development are resumed as soon as favourable conditions are restored. The term activated cysts is used to refer to cysts 'whose diapause has been terminated and which may hatch immediately or stay in a quiescent phase until the environment turns favourable'. Dormancy, dormant, and resting refer to any hypometabolic state whether intrinsic or extrinsic.

Not much is known about the physiological processes taking place during dormancy in large freshwater branchiopods. In general, the carbohydrate metabolism seems to be essential for survival and emergence in different types of drought-resistant stages of various types of organisms including plant seeds and fungal spores (Ewing & Clegg, 1969; Crowe & Clegg, 1973, 1978; Crowe *et al.*, 1984; Leopold, 1986; Crowe & Crowe, 1986; Panek *et al.*, 1986). The disaccharide trehalose, for example, has also been found to preserve membrane integrity during dehydration in *Artemia* (Slegers, 1991; Drinkwater & Clegg, 1991) and to serve as a source of carbon for glycogen and glycerol synthesis in activated cysts (Dutrieu, 1960; Clegg, 1964; Drinkwater & Clegg, 1991). The accumulation of glycerol inside the egg shell facilitates osmotic water uptake in activated cysts until the egg shell is broken (= breaking) (Clegg, 1964; Conte *et al.*, 1977; Slegers, 1991). Hatching is then supposed to proceed further by the action of hatching enzymes on the remaining egg membranes as shown in Artemia (Clegg, 1964; Sato, 1967; Slegers, 1991), and the fairy shrimp *Streptocephalus dichotomus* (Munuswamy, 1987), and is completed by the physical struggle of the first naupliar stage rupturing through the ultimate membrane.

General

That cysts of large freshwater branchiopods usually do not hatch immediately after deposition even if environmental conditions are still favourable for growth and reproduction, has led to the assumption that the onset of dormancy is rigidly programmed into embryonal development and that cysts are therefore diapausing (Belk & Cole, 1975). Several authors, however, have recorded from field and laboratory observations that a portion of the freshly deposited cysts is able to hatch after renewal or dilution of the original medium (for a review see Brendonck *et al.*, 1993). These cysts fit the definition of quiescent cysts. True diapausing cysts, do not hatch even when environmental conditions are favourable.

It seems therefore that both diapause and quiescence occur in freshwater branchiopods, and that hatching may be effected by activating diapausing cysts and exposing them to a suitable environment and by stimulating quiescent cysts with favourable conditions. In most literature available to date, however, no distinction is made between these processes. Here, an attempt is made to organize literature information

Table 1. Drought-resistance and effect of desiccation on hatching performance in large freshwater branchiopods. (part)= only a fraction of the eggs was affected by drying treatment (see text).

	Tolerate	Required	Stimulate	Author
ANOSTRACA				
Branchinecta lindahli	+		+(part)	Belk, 1973
B. mackini	+	+(part)		Brown & Carpelan, 1971
B. packardi	+		+	Prophet, 1963a
Branchinella kugenumaensis	+		+	Raj, 1961
Branchipus schaefferi	+	–		Mathias, 1934
B. stagnalis			+	Mathias, 1937
Chirocephalus diaphanus	+	–	+	Hall, 1953,1959b
	+	–	–	Mathias 1926, 1934, 1937
Eubranchipus bundyi	–			Broch, 1965
E. gelidus	+	+		Mozley, 1932
E serratus	–			Dexter & Ferguson, 1943
E. vernalis	+		+(part)	Weaver, 1943
		–		Castle, 1938; Avery, 1939
Siphonophanes grubei	–			Mossin, 1986
Streptocephalus dichotomus	+	+(part)		Sam & Krishnaswamy, 1979
		+		Bernice, 1972
			+	Bond, 1934
S. macrourus	+	–		Mitchell, 1990
S. proboscideus	+	+(part)		Brendonck *et al.*, 1993
			+	Carlisle, 1968
S. seali	–			Moore, 1957,1959b, 1967
		+(part)		Bulkowski & Meade, 1986
S. sudanicus	+	+(part)		Brendonck *et al.*, 1993
S. vitreus	+	+(part)		Carlisle, 1968; Hildrew, 1985
Tanymastix stagnalis	+	+(part)		Müller, 1918; Freiner & Grüttner, 1984; Al-Tikrity & Grainger, 1990; Grainger, 1991
NOTOSTRACA				
Lepidurus arcticus		–		Fryer (pers. comm.)
L. couesii	+	+(part)	+	Hann & Lonsberry, 1991
L. packardi	+	–		Ahl, 1991
Triops cancriformis	+	–	+	Hempel-Zawitkowska, 1971
	+		+	Fox, 1949
	+	+(part)		Takahashi, 1977b
T. granarius	+	–		Seaman *et al.*, 1991
	+	+(part)	+	Takahashi, 1977a
			+	Carlisle, 1968
T. longicaudatus	–	–	–	Scott & Grigarick, 1979
	+		+	Fry & Mulla, 1992
	+	+(part)	+	Takahashi, 1977a
'CONCHOSTRACA'				
Caenestheriella gynecia	+	–	–	Mattox & Velardo, 1950
Cyclestheria hislopi	+			Paul & Nayar, 1977
Eocyzicus klunzingeri	+	–		Brendonck *et al.*, 1993
Eulimnadia agassizii	+	+(part)		Zinn & Dexter, 1962
E. antlei	+	+(part)	+	Belk, 1972
E diversa	+	+(part)	+	Stern & Stern, 1971
Limnadia stanleyana	+	+(part)	–	Bishop, 1967a,b, 1968

Table 2. Effect of a low osmotic pressure on hatching performance in large freshwater branchiopods.

	Stimulates	Author
ANOSTRACA		
Branchinecta gigas	+	Daborn, 1975
B. mackini	+	Brown & Carpelan, 1971; Daborn, 1977
Chirocephalus diaphanus	+	Hall, 1953,1959b
Siphonophanes grubei	–	Mossin, 1986
Streptocephalus dichotomus	+	Bernice, 1972
	+	Sam & Krishnaswamy, 1979
S. seali	+	Moore, 1967
NOTOSTRACA		
Triops longicaudatus	+	Takahashi, 1977a
	+	Scott & Grigarick, 1979
'CONCHOSTRACA'		
Caenestheriella gynecia	+	Mattox & Velardo, 1950
Eocyzicus klunzingeri	+	Brendonck *et al.*, 1993
Eulimnadia antlei	+	Belk, 1972

Table 3. Effect of oxygen level on hatching performance in large freshwater branchiopods.

	High level stimulates	Low level stimulates	Author
ANOSTRACA			
Branchinecta mackini	±	–	Brown & Carpelan, 1971
Eubranchipus bundyi		+	Broch, 1965
E. holmani	+	–	Moore, 1963
Linderiella occidentalis	+		Eng *et al.*, 1990
Siphonophanes grubei	+		Mossin, 1986
Streptocephalus seali	+		Moore, 1963,1967
		–	Moore & Burn, 1968
NOTOSTRACA			
Triops cancriformis		–	Hempel-Zawitkowska, 1971
T. longicaudatus	+		Scott & Grigarick, 1979
'CONCHOSTRACA'			
Limnadia stanleyana	+		Bishop, 1967b

under different headings. All treatments with cysts in a dry state and/or related with resting conditions are considered as operating on the diapause state. Conditions that have an immediate positive effect on hatching during incubation are considered as effective on the quiescent state. The conditions required to complete successfully the entire sequence from diapause to hatching are indicated here as hatching requirements.

Hatching requirements

Activation of diapausing cysts

Desiccation

In most species, cysts tolerate drought but at least a fraction of them do not need desiccation as a prerequisite to hatching (Table 1). Only in *Streptocephalus dichotomus* cysts hatched exclusively after a dry period (Bernice, 1972). Cysts of *Eubranchipus gelidus* required to be dried and subsequently frozen in order to hatch (Mozley, 1932). In *Eubranchipus vernalis* cysts that were previously dried, hatched much earlier than undesiccated cysts (Weaver, 1943). In *Chirocephalus diaphanus*, a short period of desiccation retarded development, while an extended dry period enhanced development (Hall, 1953). In *Eulimnadia antlei* and *Branchinecta lindahli*, drying and rewetting stimulated hatching in cysts that did not hatch after a first incubation under poor conditions (Belk, 1972; Belk, 1973, respectively). On the contrary, extended periods of storage at low relative humidity were detrimental in *Streptocephalus seali* (Moore, 1967), *Triops longicaudatus* (Scott & Grigarick, 1979), and *Siphonophanes grubei* (Mossin, 1986). When embryos of *Eubranchipus bundyi* were in the prehatching phase they also no longer withstood desiccation (Broch, 1965). In several species, a fraction of the cysts need repeated drying and wetting in order to hatch, such as in *Branchinecta mackini* (Brown & Carpelan, 1971), *Eulimnadia antlei* (Belk, 1972), *Triops* spp. (Takahashi, 1977a, b), *Streptocephalus dichotomus* (Sam & Krishnaswamy, 1979), *S. vitreus* (Hildrew, 1985), *Tanymastix stagnalis* (Grainger, 1991), *S. proboscideus* and *S. sudanicus* (Brendonck *et al.*, 1993).

Hibernation

Low temperature treatment is necessary for successful hatching in *Limnadia stanleyana* (Bishop, 1967a), *Eubranchipus holmani* and *E. vernalis* (Modlin, 1982). Sometimes, hibernation has a positive influence on hatching phenology rather than on hatching percentage. In *Lepidurus couesii* the threshold temperature for hatching was less restrictive after a freezing shock (Hann & Lonsberry, 1991). In some cases hibernation stimulated hatching, as in *Triops cancriformis* where hatchability of frozen cysts was higher than in non-frozen cysts (Hempel-Zawitkowska, 1971). In *T. granarius*, hatching percentage increased with length of low-temperature treatment (Takahashi, 1977a). In *Chirocephalus diaphanus*, chilling increased hatching percentage but delayed the start of hatching (Nourisson, 1961). Cysts of *Eubranchipus gelidus* require to be dried and subsequently frozen before they will hatch (Mozley, 1932).

Resting period
The influence of length of resting period of cysts on hatching phenology has been tested in only a few species. Sometimes, a minimum period of rest (either in wet or dry condition) is required for optimal hatching. In *Streptocephalus seali*, for example, hatching was maximal after 4 months of storage in a wet condition and declined thereafter (Moore, 1967). In *S. dichotomus*, a period of 10–20 days of rest was most favourable (Bernice, 1972). Undesiccated cysts of *Triops granarius* need a rest of up to 20 days (Seaman *et al.*, 1991), while hatching of desiccated cysts first increased but decreased again when the drying period exceeded 70 days (Takahashi, 1977a). Fry & Mulla (1991) demonstrated an increase in hatch of *T. longicaudatus* with increased drying period. Hatching of dried cysts of *Streptocephalus macrourus* was maximal after 2 months but decreased thereafter (Mitchell, 1990). Hatching as a function of resting time was variable in *Triops granarius*; hatching percentage increased to a peak at 10 to 16 days, decreased to a minimum at 24 days of storage, then increased again (Takahashi, 1977a). Carlisle (1968) found that mud collected in November/December gave no hatching of Sudanese *Triops granarius*, *Streptocephalus proboscideus*, and *S. vitreus*, while samples collected at the beginning of March showed rapid hatching of all species. In *Limnadia stanleyana* Bishop (1967a) found that cysts collected from the field during the cooler months did not hatch as readily as those collected in the warmer months. Both observations may indicate that cysts first needed a period of rest during which diapause was deactivated.

Note: In several species freshly deposited cysts need a minimal period of rest before they are able to hatch (Avery, 1940; Prophet 1963a; Stern & Stern, 1971; Brown & Carpelan, 1971; Scott & Grigarick, 1979), or resist desiccation (Packard, 1875; Müller, 1918; Prophet, 1963a; Scott & Grigarick, 1979). Hatching usually does not begin until 3–7 days after ejection from the ovisac. This period, however, probably corresponds with additional embryonal development and has nothing to do with diapause.

Triggering of quiescent cysts

Reduced water level
In *Eubranchipus vernalis* (Castle, 1938), *Chirocephalus diaphanus* (Hall, 1959b, c), and *Streptocephalus seali* (Moore, 1951), hatching was positively affected by a reduction of the water level.

Low osmotic pressure
Where tested, hatching could in most species be invoked or augmented by renewal or dilution of the medium (Table 2). Usually, cysts did not hatch in habitat water or original culture medium, but did so after dilution or transfer into demineralized or distilled water. Some species, such as *Branchinecta lindahli*, however, hatch at a wide range of ionic strengths (Horne, 1967). Only in exceptional cases was some hatching observed in the original medium, as in *Branchipodopsis hodgsoni* by Sars (1898), *Chirocephalus diaphanus* by Hall (1959a), *Lepidurus arcticus* by Fryer (pers. comm.). In *Streptocephalus seali*, a few cysts hatched in the original medium after reduction of the volume of water (Moore, 1951) or after removal of adults (Moore, 1957).

Oxygen pressure
In most cases, low oxygen tensions inhibited or strongly reduced hatching (Table 3). Above a certain threshold level, generally no further influence could be detected. Only Broch (1965) suggested that lack of oxygen was the final stimulus for hatching in *Eubranchipus bundyi*; a result, however, put in serious question by Mossin (1986) who found that in *Siphonophanes grubei* final hatching was stimulated by an elevated pCO_2 taking effect only when sufficient oxygen was available. In *Triops longicaudatus*, cysts failed to hatch when covered with soil (Scott & Grigarick, 1979). This might be due to low oxygen levels or lack of light. It was assumed that exposure of cysts of *Triops cancriformis* to anaerobic conditions induced diapause (Hempel-Zawitkowska, 1971).

Table 4. Optimum incubation temperature for hatching in large freshwater branchiopods.

	Range	Optimum	Author
ANOSTRACA			
Branchinecta lindahli	5–25	15	Belk, 1973, 1977
B. mackini	1–32		Eng *et al.*, 1990
B. packardi	5–20	5–10	Belk, 1977
B. paludosa	5–10	5	Belk, 1977
Chirocephalus diaphanus	see text	see text	Nourisson, 1961
Eubranchipus bundyi	see text	see text	Broch, 1965
	5–10	5	Belk, 1977
E. serratus	6–15		Prophet, 1963b
	5–15	5	Belk, 1977
Linderiella occidentalis		10	Eng *et al.*, 1990
Siphonophanes grubei	see text	see text	Mossin, 1986
Streptocephalus dichotomus		30	Bernice, 1972
S. dorothae	15–35	15–30	Belk, 1977
S. mackini	15–40	25–35	Belk, 1977
S. macrourus		14–20	Mitchell, 1990
S. seali	10–28	21	Moore, 1963, 1967
	10–32	20	Prophet, 1963b
S. texanus	15–32	22–30	Home, 1967, 1971
Tanymastix stagnalis		10–15	Al-Tikrity & Grainger, 1990; Grainger, 1991
Thamnocephalus mexicanus	20–35	20–25	Belk, 1977
T. platyurus	17–32		Prophet, 1963b
	15–35	20–35	Belk, 1977
NOTOSTRACA			
Lepidurus couesii		20	Hann & Lonsberry, 1991
L. packardi		15	Ahl, 1991
Triops cancriformis	12–25		Gaschott, 1928
T. longicaudatus		22–30	Home, 1971
	14–29	22	Scott & Grigarick, 1979
'CONCHOSTRACA'			
Caenestheriella gynecia	13–38	28	Mattox & Velardo, 1950
C. setosa		25	Sublette and Sublette, 1967; Belk & Belk, 1975;
Eocyzicus spinifer	28–30		Khalaf & Al-Jaafery, 1985
Limnadia stanleyana		20	Bishop, 1967a

In *Streptocephalus macrourus* oxygen levels are neither stimulating nor inhibiting, and nauplii still hatched at dissolved oxygen tensions of less than 0.5 mg l^{-1} (Mitchell, 1990). In *Branchinecta mackini*, the effect of oxygen varied with salinity; While sufficient oxygen was necessary at low salinity, high oxygen concentrations became increasingly inhibitory at high salinities (Brown & Carpelan, 1971).

Low pH

A pH of 5–6 stimulated hatching in *Eubranchipus vernalis* (Dexter & Kuehnle, 1948), *Siphonophanes grubei* (Mossin, 1986), *Triops longicaudatus* (Scott & Grigarick, 1979), and *Streptocephalus proboscideus* (Brendonck, 1992).

Table 5. Quantitative hatching characteristics of large freshwater branchiopods.

	Start	Spread	Peak	%	Author
ANOSTRACA					
Branchinecta lindahli	d2			69–94	Belk, 1973, 1977
B. mackini	d1	>38d	d1/2	70–90	Brown & Carpelan, 1971
		>14d			Daborn, 1977
B. packardi	d1		d1/2		Prophet, 1963a
	d3	9d			Prophet, 1963a
				70–80	Belk, 1977
B. paludosa				80	Belk, 1977
Branchipus stagnalis	d5/6				Mathias, 1937
Chirocephalus diaphanus	d7–14	>8d	d2–3	20–90	Hall, 1953,1959a,b
	d12–15	15–20d	d1/2	80–100	Nourisson, 1961
	d13–17				Mathias, 1926, 1937
	d6–10	3d	d2	80	Hall & MacDonald, 1975
Eubranchipus bundyi		2d		70–100	Broch, 1965
				70–80	Belk, 1977
E. holmani	d2				Moore, 1959a
E. oregonus	d3	>35d			Coopey, 1950
E. serratus	d14				Prophet, 1963a
				70–90	Belk, 1977
E. vernalis	d12				Weaver, 1943
Streptocephalus dichotomus	d1	1–4d	d1	max: 100	Bernice, 1972
	d1	3d	d1	78	Sam & Krishnaswamy, 1979
S. dorothae				20–40	Belk, 1977
S. mackini				50–60	Belk, 1977
S. macrourus	d2	6d	d1/2	60–90	Mitchell, 1990
S. proboscideus	d3				Carlisle, 1968
	d1		d2	10–60	Brendonck, 1991,1992; Brendonck *et al.*, 1993
S. seali	d1–2				Moore, 1955, 1967; Anderson & Hsu, 1990
	d1	2d	d1	max: >90	Moore, 1959b
				80–90	Moore, 1967
	d1	7–8d	d2	max: 71	Moore, 1957
				10–32	Moore, 1963; Prophet, 1963a
	d1	>6d		5.6	Bulkowski & Meade, 1986
S. sudanicus	d1–2			10–30	Brendonck *et al.*, 1993
S. texanus	d1		d1/2		Prophet, 1963a
S. vitreus	d1	3d	d2	5/6 (0–25)	Hildrew, 1985
	d4				Carlisle, 1968
Tanymastix stagnalis				<1	Freiner & Grüttner, 1984
	d2/3	>20d			Maier & Tessenow, 1983
	d1			70	Müller, 1918
Thamnocephalus mexicanus				60–80	Belk, 1977
T. platyurus				50–60	Belk, 1977

Table 5. Continued.

	Start	Spread	Peak	%	Author
NOTOSTRACA					
Lepidurus apus	d5			44–67	Burmeister, 1982
L. couesii	d3	3–5d	d2	60–80	Hann & Lonsberry, 1991
L. packardi	d9	8–20d	d3/4	64–96	Ahl, 1991
Triops cancriformis	d9–11	>30d		70–80	Hempel-Zawitkowska, 1971
T. granarius	d8	>100d		60–75	Takahashi, 1977a
	d1–3			<50	Seaman *et al.*, 1991
	d2				Carlisle, 1968
T. longicaudatus	d1	1–2d		70–90	Scott & Grigarick, 1979
	d8	>100d		60–90	Takahashi, 1977a
'CONCHOSTRACA'					
Caenestheriella gynecia	d2–3				Mattox & Velardo, 1950
C. setosa				60–80	Belk & Belk, 1975; Sublette & Sublette, 1967
Eocyzicus klunzingeri	d1			100	Brendonck *et al.*, 1993
Eulimnadia antlei	d1		d1	60–90	Belk, 1972
E. diversa	d3				Stern & Stern, 1971
Limnadia stanleyana	d2–3	>50d	d6–11	40–60	Bishop, 1967a,b; 1968

CO_2

An elevated free pCO_2 was the ultimate hatching stimulus for dormant prehatched nauplii of *Siphonophanes grubei* (Mossin, 1986).

Temperature

Table 4 gives information on hatching temperatures. In some species there is no optimal temperature range for hatching but an optimal regime of successive temperature conditions depending on the developmental stage. In *Eubranchipus bundyi* and *Siphonophanes grubei*, for example, cysts first require a temperature of >15 °C for at least 21 days before development at < 10 °C can proceed to dormant prehatched nauplii (Broch, 1965; Mossin, 1986, respectively). In *Streptocephalus seali* and *Chirocephalus diaphanus*, hatching could be further stimulated by fluctuating water temperatures within the optimal temperature range (Moore, 1967; Nourisson, 1961, respectively). In *C. diaphanus*, maximum hatching was obtained when cysts were exposed daily for 9 hrs at 4 °C and for the rest of the day at 17 °C (Nourisson, 1961). In *S. seali*, fluctuating temperatures within 19–23 °C increased hatching in comparison with incubation at a constant temperature within this range (Moore, 1967).

Light

Light was indispensable for hatching in *Eulimnadia antlei* (Belk, 1972), *Limnadia stanleyana* (Bishop, 1967a, b), *Triops longicaudatus* (Scott & Grigarick, 1979), and *Streptocephalus macrourus* (Mitchell, 1990), or significantly increased hatching performance in *Triops cancriformis* (Takahashi, 1977a), *T. granarius* (Takahashi, 1977a; Seaman *et al.*, 1991), and *T. longicaudatus* (Takahashi, 1977a). That cysts of *Triops longicaudatus* did not hatch when covered with soil (Scott & Grigarick, 1979), could, however, also be attributed to low oxygen levels (see above).

Hatching phenology

Variability

In many species, hatching characteristics (rate, percentage, etc.) and/or requirements are variable (Moore, 1957, 1959b, 1967; Zinn & Dexter, 1962; Bishop, 1967b; Belk, 1977; Takahashi, 1977a; Modlin, 1982; Hann & Lonsberry, 1991; Seaman *et al.*, 1991), even within the same brood (Brown & Carpelan, 1971; Scott & Grigarick, 1979, Hildrew, 1985).

Generation carry-over

In most species, only a fraction of the cyst pool hatches at any subsequent inundation (Table 1). In *Streptocephalus vitreus* hatching still occurred after 9 subsequent drying/rewetting cycles (Hildrew, 1985).

Quantitative characteristics

In most species, hatching starts a few days after inundation (Table 5). Only in *Chirocephalus diaphanus* (Hall, 1953, 1959a, b, c; Nourisson, 1961) and *Triops cancriformis* (Hempel-Zawitkowska, 1971) did it start after more than a week. Absolute spread of hatching has only been assessed in a few species. Although the peak generally occurs within a few days after the start of hatching, cysts continue to hatch at low percentage in some species. In *Eubranchipus oregonus* (Coopey, 1950) and *Limnadia stanleyana* (Bishop, 1967a, b), hatching continued after more than a month. Maximum hatching percentages were generally between 50 and 100%. Only in *Triops granarius* (Seaman *et al.*, 1991), *Limnadia stanleyana* (Bishop, 1967a, b), *Streptocephalus vitreus* (Hildrew, 1985), *Streptocephalus proboscideus* (Brendonck, 1991; Brendonck *et al.*, 1993), *Streptocephalus sudanicus* (Brendonck *et al.*, 1993) hatching was generally below 50%. All these species occur in sub-tropical/desert climate.

Discussion

General

High variability in hatching behaviour and lack of background ecological information of the test species makes it difficult to generalise or to correlate hatching characteristics with environmental conditions. This variability has led to contradictory classifications of types of resting eggs (e.g. Bond, 1934; Mathias, 1937), of which most proved to be irrelevant. It is recommended for future research on dormancy, either in field or laboratory, to include ample information on the habitat of target species. Laboratory experiments should be conducted under accurately defined conditions to increase repeatability of results and reliability of interpretations.

Some of the cysts of a single brood generally only give rise to a next generation after at least one cycle of drying and refilling (Table 1), or after dilution. Additional research is needed to determine whether cysts which fail to hatch on any one occasion, are in diapause, activated but requiring another or more intense trigger, were of poor quality or dead, or needed further embryonal development. It is therefore still not clear to what extent dormancy in freshwater phyllopods is controlled endogenously (diapause), or exogenously (quiescence).

From this literature review and personal experience, it is suspected that both processes occur even within the same batch of eggs. This was also assumed by Belk & Cole (1975) who also stated that the portion of quiescent cysts becomes more important in more unpredictable environments such as desert regions. Generally, the occurrence of a portion of quiescent cysts is an effective means of swift response to improving conditions in an unreliable environment. It gives the species a quick start in the race against time, to mature and reproduce before the pool dries out. The fraction of diapausing cysts, on the other hand, is expected to be more prominent in more predictable environments.

Activation of diapausing cysts

Conventional treatments to deactivate diapause, particularly in the brine shrimp *Artemia*, are resting, drying, and/or hibernation (Lavens & Sorgeloos, 1987; Drinkwater & Clegg, 1991). Based on the effectiveness of these treatments, however, there is not much unequivocal evidence of diapause being a general process controlling responsiveness of cysts in large freshwater branchiopods.

In the tropical *Streptocephalus dichotomus*, cysts hatched exclusively after a dry period (Bernice, 1972), which may be an indication of true diapause. In most species, however, a fraction of the cysts need several drying/rewetting cycles (Table 1). Whether these observations are to be explained by activation by resting or drying/rewetting (diapause) or by the use of an incubation medium with different characteristics (quiescence) is, however, not clear. This generation carry-over of a fraction of the cyst population, was found by Takahashi (1977b) to stabilize and increase the population density of a notostracan species under unstable environmental conditions. According to Cohen (1966, 1968), only when the variance of the production of propagules of an annual organism becomes large in relation to the mean yield, does it become advantageous to postpone (by dormancy) the hatching of some fraction of them.

The positive effect of hibernation of cysts on both hatching percentage and rate in some species may indicate that diapause indeed exists, at least in a portion of the cysts.

In few cases only, cysts first need to rest for a certain period either in a wet or dry condition before they become responsive, while in most species there appears to be an optimal storage period (e.g. Moore, 1967; Bernice, 1972; Takahashi, 1977a; Mitchell, 1990; Seaman *et al.*, 1991). In *Streptocephalus proboscideus*, contrasting patterns were displayed by different broods (Brendonck, 1992). The period of rest required by freshly deposited cysts to hatch or resist desiccation as observed in some species (Avery, 1940; Prophet, 1963a; Brown & Carpelan, 1971; Scott & Grigarick, 1979) probably indicates the need for a minimum period of embryonal development before cysts become responsive and is probably not related to diapause.

Stimulation of quiescent cysts

Hatching sometimes takes place shortly after deposition in the original medium (Moore, 1951; Moore, 1957; Hall, 1959a; Brown & Carpelan, 1971), but in most cases no hatching occurs in the culture medium and a specific trigger seems to be required. It is generally assumed that hatching is effected by a combination of environmental conditions that relates well to the climatic region and type of habitat. With this strategy, the best synchronization of life-cycle with the temporal heterogeneity of the habitat is obtained (Broch, 1965; Brown & Carpelan, 1971; Belk & Cole, 1975).

The occurrence of intra-specific differences in hatching results and requirements, supports the idea of local adaptation. Hatching requirements are also often stated as determining colonization potentials and distribution patterns (Horne, 1967; Belk & Belk, 1975). Species-specific differences in requirements for embryonic development and egg hatching may furthermore explain, to some extent, year to year discontinuous occurrence patterns in anostracan populations as observed in different regions (Dexter, 1946, 1953; Dexter & Kuehnle, 1948; Prophet, 1963a; Broch, 1965; Horne, 1967, 1971; Brown & Carpelan, 1971; Donald, 1983).

Important environmental variables determining hatching rate and percentage of activated cysts are temperature (Table 4), photoperiod and illumination, oxygen (Table 3), and osmotic pressure (Table 2). According to Brown & Carpelan (1971), temperature and oxygen level are the most important variables in humid climates, while osmotic pressure and oxygen are more relevant in arid regions. To date, however, there is insufficient reliable information to confirm this statement. When there is sufficient light and oxygen hatching almost always seems to be invoked by a lowering of the osmotic level or renewing of the medium (Table 2). Temperature may influence the osmotic process during breaking or intra-cystic enzymatic activity (see Brendonck *et al.*, 1996). Light may be important to activate a supposed embryonal light receptor (Drinkwater & Clegg, 1991 and references therein), while oxygen is required for metabolic activities. Quantitative characteristics of the incubation medium, such as hardness, may, in turn, affect the amount of glycerol that needs to be built up to reach the critical intracystic osmotic pressure required to break the egg shell. The most rapid hatching rates noted seem to be at low hardness levels (as in distilled or deionized water).

Hatching phenology

Hatching is often highly variable, even within the same batch of cysts, and is thought to be an adaptation to ensure the persistence of populations in a variable and unpredictable environment. This variability may result from within clutch balanced polymorphisms in response to environmental cues (Brendonck & Persoone, 1993). Dissimilar requirements for optimal hatching in a mixed propagule-pool may contribute to changes in population density, species composition, and genetic diversity (Korpelainen, 1986), depending on prevailing conditions (Carvalho & Wolf, 1989).

In many species, hatching extends over several days or even weeks even under favourable conditions. The highest peak, however, is generally on the first or second day of hatching (Table 5). In species from arid regions, however, a sharp increase in salinity of the pond may prevent delayed hatching (Brown & Carpelan, 1971).

No general patterns can be drawn for hatching percentages in relation to environmental conditions. However, most species with relatively low hatching yields (<60%) occur in sub-tropical/desert regions. Poor hatching may reflect low chance for successful production of resting eggs due to frequent early drying of the ponds. According to Cohen (1966, 1968), the fraction of resting eggs that hatches yearly should be approximately equal to the probability of a large production of these structures, and the fraction which does not hatch should be almost equal to the probability of total or near total failure to produce propagules.

The low and erratic hatching results observed in many freshwater branchiopods, in comparison with *Artemia*, may relate to the different ecology of these species, *Artemia* generally lives in predictable environments.

Conclusions

Survival in unstable environments with a high probability of total reproductive failure, necessitates a fast hatching response to favourable conditions at low conductivity, a spread of risk among the progeny by hatching of a fraction of the cyst bank that correlates with chances for successful reproduction, and the ability of propagules to survive for many generations in the soil. Together with early maturation, high fecundity, obligate oviparity, and the absence of an asexual life cycle, these characteristics of resting eggs are thought to be highly adaptive to the temporary environment and explain the success of the group (see also Brendonck & Persoone, 1993).

Study of the phenology of the hatching process in different types of ponds from various climates may give more insight into the adaptation of freshwater branchiopods to their specific habitat. They have been able to persist from the Palaeozoic to today because of this fine-tuned strategy. It will surely take several more decades before the controlling factors in diapause and diapause deactivation are discovered.

Acknowledgments

The author is a Postdoctoral Fellow of the Belgian National Fund for Scientific Research (N.F.W.O.).

References

Ahl, J. S. B., 1991. Factors affecting contributions of the tadpole shrimp, *Lepidurus packardi*, to its oversummering egg reserves. In D. Belk, H. J. Dumont and N. Munuswamy (eds), Studies on Large Branchiopod Biology and Aquaculture. Hydrobiologia 212 (Dev. Hydrobiol. 64): 137–144.

Al-Tikrity, M. R. & J. N. R. Grainger, 1990. The effect of temperature and other factors on the hatching of the resting eggs of *Tanymastix stagnalis* (L.) (Crustacea, Anostraca). J. theor. Biol. 15: 87–90.

Anderson, G. & S-Y. Hsu, 1990. Growth and maturation of a North American fairy shrimp, *Streptocephalus seali* (Crustacea: Anostraca): a laboratory study. Freshwat. Biol. 24: 429–442.

Avery, J. L., 1940. Studies on the egg-laying habits of the fairy shrimp. J. Wash. Acad. Sci. 30: 31–33.

Belk, D., 1972. The biology and ecology of *Eulimnadia antlei* Mackin (Conchostraca). S. West Nat. 16: 297–305.

Belk, D., 1973. Suggestion of timing mechanism inhibiting hatching after the first day of wetting in *Branchinecta lindahli* Packardi eggs. Am. Zool. 13: 1339.

Belk, D., 1977. Evolution of egg size strategies in fairy shrimps. S. West Nat. 22: 99–105.

Belk, D. & G. A. Cole, 1975. Adaptational biology of desert temporary-pond inhabitants. In N. F. Hadley (ed.), Environmental physiology of desert organisms. Dowden, Hutchinson and Ross, Inc., Stroudsburg, Pa.: 207–226.

Belk, D. & M. S. Belk, 1975. Hatching temperatures and new distributional records for *Caenestheriella setosa* (Crustacea, Conchostraca). S. West Nat. 20: 409–420.

Bernice, R., 1972. Ecological studies on *Streptocephalus dichotomus* Baird. Hydrobiologia 39: 217–240.

Bishop, J. A., 1967a. Seasonal occurrence of a branchiopod crustacean, *Limnadia stanleyana* King (Conchostraca) in eastern Australia. J. anim. Ecol. 36: 77–95.

Bishop, J. A., 1967b. Some adaptations of *Limnadia stanleyana* King (Crustacea: Branchiopoda: Conchostraca) to a temporary freshwater environment. J. anim. Ecol. 36: 599–609.

Bishop, J. A., 1968. Resistance of *Limnadia stanleyana* King (Branchiopoda, Conchostraca) to desiccation. Crustaceana 14: 35–38.

Bond, R. M., 1934. Report on phyllopod Crustacea, including a revision of the Anostraca of the Indian Empire. Mem. Conn. Acad. Arts Sci. 10: 29–62.

Brendonck, L., 1991. Contributions to the study of the reproductive biology of *Streptocephalus proboscideus* (Anostraca, Streptocephalidae). Crustaceana 60: 145–162.

Brendonck, L., 1992. Study of the biology of large freshwater branchiopods, with special reference to the fairy shrimp *Streptocephalus proboscideus* (Frauenf.) (Crustacea: Branchiopoda: Anostraca). Ph.D. thesis. University of Ghent, 529 pp.

Brendonck, L. & G. Persoone, 1993. Biological, ecological characteristics of large freshwater branchiopods from endorheic regions and consequences for their use in cyst-based toxicity tests. In A. M. V. M. Soares and P. Calow (eds), Progress in standardization of aquatic toxicity tests. Lewis Publishers, USA: 7–35.

Brendonck, L., D. M. Centeno & G. Persoone, 1993. Fecundity and resting egg characteristics of some subtropical fairy shrimp and clam shrimp species (Crustacea: Branchiopoda), reared under laboratory conditions. Arch. Hydrobiol. 126: 445–459.

Brendonck, L., M. D. Centeno & G. Persoone, 1996. The influence of processing and temperature conditions on hatching of resting eggs of *Streptocephalus proboscideus* (Crustacea: Branchiopoda: Anostraca). Hydrobiologia 320 (Dev. Hydrobiol. 114): 99–105.

Broch, E. S., 1965. Mechanism of adaptation of the fairy shrimp *Chirocephalopsis bundyi* Forbes to the temporary pond. Cornell Univ., Cornell Agric. Exp. Stn Mem. 392: 1–48.

Brown, L. R. & L. H. Carpelan, 1971. Egg hatching and life history of a fairy shrimp *Branchinecta mackini* Dexter (Crustacea: Anostraca) in a Mohave desert playa (Rabbit Dry Lake). Ecology 52: 41–54.

Bulkowski, L. & J. W. Meade, 1986. Chemical decapsulation of fairy shrimp *Streptocephalus seali* Ryder (Anostraca). Crustaceana 51: 212–215.

Burmeister, E.-G., 1982. Ein Beitrag zur Biologie und Populationsstruktur von *Lepidurus apus* L. Spixiana 5: 193–209.

Carlisle, D. B., 1968. *Triops* (Entomostraca) eggs killed only by boiling. Science 161: 279.

Carvalho, G. R. & H. G. Wolf, 1989. Resting eggs of lake-*Daphnia*. I. Distribution, abundance and hatching of eggs collected from various depths in lake sediments. Freshwat. Biol. 22: 459–470.

Castle, W. A., 1938. Hatching of the eggs of the fairy shrimp. Science 87: 531.

Clegg, J. S., 1964. The control of emergence and metabolism by external osmotic pressure and the role of free glycerol in developing cysts of *Artemia salina*. J. exp. Biol. 41: 879–892.

Cohen, D., 1966. Optimizing reproduction in a randomly varying environment. J. theor. Biol. 12: 119–129.

Cohen, D., 1968. A general model of optimal reproduction in a randomly varying environment. Ecology 56: 219–228.

Conte, F. P., P. C. Droukas & R. D. Ewing, 1977. Development of sodium regulation and de novo synthesis of $NA^{+}K$-activated ATPase in larval brine shrimp *Artemia salina*. J. exp. Zool. 202: 339–361.

Coopey, R. W., 1950. The life history of the fairy shrimp *Eubranchipus oregonus*. Trans. Am. microsc. Soc. 69: 125–132.

Crowe, J. H. & J. S. Clegg, 1973. Anhydrobiosis. Dowden, Hutchinson and Ross, Inc., Stroudsburg, 477 pp.

Crowe, J. H. & J. S. Clegg, 1978. Dry biological systems. Academic Press, New York, USA, 357 pp.

Crowe, J. H. & L. M. Crowe, 1986. Stabilization of membranes in anhydrobiotic organisms. In Leopold, A. C. (ed.), Membranes, metabolism, and dry organisms. Comstock Publishing Associates, Cornell University Press, Ithaca, NY, USA: 188–209.

Crowe, J. H., L. M. Crowe & D. Chapman, 1984. Preservation of membranes in anhydrobiotic organisms: the role of trehalose. Science 223: 701.

Daborn, G. R., 1975. Life history and energy relations of the giant fairy shrimp *Branchinecta gigas* Lynch 1937 (Crustacea: Anostraca). Ecology 56: 1025–1039.

Daborn, G. R., 1977. The life history of *Branchinecta mackini* Dexter (Crustacea: Anostraca) in an argillotrophic lake of Alberta. Can. J. Zool. 55: 161–168.

Dexter, R. W., 1946. Further studies on the life history and distribution of *Eubranchipus vernalis* (Verrill). Ohio J. Sci. 46: 31–44.

Dexter, R. W., 1953. Studies on North American fairy shrimps with the description of two new species. Am. Midl. Nat. 49: 751–771.

Dexter, R. W. & C. Kuehnle, 1948. Fairy shrimp populations of northeastern Ohio in the seasons of 1945 and 1946. Ohio J. Sci. 48: 15–26.

Dexter, R. W. & M. S. Ferguson, 1943. Life history and distributional studies on *Eubranchipus serratus* Forbes (1876). Am. Midl. Nat. 29: 210–222.

Donald, D. B., 1983. Erratic occurrence of anostracans in a temporary pond: colonization and extinction or adaptation to variations in annual weather. Can. J. Zool. 61: 1492–1498.

Drinkwater, L. E. & J. S. Clegg, 1991. Experimental biology of cyst diapause. In R. A. Browne, P. Sorgeloos and C. N. A. Trotman (eds), Artemia Biology. CRC Press, USA: 93–117.

Dutrieu, J., 1960. Observations biochimiques et physiologiques sur le développement d'*Artemia salina* Leach. Archs Zool. exp. gén. 99: 1–134.

Eng, L. L., D. Belk & C. H. Eriksen, 1990. Californian Anostraca: distribution, habitat, and status. J. Crust. Biol. 10: 247–277.

Ewing, R. D. & J. S. Clegg, 1969. Lactate dehydrogenase activity and anaerobic metabolism during embryonic development in *Artemia salina*. Comp. Biochem. Physiol. 31: 297–307.

Fox, H. M., 1949. On *Apus*: its rediscovery in Britain, nomenclature and habits. Proc. zool. Soc. Lond. 119: 693–702.

Freiner, D. & O. Grüttner, 1984. Der eichener Kiemenfuskrebs. Natur und Museum 114: 273–286.

Fry, L. L. & M. R. Mulla, 1992. Effect of drying period and soil moisture on egg hatch of the tadpole shrimp (Notostraca: Triopsidae). J. econ. ent. 85: 65–69.

Fryer, G., 1986. Enemy-free space: a new name for an ancient ecological concept. Biol. J. linn. Soc. 27: 287–292.

Gaschott, O., 1928. Beobachtungen und Versuche an den Augsburger Euphyllopoden. Ber. Naturwiss. Ver Schwaben U. Neuberg 46: 2–6.

Grainger, J. N. R., 1991. The biology of *Tanymastix stagnalis* (L.) and its survival in large and small temporary water bodies in Ireland. In D. Belk, H. J. Dumont and N. Munuswamy (eds), Studies on Large Branchiopod Biology and Aquaculture. Hydrobiologia 212 (Dev. Hydrobiol. 64): 77–82.

Hall, R. E., 1953. Observations on the hatching of eggs of *Chirocephalus diaphanus* Prévost. Proc. zool. Soc. Lond. 123: 95–109.

Hall, R. E., 1959a. The development of eggs of *Chirocephlaus diaphanus* Prevost in relation to depth of water. Hydrobiologia 14: 79–84.

Hall, R. E., 1959b. Delayed development of eggs of *Chirocephalus diaphanus* Prevost. Hydrobiologia 13: 160–169.

Hall, R. E., 1959c. The development of eggs of *Chirocephalus diaphanus* Prévost at a low temperature. Hydrobiologia 13: 156–159.

Hall, R. E. & L. J. MacDonald, 1975. Hatching of the Anostracan branchiopod *Chirocephalus diaphanus* Prévost. I. Osmotic processes and the possible role of glycerol. Hydrobiologia 46: 369–375.

Hann, B. J. & B. Lonsberry, 1991. Influence of temperature on hatching of eggs of *Lepidurus couesii* (Crustacea, Notostraca). In D. Belk, H. J. Dumont and N. Munuswamy (eds), Studies on Large Branchiopod Biology and Aquaculture. Hydrobiologia 212 (Dev. Hydrobiol. 64): 61–66.

Hempel-Zawitkowska, J., 1971. Resistance of eggs of *Triops cancriformis* (Bosc.) to low temperatures as related to several chosen environmental factors. Polsk. Arch. Hydrobiol. 18: 295–302.

Hildrew, A. G., 1985. A quantitative study of the life history of a fairy shrimp (Branchiopoda: Anostraca) in relation to the temporary nature of its habitat, a Kenyan rainpool. J. anim. Ecol. 54: 99–110.

Horne, F. R., 1967. Effects of physical-chemical factors on the distribution and occurrence of some southeastern Wyoming phyllopods. Ecology 48: 472–477.

Horne, F. R., 1971. Some effects of temperature and oxygen concentration on phyllopod ecology. Ecology 52: 343–347.

Khalaf, A. N. & A. R. Al-Jaafery, 1985. Hatching conditions for *Eocyzicus spinifer* (Conchostraca: Crustacea) from Iraq. Hydrobiologia 123: 263–264.

Korpelainen, H., 1986. The effect of diapause on the genetic structure of *Daphnia magna* populations. Z. zool. Syst. Evolut.-forsch. 24: 291–299.

Longhurst, A. R., 1955. A review of the Notostraca. Bulletin of the British Museum (Natural History), Zoology 3: 1–57.

Lavens, P. & P. Sorgeloos, 1987. The cryptobiotic state of *Artemia* cysts, its diapause deactivation and hatching: a review. In P. Sorgeloos, D. A. Bengtson, W. Decleir, and E. Jaspers (eds), Artemia Research and its Applications. 3. Ecology, culturing, use in aquaculture, 3. Universa Press, Wetteren: 27–63.

Leopold, A. C., 1986. Introduction. In Leopold, A. C. (ed.), Membranes, metabolism, and dry organisms. Comstock Publishing Associates, Cornell University Press, Ithaca and London: 17–21.

Maier, G. & U. Tessenow, 1983. *Tanymastix stagnalis*: Vorkommen im Hannoverschen Wendland und Befunde zur Larvalentwicklung (Crust., Anostraca). Abh. Naturwiss. Ver. Hamburg (N.F.) 25: 351–355.

Mathias, P., 1926. Sur la biologie d'un Crustacé Phyllopode (*Chirocephalus diaphanus* Prevost). C. r. Soc. biol. 94: 1193–1194.

Mathias, P., 1934. Sur le développement de l'oeuf des crustacés phyllopodes. Bull. Soc. zool. Fr. 54: 342–344.

Mathias, P., 1937. Biologie des Crustacés Phyllopodes. In Hermann, (ed.), Activités Scientifiques et Industrielles. Bibliothèque de la Société Philomathique de Paris: 1–107.

Mattox, N. T. & J. T. Velardo, 1950. Effect of temperature on the development of the eggs of a conchostracan phyllopod, *Caenestheriella gynecia*. Ecology 31: 497–506.

Mitchell, S. A., 1990. Factors affecting the hatching of *Streptocephalus macrourus* Daday (Crustacea; Eubranchiopoda) eggs. Hydrobiologia 194: 13–22.

Modlin, R. F., 1982. A comparison of two *Eubranchipus* Species (Crustacea: Anostraca). Am. Midl. Nat. 107: 107–113.

Moore, W. G., 1951. A fairy shrimp new to Louisiana. Am. Midl. Nat. 46: 255.

Moore, W. G., 1955. The life history of the spiny-tailed fairy shrimp in Louisiana. Ecology 36: 176–184.

Moore, W. G., 1957. Studies on the laboratory culture of Anostraca. Trans. am. micr. Soc. 76: 159–173.

Moore, W. G., 1959a. Observations on the biology of the fairy shrimp *Eubranchipus holmani*. Ecology 40: 398–403.

Moore, W. G., 1959b. Production and viability of eggs in laboratory cultured fairy shrimp. Proc. La. Acad. Sci. 22: 53–62.

Moore, W. G., 1963. Some interspecies relationships in Anostraca populations of certain Louisiana ponds. Ecology 44: 131–139.

Moore, W. G., 1967. Factors affecting egg hatching in *Streptocephalus seali* (Branchiopoda: Anostraca). In Proc. Symp. Crustacea, Part 2: 724–735.

Moore, W. G. & A. Burn, 1968. Lethal oxygen thresholds for certain temporary pond invertebrates and their application to field situations. Ecology 49: 349–351.

Mossin, J., 1986. Physicochemical factors inducing embryonic development and spring hatching of the european fairy shrimp *Siphonophanes grubei* (Dybowsky) (Crustacea: Anostraca). J. crust. Biol. 6: 693–704.

Mozley, A., 1932. A biological study of a temporary pond in western Canada. Am. Nat. 66: 235–249.

Müller, R., 1918. Zur Biologie von Tanymastix lacunae Guérin. Biol. Zentralblatt (Leipzig) 38: 257–268.

Munuswamy, N., 1987. A note on hatching and decapsulation in *Streptocephalus dichotomus* Baird, 1860 (Anostraca). Crustaceana 53: 310–313.

Nourisson, M., 1961. Influence d'un thermopériodisme journalier sur le développement des oeufs non assechés de *Chirocephalus stagnalis* Shaw (Crustacé Phyllopode). C. r. hebd. Acad. Sci., Paris 253: 1870–1872.

Packard, A. S., Jr. 1875. Life histories of the crustacea and insects. Am. Nat. 9: 583–624.

Panek, A. C., E. Bernardes & A. D. Panek, 1986. Does trehalose play a role in yeast cells under stress? In Leopold, A. C. (ed.), Membranes, metabolism, and dry organisms. Comstock Publishing Associates, Cornell University Press, Ithaca and London, 7: 123–142.

Paul, M. A. & K. G. Nayar, 1977. Studies on a natural population of *Cyclestheria hislopi* (Baird) (Conchostraca: Crustacea). Hydrobiologia 53: 173–179.

Prophet, C. W., 1963a. Some factors influencing hatching of anostracan eggs. Trans. Kans. Acad. Sci. 66: 150–159.

Prophet, C. W., 1963b. Physico-chemical characteristics of habitats and seasonal occurrence of some Anostraca in Oklahoma and Kansas. Ecology 44: 798–801.

Raj, P. J. S., 1961. Morphology and distribution of *Branchinella kugenumaensis* (Ishikawa), var. madurai Raj. Ohio J. Science 61: 257–262.

Sam, S. T. & S. Krishnaswamy, 1979. Effect of osmomolarity of the medium upon hatching of undried eggs of *Streptocephalus dichotomus* Baird (Anostraca: Crustacea). Arch. Hydrobiol. 86: 125–130.

Sars, G. O., 1898. On some South-African Phyllopoda raised from dried mud. Arch. Math. Naturv. Kristiania B 20: 1–43.

Sato, N. L., 1967. Enzymatic contribution to the excystment of *Artemia salina*. Scient. Rep. Tohoku Univ. Ser. 4 (Biol.) 33: 319–327.

Scott, S. R. & A. Grigarick, 1979. Laboratory studies of factors affecting egg hatch of *Triops longicaudatus* (Leconte) (Notostraca: Triopsadae). Hydrobiologia 63: 145–152.

Seaman, M. T., D. J. Kok, B. J. von Schlichting & A. J. Kruger, 1991. Natural growth and reproduction in *Triops granarius* (Lucas) (Crustacea: Notostraca). In D. Belk, H. J. Dumont and N. Munuswamy (eds), Studies on Large Branchiopod Biology and Aquaculture. Hydrobiologia 212 (Dev. Hydrobiol. 64): 87–94.

Slegers, H., 1991. 3. Enzyme activities through development: a synthesis of the activity and control of the various enzymes as the embryo matures. In Browne, R. A., P. Sorgeloos and C. N. A. Trotman (eds), Artemia Biology. CRC Press, USA: 37–74.

Stern, D. H. & M. S. Stern, 1971. Morphology and culture of *Eulimnadia diversa* (Crustacea, Conchostraca) found in Louisiana. Trans. Am. micr. Soc. Menaska 90: 483–486.

Sublette, J. E. & M. S. Sublette, 1967. The limnology of playa lakes on the Llando Estacado, New Mexico and Texas. S. West Nat. 12: 369–406.

Takahashi, F., 1977a. Pioneer life of the Tadpole Shrimps, *Triops* spp. (Notostraca: Triopsidae). Appl. Ent. Zool. 12: 104–117.

Takahashi, F., 1977b. Generation carryover of a fraction of population members as an animal adaptation to unstable environmental conditions. Res. Popul. Ecol. 18: 235–242.

Weaver, C. R., 1943. Observations on the life cycle of the fairy shrimp *Eubranchipus vernalis*. Ecology 24: 500–502.

Zinn, D. J. & R. W. Dexter. 1962. Reappearance of *Eulimnadia agassizii* with notes on its biology and life history. Science 137: 676–677.

Hydrobiologia **320**: 99–105, 1996.
V. R. Alekseev & G. Fryer (eds), Diapause in the Crustacea.

The influence of processing and temperature conditions on hatching of resting eggs of *Streptocephalus proboscideus* (Crustacea: Branchiopoda: Anostraca)

Luc Brendonck[1], Maria D. Centeno[2] & Guido Persoone[2]
[1] *Royal Belgian Institute of Natural Sciences, Freshwater Biology, Vautierstraat 29, B-1040 Brussels, Belgium*
[2] *Laboratory for Biological Research in Aquatic Pollution, University of Ghent, J. Plateaustraat 22, B-9000 Gent, Belgium*

Key words: diapause, quiescence, hatching, resting eggs, Anostraca, processing, incubation

Abstract

In large freshwater branchiopods, erratic hatching success of resting eggs is a major obstacle to various applications. Lack of knowledge of the diapause-regulating processes makes control of hatching difficult. In the Sudanese fairy shrimp *Streptocephalus proboscideus* cysts are considered to include a diapausing and a quiescent fraction. To effect hatching, diapausing cysts have to be activated, while the quiescent portion has to be triggered by suitable environmental conditions. Of several attempts to control hatching with varying production, processing, and incubation conditions, only a few treatments proved consistently successful. Cysts were produced in an indoor culture system under controlled conditions and were harvested, washed, and dried according to defined procedures before processing (if any) and incubation. Hatchability (first-day and cumulative) was consistently higher in chemically decapsulated than in untreated (non-decapsulated) cysts at 25 °C. At a temperature of 28 °C, hatching was comparable in untreated and in decapsulated cysts and was significantly higher than at 25 °C. Pre-treatment with 7.5% NaOCl for 5–10 minutes, resulted in higher hatching than other decapsulation procedures and durations. It is believed that the decapsulation and temperature treatments were only effective in triggering quiescent cysts but did not activate the diapausing fraction.

Introduction

Large ('non-cladoceran') freshwater branchiopods (fairy shrimps, clam shrimps, tadpole shrimps) live in astatic water bodies that dry out or freeze periodically, or show striking changes in the water level. No trigger is required for resting egg (cyst) production. By contrast, the halobiontic *Artemia* needs a stimulus to switch from ovoviviparous to oviparous reproduction, and rotifers and 'cladocerans' have distinct life-cycle phases and only produce resting stages in response to specific conditions prior to the onset of environmental adversity. In particular large branchiopods of (semi-) arid regions are characterized by early maturation and high fecundity rates as an adaptation to their ephemeral and unpredictable habitat. In addition, cysts are resistant to environmental extremes and often have a long-term viability (Brendonck & Persoone, 1993 and references therein).

Despite the above advantages, a major constraint in the use of freshwater branchiopods in aquaculture and aquariology is the asynchronous, unpredictable, and often population-dependent nature of hatching of their resting eggs. These are also a major drawback to the use of cyst-based tests in aquatic toxicology since sufficient numbers of test organisms of uniform age are required at a predictable time after incubation of the cysts (see Centeno *et al.*, 1993). The erratic hatching pattern is thought to be an adaptation to ensure the persistence of natural populations, even if habitat factors eliminate all individuals before they have an opportunity to reproduce.

That cysts do not hatch immediately after their deposition, even if environmental conditions are still favourable for growth and reproduction, may imply that the onset of dormancy is rigidly programmed in their development and as such are diapausing. Several authors, however, have recorded that a portion of freshly deposited cysts hatches after dilution or refreshment of the medium (reviewed in Brendonck *et al.*, 1993; Brendonck, 1996). This fraction is considered quiescent, since it responds directly to favourable environmental conditions. To date, the actual processes taking place during diapause and post-diapause development of large branchiopod resting eggs and the conditions required to terminate diapause ('activation') are not understood. To stimulate and control hatching one can attempt to activate diapausing cysts and/or find a suitable trigger for the quiescent portion. A review of diapause terminating techniques for the cysts of several crustaceans and rotifers is presented in Lavens & Sorgeloos (1987).

Because of the continuous availability of large numbers of cysts and the economic importance of brine shrimp in aquaculture, most work on the biological and biochemical aspects of dormancy and hatching has been concentrated on *Artemia* (for a review see Drinkwater & Clegg, 1991). From the plethora of studies on hatching in large freshwater branchiopods, it may be concluded that lack of background information on the test material has often led to misinterpretations and low reproducibility of results (see Brendonck, 1996).

The Sudanese fairy shrimp *Streptocephalus proboscideus* is very fecund (Brendonck, 1991; Brendonck *et al.*, 1993) and easy to culture (Brendonck *et al.*, 1990), making it suitable for various applications (Brendonck & Persoone, 1993). A resting egg-based acute toxicity test was recently developed with this species (Centeno *et al.*, 1993). More knowledge of hatching stimulating and inhibiting factors, however, is required to better control hatching in order to standardize fully the test protocol. The development of a laboratory culture system for the controlled production of resting eggs (Brendonck *et al.*, 1990) provided the experimental tool to study the influence of environmental factors on the hatchability of cysts. In a study of the effects of various production, processing, and incubation conditions, hatching was highly variable in almost all experiments, even when only one batch of cysts produced under controlled conditions was used (see Brendonck, 1992). Here we present only those results that proved to be both successful and repeatable, i.e. decapsulation and increased incubation temperature.

Materials and methods

General

S. proboscideus was obtained from continuous cultures kept in the laboratory since 1986. The starting material was resting eggs collected from a rain pool in Al Gedid, south of Khartoum, Sudan.

Unless otherwise stated, all cysts were produced in closed recirculation units (described in Brendonck *et al.*, 1990) at a temperature of 25 °C and a photoperiod of 12L/12D by parents kept at a density of $50\,l^{-1}$ and fed on the micro-alga *Selenastrum capricornutum*.

To minimize the influence of uncontrolled variables in the collection, processing, and storage of cysts, a standard method was adopted for all experiments. Resting eggs were harvested weekly according to the procedure presented in Brendonck *et al.* (1990). Cysts and debris were washed over a series of sieves with mesh sizes of 600, 200, and 112 μm, respectively. Resting eggs and debris of the same size range were collected from the 200 μm-sieve. The remaining debris was removed by repeated decantations using tap water. The end-product consisted of clean resting eggs and some mandible exuviae (which often contain one resting egg) of similar or higher density than the cysts. Cysts were then filtered under vacuum and dry-stored at 25 °C in dimmed light and 40% relative humidity.

In several experiments, cysts were decapsulated prior to incubation in a solution of 7.5 ml NaOCl and 2.5 ml NaOH in 100 ml deionized water (= solution a), or in a 7.5% NaOCl solution (= solution b). The chemical activity of the pure NaOCl ranged from 9.4 to 7.5% depending on the period of preparation. For decapsulation, cysts were placed on a screen submerged in the decapsulation solution which was stirred by air-bubbling. After decapsulation, they were washed in a stream of tap water for several minutes to remove all traces of hypochlorite. In some experiments, decapsulated washed cysts were treated further with a 0.1N HCl solution or a 0.1% sodium thiosulphate ($Na_2S_2O_3$) solution, in order to deactivate chlorine residues adsorbed on the decapsulated cysts (Sorgeloos *et al.*, 1986).

Cysts were generally incubated in 6 × 10 ml covered polystyrene multi-well plates (water depth ± 1 cm) at 25 °C and with continuous illumination at 2000 lux, provided by white fluorescent bulbs. Artificial freshwater (standard reference water (SRW)), according to the formula of Freeman (1953) was used as the incubation medium. Each well (= 1 replicate) contained 45–55 cysts. The number (x) of replicates performed for each treatment is indicated hereunder as: (n = x). Plates were checked daily and hatched nauplii counted and removed. For each experiment, beside cumulative hatching results, first-day hatching percentages have also been considered, since the latter gives an indication of the hatching rate.

Decapsulation procedures

Different decapsulation durations and procedures were compared with the same batch of cysts. For practical reasons, not all treatments could be performed concurrently; so cysts used in different treatments were stored for different durations. In previous work, however, no constant relation was found between storage duration and hatching characteristics (Brendonck, 1992). Treatments were as follows:

A. Cysts stored for 56 days were decapsulated in solution a. Cysts were decapsulated for 5, 10, 15, 20 and 25 minutes, respectively; B. Cysts stored for 72 days were hydrated for two hours in SRW prior to decapsulation in solution a. Decapsulation times as above; C. Cysts stored for 77 days were decapsulated in solution a, but were post-treated with 0.1 N HCl. Decapsulation times as above; D. Cysts stored for 30 days were decapsulated in solution a, but were post-treated with 0.1% $Na_2S_2O_3$. Decapsulation times as above; E. Cysts stored for 41 days were hydrated for two hours in SRW prior to decapsulation. Cysts were decapsulated in solution a, but were post-treated with 0.1 N HCl. Decapsulation times as above; F. Cysts stored for 42 days were hydrated for two hours in SRW prior to decapsulation. Cysts were decapsulated in solution a, but were post-treated with 0.1% $Na_2S_2O_3$. Decapsulation times as above; G. Cysts stored for 84 days were decapsulated in solution b. Decapsulation times as above.

For each decapsulation time in each treatment six replicates were used (n = 6). In all experiments hatching was scored daily for six days.

Temperature conditions

One part of a batch of cysts stored for three months was decapsulated for 7.5 minutes in solution b while the other part was not processed. One series of both decapsulated and untreated cysts was incubated at 25 °C and a second at 28 °C. A temperature of 28 °C corresponds to the average temperature in the natural habitat (see Rzóska, 1961) and with the comfort zone for culturing under laboratory conditions (De Walsche *et al.*, 1991). For each combination of processing type and temperature three replicates were used (n = 3). Hatching was scored daily for six days.

Statistics

Hatching data were statistically analyzed for one dependent (hatching at the first day or cumulative hatching) and one independent variable (each separate treatment) at a time. To test for the assumptions of normality, a Kolmogorov-Smirnov goodness of fit test and Bartlett's test for homogeneity of variance were performed. When only two samples were involved, an independent samples t-test, a one-way-ANOVA, or a Mann-Whitney test was applied, depending on the distribution of the respective dependent variable. When more than two samples were compared, a Kruskal-Wallis test or a one-way ANOVA were used to establish whether there was a significant overall difference in one of the dependent variables for a number of samples differentiated on a particular independent variable. A one-way ANOVA extended with a Duncan-range-test was performed to assess which pairs of treatments had significantly different means (at the 0.05 level) for a defined dependent variable (Sokal & Rohlf, 1981).

Terminology related to dormancy follows Lavens & Sorgeloos (1987), Drinkwater & Clegg (1991), and Brendonck (1996).

Results

Decapsulation procedures

Pre-hydration resulted in an increase of first-day and cumulative hatching results only for the untreated cysts. In pre-hydrated cysts, no significant difference (P>0.05) was found between decapsulation procedures in their effect on first-day and cumulative hatching results (Fig. 1A, B). In non-pre-hydrated cysts, a significant overall difference was measured between decapsulation procedures for both first-day (P<0.001) and cumulative (P<0.001) hatching percentages (Fig. 1A, B). First-day hatching was most successful with cysts decapsulated in solution b (12.5%), and those decapsulated in solution a and post-treated with HCl (9.7%). Cumulative hatching values were highest for cysts decapsulated in solution b (21.3%).

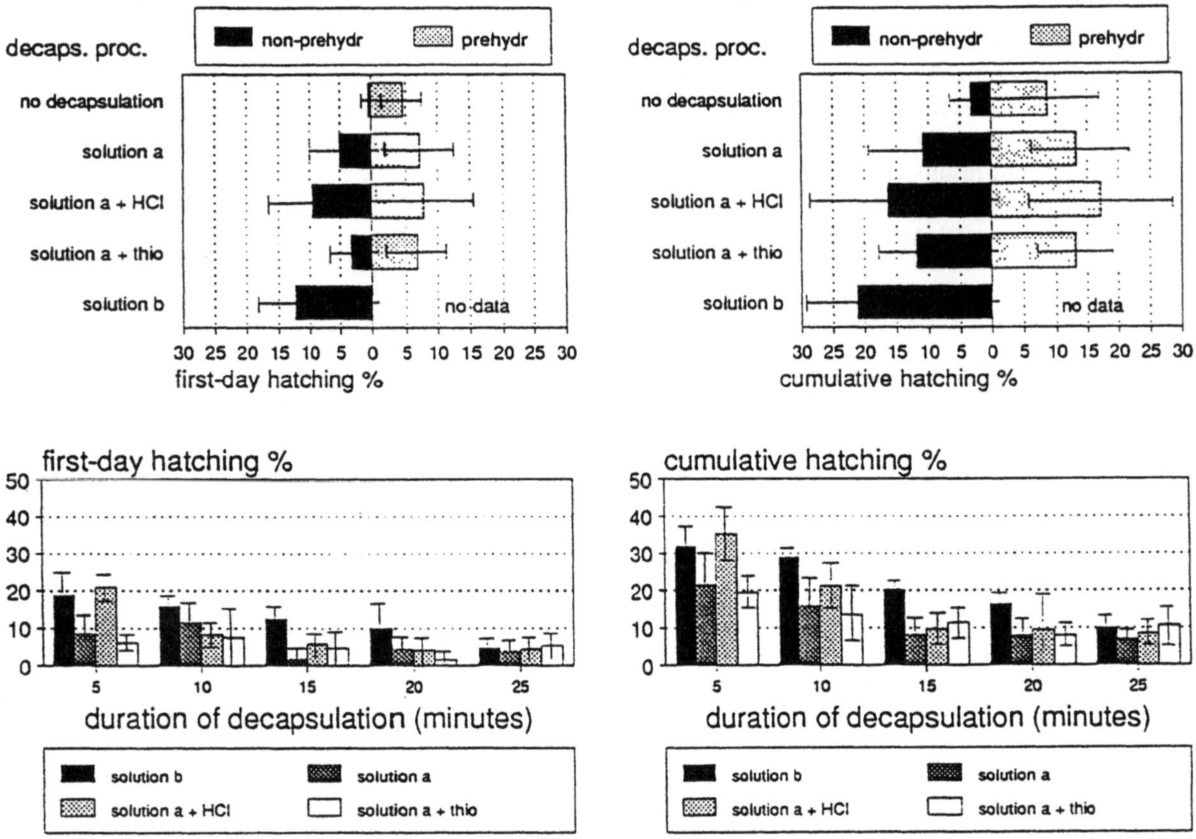

Fig. 1. Decapsulation procedures (see also materials and methods): A. First-day and B. cumulative (6 days) hatching percentages of resting eggs processed differently; C. First-day and D. cumulative (6 days) hatching percentages of both prehydrated and non-pre-hydrated resting eggs, decapsulated differently for increasing periods and post-treated differently. solution a = 7.5 ml NaOCl + 2.5 ml NaOH + 90 ml deionized water; solution b = 7.5 ml NaOCl + 92.5 ml deionized water. decaps. proc. = decapsulation procedure; prehydr. = pre-hydrated; thio = sodium thiosulphate.

For both the pre-hydrated and the non-pre-hydrated series, non-decapsulated cysts showed the lowest values for first-day and cumulative hatching (Fig. 1A, B). For all decapsulation procedures, there was a critical influence of the duration of decapsulation on the hatching success (Fig. 1C, D). A decapsulation time of from five to 10 minutes generally resulted in the highest first-day and cumulative hatching percentages.

Temperature conditions

For decapsulated cysts, only cumulative hatching significantly ($P<0.05$) differed between cysts incubated at different temperatures with higher hatching at 28 °C (Fig. 2A, B). Temperature, however, had a significant effect on both first-day ($P<0.001$) and cumulative hatching percentages ($P<0.001$) of untreated cysts, with ten-fold higher values at 28 °C (Fig. 2A, B).

For cysts incubated at 28 °C, no significant difference ($P>0.05$) was observed in hatching results between decapsulated and non-decapsulated cysts (Fig. 2A, B).

For cysts incubated at 25 °C, a significant difference was observed between decapsulated and untreated cysts for both first-day ($P<0.01$) and cumulative ($P<0.01$) hatching percentages (Fig. 2A, B). Much higher hatching results were obtained with decapsulated cysts.

Discussion

Apart from a few cysts which sunk two to three hours after incubation, most dried, non-processed *S. proboscideus* cysts remained floating throughout each experiment, as do dried cysts of *Artemia* (Myint,

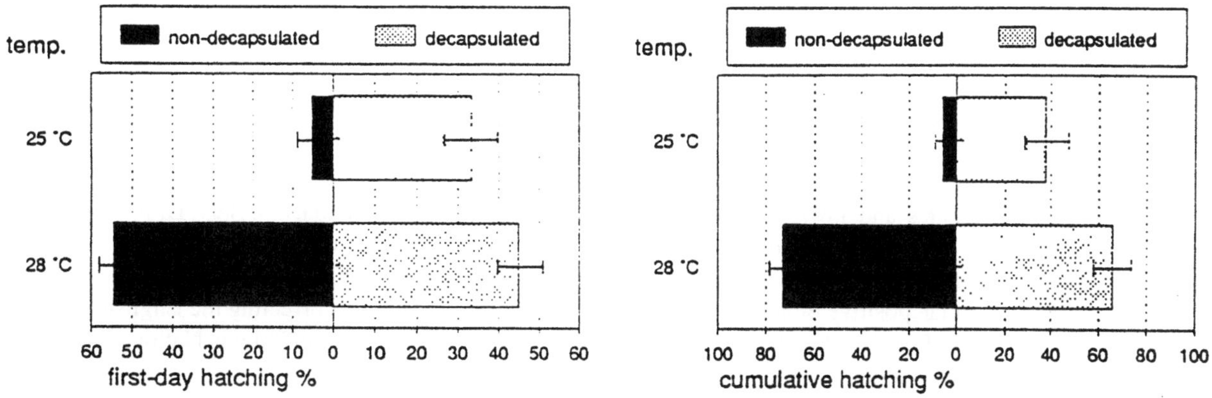

Fig. 2. Temperature conditions (see also materials and methods): A. First-day and B. cumulative (5 days) hatching percentages of decapsulated and untreated eggs and incubated at 25 °C or 28 °C. temp. = incubation temperature.

1956) and *Streptocephalus seali* (Bulkowski & Meade, 1986). A reduction in egg density during drying may result from dehydration and replacement of water by air in the alveolar network of the tertiary envelope. In addition, a hydrophobous film in the alveolar egg layer probably hampers the entrance of water. A treatment of 2–3 min with hypochlorite did not affect the external structure of *S. proboscideus* cysts, but caused all cysts to sink.

Despite all precautions, hatching was very variable in almost all experiments, even within the same batch of cysts. Moore (1957, 1959, 1967) also emphasizes the considerable variation in hatchability of *S. seali* cysts produced under so-called identical culture conditions. Scott & Grigarick (1979) also noted great variability in hatching of similarly treated cysts of the same brood of the notostracan *Triops longicaudatus*.

S. proboscideus cysts, decapsulated for 5–10 minutes, gave consistently better hatching results than untreated cysts when incubated at 25 °C (Fig. 2A, B). Since this is the standard temperature used during toxicity testing with *S. proboscideus* (Centeno *et al.*, 1993) this decapsulation method may be incorporated in the standard processing procedure of cysts to increase both rate and percentage of hatching. At 28 °C, however, decapsulated and untreated cysts behaved similarly (Fig. 2A, B).

Although decapsulation in fairy shrimps has been studied for its effect on egg morphology in different species, its effect on hatching has only been assessed in *S. seali* (Bulkowski & Meade, 1986) where the hatching percentage of decapsulated cysts (25%) was higher than in the control (6%). Decapsulation also increased hatching success in *Artemia* (Lavens & Sorgeloos, 1987). The positive effect of decapsulation may be explained by the 'trehalose-glycerol hyperosmotic regulatory system'-theory (Clegg, 1964; Conte *et al.*, 1977). By the (partial) removal of the tertiary egg envelope, 'breaking' may become possible at lower intra-cystic glycerol concentrations than those required in untreated cysts. In *Artemia*, under certain circumstances, hypochlorite may have a direct physiological effect and activate diapausing embryos (Lavens & Sorgeloos, 1987; Drinkwater & Clegg, 1991). Removal of the egg shell which is impregnated with haematine, may make branchiopod resting stages more photosensitive (Pancella & Stross, 1963; Belk, 1970; Vanhaecke *et al.*, 1981; Van Der Linden *et al.*, 1986) and thus more easily 'activated' by light stimuli.

At an egg density of 50 cysts per 10 ml, a post-decapsulation deactivation treatment with 0.1 N HCl or 0.1% $Na_2S_2O_3$ did not improve hatching (Fig. 1A–D). Best results were obtained when cysts were immersed for 5–10 minutes in 7.5% NaOCl. Since the hypochlorite solution was sufficiently basic (pH>10), no addition of NaOH was required as prescribed by Sorgeloos *et al.* (1986).

Rehydration of dried cysts before they were decapsulated did not increase the positive effect of decapsulation (Fig. 1A, B), though it does in *Artemia*. This difference may be related to the collapsed configuration of dehydrated *Artemia* cysts which first need to be rehydrated to resume their spherical shape, a prerequisite for a uniform decapsulation. *S. proboscideus* cysts, on the other hand, have a rigid tertiary egg envelope (Brendonck, 1990) which is not obviously affected during dehydration.

Prolonged decapsulation durations (>10 min), led to a gradual decrease in hatching success, even with deactivation of possible chemical residues (Fig. 1C, D). A similar effect was found in *S. seali* by Bulkowski & Meade (1986). This may be the result of damage to the outer cuticular egg membrane or the embryonic cuticle.

At higher incubation temperatures, increased mobility of water molecules and lower surface tension in the wells may have facilitated the entry of water into the unprocessed floating *S. proboscideus* cysts, resulting in a similar positive effect on hatching as with decapsulation (Fig. 2A, B). More successful hatching at higher temperatures may also result from temperature-related intra-cystic enzymatic activity. Temperature is furthermore one of the most important environmental variables for stimulating quiescent cysts to hatch (Drinkwater & Clegg, 1991). The effect of temperature on hatching has been studied in several taxa of large branchiopods: anostracans (Mathias, 1937; Hall, 1953; Nourisson, 1961; Moore, 1963, 1967; Prophet, 1963a, b; Bernice, 1972; Jones, 1972; Wiggins *et al.*, 1980; Mitchell, 1990), clam shrimps (Mattox, 1946, 1950; Mattox & Velardo, 1950; Belk & Belk, 1975; Khalaf & Al-Jaafery, 1985), and tadpole shrimps (Fox, 1949; Horne, 1967, 1971; Takahashi, 1977; Scott & Grigarick, 1979). For most species, optimal temperatures for hatching were recorded and some relationship with prevailing climate has been suggested (Prophet, 1963a; Moore, 1967; Bernice, 1972). Khalaf & Al-Jaafery (1985), however, showed that laboratory results of thermal influence on hatching in the clam shrimp *Eocyzicus spinifer* did not fully correspond with field observations. Since the more succesfull temperature of 28 °C is close to the average of the natural *S. proboscideus* habitat in Sudan (Rzóska, 1961), it is supposed that it stimulated a larger portion of quiescent cysts.

Conclusion

Decapsulation enhances both hatching rate and cumulative percentage at the temperature used for toxicity testing (25 °C). Partly deshelling of the cysts probably enables hatching at lower intra-cystic osmotic pressures than in untreated cysts.

Faster and increased hatching occurs at 28 °C (close to the natural average) than at 25 °C. The elevated temperature probably triggers more quiescent cysts and/or enhances imbibition of water by rehydrating cysts.

Decapsulation and temperature probably only affect the quiescent fraction of the cysts. Further research is required to find out how diapause can be broken in the non-responsive fraction.

Acknowledgments

The authors are grateful to Dr D. Belk, Dr P. Sorgeloos, and Dr L. De Meester for their invaluable comments on earlier drafts of this paper. Thanks are also due to Mrs Espiritu for kindly correcting the English. L.B. is a Postdoctoral Fellow of the Belgian National Fund for Scientific Research (N.F.W.O.).

References

Belk, D., 1970. Functions of the conchostracan egg shell. Crustaceana 19: 105–106.

Belk, D. & M. S. Belk, 1975. Hatching temperatures and new distributional records for *Caenestheriella Setosa* (Crustacea, Conchostraca). S. West Nat. 20: 409–420.

Bernice, R., 1972. Hatching and postembryonic development of *Streptocephalus dichotomus* Baird (Crustacea: Anostraca). Hydrobiologia 40: 251–278.

Brendonck, L., 1990. Redescription of the fairy shrimp *Streptocephalus proboscideus* (Frauenfeld, 1873) (Crustacea, Anostraca). Bull. k. belg. Inst. Nat. Wet., Biol. 59 (1989): 49–57.

Brendonck, L., 1991. Contributions to the study of the reproductive biology of *Streptocephalus proboscideus* (Anostraca, Streptocephalidae). Crustaceana 60: 145–162.

Brendonck, L., 1992. Study of the biology of large freshwater branchiopods, with special reference to the fairy shrimp *Streptocephalus proboscideus* (Frauenf.) (Crustacea: Branchiopoda: Anostraca). Ph.D. thesis. University of Ghent, 529 pp.

Brendonck, L., 1996. Diapause, quiescence, hatching requirements: what we can learn from large freshwater branchiopods (Crustacea: Branchiopoda: Anostraca, Notostraca, Conchostraca). Hydrobiologia 320 (Dev. Hydrobiol. 114): 85–98.

Brendonck, L. & G. Persoone, 1993. 2. Biological/ecological characteristics of large freshwater branchiopods from endorheic regions and consequences for their use in cyst-based toxicity tests. In A. M. V. M. Soares & P. Calow (eds), Progress in standardization of aquatic toxicity tests. Lewis Publishers, Boca Raton, USA: 7–35.

Brendonck, L., M. D. Centeno & G. Persoone, 1993. Fecundity and resting egg characteristics of some subtropical fairy shrimp and clam shrimp species (Crustacea: Branchiopoda), reared under laboratory conditions. Archiv. Hydrobiol. 4: 445–459.

Brendonck, L., G. Uyttersprot & G. Persoone, 1990. A culture system for fairy shrimps (Crustacea: Anostraca). Aquac. Engng 9: 267–283.

Bulkowski, L. & J. W. Meade, 1986. Chemical decapsulation of fairy shrimp *Streptocephalus seali* Ryder (Anostraca). Crustaceana 51: 212–215.

Centeno, M. D., L. Brendonck & G. Persoone, 1993. Cyst-based toxicity tests. 3. Development and standardization of an acute toxicity test with the fairy shrimp *Streptocephalus proboscideus* (Crustacea: Branchiopoda: Anostraca). In A. M. V. M. Soares & P. Calow (eds), Progress in standardization of aquatic toxicity tests. Lewis Publishers, Boca Raton, USA: 37–55.

Clegg, J. S., 1964. The control of emergence and metabolism by external osmotic pressure and the role of free glycerol in developing cysts of *Artemia salina*. J. exp. Biol. 41: 879–892.

Conte, F. P., P. C. Droukas & R. D. Ewing, 1977. Development of sodium regulation and de novo synthesis of $NA^{+}K$-activated ATPase in larval brine shrimp *Artemia Salina*. J. exp. Zool. 202: 339–361.

Crowe, J. H., L. M. Crowe, L. Drinkwater & W. B. Busa, 1987. Intracellular pH and anhydrobiosis in *Artemia* cysts. In W. Decleir, L. Moens, H. Slegers, E. Jaspers & P. Sorgeloos (eds), Artemia Research and its Applications, 2. Universa Press, Wetteren: 19–40.

De Walsche, C., J. Mertens & H. J. Dumont, 1991. Observations on temperature optimum, cyst production, and survival of *Streptocephalus proboscideus* (Frauenfeld, 1873) (Crustacea: Anostraca), fed different diets. In: D. Belk, H. J. Dumont and N. Munuswamy (eds), Studies on Large Branchiopod Biology and Aquaculture. Hydrobiologia 212 (Dev. Hydrobiol. 64): 21–26.

Drinkwater, L. E. & J. S. Clegg, 1991. 5. Experimental biology of cyst diapause. In R. A. Browne, P. Sorgeloos & C. N. A. Trotman (eds), Artemia Biology. CRC Press, USA: 93–117.

Fox, H. M., 1949. On *Apus*: its rediscovery in Britain, nomenclature and habits. Proc. zool. Soc. Lond. 119: 693–702.

Freeman, L., 1953. A standardized method for determining toxicity of pure compounds to fish. Sewage ind. Wastes 25: 845–848.

Hall, R. E., 1953. Observations on the hatching of eggs of *Chirocephalus diaphanus* Prévost. Proc. zool. Soc. Lond. 123: 95–109.

Horne, F.R., 1967. Effects of physical-chemical factors on the distribution and occurrence of some southeastern Wyoming phyllopods. Ecology 48: 472–477.

Horne, F. R., 1971. Some effects of temperature and oxygen concentration on phyllopod ecology. Ecology 52: 343–347.

Jones, A. J., 1972. An inexpensive apparatus for the large scale hatching of *Artemia salina* L. J. Cons. int. Explor. Mer 34: 351–356.

Khalaf, A. N. & A. R. Al-Jaafery, 1985. Hatching conditions for *Eocyzicus spinifer* (Conchostraca: Crustacea) from Iraq. Hydrobiologia 123: 263–264.

Lavens, P. & P. Sorgeloos, 1987. The cryptobiotic state of *Artemia* cysts, its diapause deactivation and hatching: a review. In P. Sorgeloos, D. A. Bengtson, W. Decleir & E. Jaspers (eds), Artemia Research and its Applications, 3. Universa Press, Wetteren: 27–63.

Mathias, P., 1937. Biologie des Crustacés Phyllopodes. In Hermann and Cie (eds), Actual. scient. ind. Bibliothèque de la Société philomathique de Paris, 447: 1–107.

Mattox, N. T., 1946. Influence of temperature on hatching of eggs of a conchostracan phyllopod. Anat. Rec. 94: 352–353.

Mattox, N. T., 1950. Notes on the life history and description of a new species of conchostracan phyllopod, *Caenestheriella gynecia*. Trans. am. microsc. Soc. 69: 50–53.

Mattox, N. T. & J. T. Velardo, 1950. Effect of temperature on the development of the eggs of a conchostracan phyllopod, *Caenestheriella gynecia*. Ecology 31: 497–506.

Mitchell, S. A., 1990. Factors affecting the hatching of *Streptocephalus macrourus* Daday (Crustacea; Eubranchiopoda) eggs. Hydrobiologia 194: 13–22.

Moore, W. G., 1957. Studies on the laboratory culture of Anostraca. Trans. am. microsc. Soc. 76: 159–173.

Moore, W. G., 1959. Production and viability of eggs in laboratory cultured fairy shrimp. Proc. La Acad. Sci. 22: 53–62.

Moore, W. G., 1963. Some interspecies relationships in Anostraca populations of certain Louisiana ponds. Ecology 44: 131–139.

Moore, W. G., 1967. Factors affecting egg hatching in *Streptocephalus seali* (Branchiopoda: Anostraca). In Proc. Symp. Crustacea, 2: 724–735.

Myint, T., 1956. New details of excystment of *Artemia salina* leach. La Acad. Sci. 19: 24–28.

Nourisson, M., 1961. Influence d'un thermopériodisme journalier sur le développement des oeufs non assechés de *Chirocephalus stagnalis* Shaw (Crustacé Phyllopode). C. r. hebd. Acad. Sci., Paris 253: 1870–1872.

Pancella, J. R. & R. G. Stross, 1963. Light induced hatching of *Daphnia* resting eggs. Chesapeake Sci. 4: 135–140.

Prophet, C. W., 1963a. Physico-chemical characteristics of habitats and seasonal occurrence of some Anostraca in Oklahoma and Kansas. Ecology 44: 798–801.

Prophet, C. W., 1963b. Some factors influencing hatching of anostracan eggs. Trans. Kans. Acad. Sci. 66: 150–159.

Rzóska, J., 1961. Observations on tropical rainpools and general remarks on temporary waters. Hydrobiologia 17: 265–286.

Scott, S. R. & A. Grigarick, 1979. Laboratory studies of factors affecting egg hatch of *Triops Longicaudatus* (Leconte) (Notostraca: Triopsidae). Hydrobiologia 63: 145–152.

Sokal, R. R. & F. J. Rohlf, 1981. Biometry. The principles and practice of statistics in biological research. 2nd edn. W. H. Freeman and Co., San Francisco, 859 pp.

Sorgeloos, P., P. Lavens, Ph. Léger, W. Tackaert & D. Versichele, 1986. Manual for the culture and use of Brine Shrimp Artemia in aquaculture. Manual prepared for the Belgian administration for development cooperation and the food and agriculture organization of the United Nations. Artemia Reference Center, State University of Ghent, Faculty of Agriculture, Ghent.

Takahashi, F., 1977. High water temperature prevents egg hatching in *Triops granarius* (Notostraca: Triopsidae). Appl. Ent. Zool. 12: 196–197.

Van Der Linden, A., I. Vankerckhoven, R. Caubergs, W. Decleir, 1986. Action spectroscopy of light-induced hatching of *Artemia* cysts (Branchiopoda: Crustacea). Mar. Biol. (Berlin) 91: 239–243.

Vanhaecke, P., A. Cooreman & P. Sorgeloos, 1981. International study on *Artemia*. 15. Effect of light intensity on hatching rate of *Artemia* cysts from different geographic origin. Mar. Ecol. Prog. Ser. 5: 111–114.

Wiggins, G. B., R. J. Mackay & I. M. Smith, 1980. Evolutionary and ecological strategies of animals in annual temporary pools. Arch. Hydrobiol. (suppl.) 58: 97–206.

Hydrobiologia **320**: 107–117, 1996.
V. R. Alekseev & G. Fryer (eds), Diapause in the Crustacea.

Significance of photoperiodism and diapause control in the multicycle Crustacean *Daphnia pulex* Leydig

R. G. Stross
Dept. of Biological Sciences, University at Albany (SUNY), Albany, NY 12222, USA

Key words: cell-/life cycle regulation, photoperiodism, diapause control, zonation, *Daphnia, Nitella*

Abstract

Eukaryote cells traverse the cell cycle, or become growth-arrested at one of two blockpoints, as in an engine of energy conversion. Diapause describes that growth arrest in the cell cycle(s), of single and multicellular individuals. Its temporal position in an annual cycle is an evolved response to the external environment, which is also responsible for sending the necessary signal(s) to receptors in the cell, or to the neural integrator of multicellular individuals.

The 'diapause syndrome', common to protists and Crustacean waterfleas, includes the cell cycle arrest, as well as the induction of mating types, gametogenesis, packaging of the diapaused stage, and the duration of the diapause. The cell-/ life-cycle may be interrupted in a variety of seasonal patterns, which are triggered by one or more identified external signals and are integrated by cell and circadian clocks. *Daphnia pulex* displays four distinct seasonal patterns of embryonic diapause in four putative genotypes.

The ecological implications of diapause in life cycles may be sought in such questions as 1, environmental control of temporal and spatial boundaries in sessile populations; 2, application of knowledge of diapause/dormancy cycles to redesigning ecological communities; and 3, integration of mechanistic and biosystematic approaches in ecology.

Introduction

Diapause in perspective is an elaboration of cell cycle control. Such control infers that diapause is a 'seme' much as '4-leggedness'. It may be modified as to duration, or the stage of occurrence, or season of expression, but not eliminated from a genome. Its basis lies in the cell cycle engine, and its control may be 'upstream' of gametogenesis.

Diapause (and dormancy) control is relevant to ecology in the subject of boundary control, spatial as well as temporal. Cycling in and out of growth arrest involves a signal external to the individual, which needs to be of a threshold intensity. Dependency creates the possibility that a spatial boundary will exist at a point in the external environment where the gradient in signal strength conforms to the threshold. Germination of oospores of *Chara zeylanica* in the primary cycle of dormancy may be prevented by short-day photoperiods, or thermoperiods. Although not a boundary species, such controls on germination may explain sharply delineated boundaries that may exists in *Nitella* meadows and at 'treelines'.

Life cycle control is an elementary consideration in the selection of populations with which to restructure ecosystems. Arthropod life cycles revolve in phase with the annual cycle of environment because the photoperiod reaction regulates the diapause interval (Danilevskii, 1965). The expression of photoperiodism may be modified, even dependent, on variables other than the daylength signal (Zaslavski, 1972). The diapause is flexible both as to stage of life cycle and the season of occurrence. Flexibility is most apparent in the life cycles of aquatic arthropods because active growth in all seasons is possible.

The purpose of this paper is to compare the control systems that govern growth arrests in the cell cycle with those that control diapause cycle in arthropods. Such information becomes the basis for asking fundamental questions in ecology.

Diapause is the programmed growth arrest in the development of individual cells and multicellular arthropods alike. Its roots are manifest in the progression of the cell cycle as it unfolds life's three fundamental processes: growth, proliferation and differentiation. The cell cycle behaves like an engine, releasing potential- to kinetic energy (Murray & Hunt, 1993). At a phase of the revolving cell cycle, corresponding to some point in the cycle of earth's rotation, growing cells transform potential to kinetic energy in order to replicate and separate a second self.

Characteristics of the cell cycle mimic the previous cycle, or they assume a differentiated form. For example, cells continue to cycle, or they enter a 'diapause'. If the latter, the cell cycle becomes arrested, and it remains arrested at an established blockpoint in the cycle. The arrest is usually in response to signals external to the cell, which both initiate and influence duration of the arrest.

The cell cycle

Four basic phases of the cell cycle in eukaryotes are identified by the onset of synthesis of nuclear DNA (S phase) and, secondly, by the division of the nucleus (M-phase) followed by cytokinesis. The interval (gap) separating the birth of daughter cells and the beginning of S-phase is G1, the second gap between S and M is labelled G2.

A cell clock controls the speed of revolution through the G1, S, G2 and M-phases of the cycle. Cells of yeasts, including the budding yeast, *Saccharomyces cerevisiae* complete a cycle without apparent influence of a circadian, or other biological clock (Abe *et al.*, 1984). At any given temperature, cycle length is a function of nutrition and taxon; at 27 ° budding yeasts complete the cycle in 90 minutes, fission and torula yeasts take longer (Abe *et al.*, 1984). Progress along the cell cycle is monitored by an elaborate signalling system, involving signal receptors in the cell membrane and intracellular protein messengers that interact to inhibit, modulate, or amplify the signals sent to the nuclear membrane. The signaling system induces, or modifies gene expression, as in the mating reaction, or it initiates replication of DNA.

Revolution of the cell division cycle in eukaryotes is blocked at key locations in the cycle. A major blockpoint, 'START', is positioned in late G1. A second blockpoint, (G2/M), lies at the boundary of G2 and M. Kineses such as attend the synthesis of chromosomes is caused by strategic phosphorylation of certain amino acids at key locations of proteins. Control of movement is effected through the opposing actions of kinases and phosphatases. The blockpoint START is traversed, and the 'S' phase begins when a kinase protein (p34 CDC28) is joined by another protein, a G1 cyclin to form the maturation promoting factor (MPF; reviewed by Murray & Hunt, 1993).

To traverse the START block, the growing cell must reach threshold size. In a nourishing (eutrophic) environment, free of cell-cell contact, yeast cells in the haploid state cycle through G1 into S in 'routine' fashion (Fig. 1). In oligotrophy, the cell cycle may be arrested in G1, the result of feedbacks between external signals and membrane receptors of the cell. Certain nutrient ions may prevent synthesis of cyclin, and the cell cycle is arrested in G1. The cell cycle is also arrested when a mating pheromone triggers the appropriate membrane receptor during G1. Signal receptors for pheromone and nutrition connect to the START blockpoint via the proteins they induce, which are synthesized at well defined transition points in the cell cycle (Sherlock & Rosamond, 1993). The protein product of the gene FAR (1), which is activated in response to (negative) external signals, blocks synthesis of MPF and appears to be the master regulator of cell division (Peter & Hershkowitz, 1994).

Growth arrests

The growth arrest in G1 triggered by mating pheromone prevents chromosomes from unravelling until conjugation is complete. After mating, the diploid cell continues to cycle if nutrition is sufficient, or it undergoes meiosis and gives rise to four spores, which enter growth arrest (Fig. 1). In point of fact, the growth arrest caused by mating pheromone is upstream of nutrient deficiency and sexual reproduction, as shown by genetic analysis of the proteins involved (reviewed by Sherlock & Rosamond, 1993). Thus, selection for traits involving diapause may be both independent of and primary to traits asssociated with sexual reproduction.

The dynamics of a population may be explained by the life cycle interacting with the environment. In the budding yeast *S. cerevisiae*, the dynamics reveals 1, proportions of four types of cells: diploid cell, or haploid cell of one of two mating types, or zygote; and 2, the 4-dimensioned traverse of the cell division cycle (as reviewed by Hershkowitz, 1988). Feedbacks between environment and cells during the course of the cell cycle determine the population's unique characteristics, as defined in any text of general ecology. Such characteristics appear to be independent of circadian clocks, however.

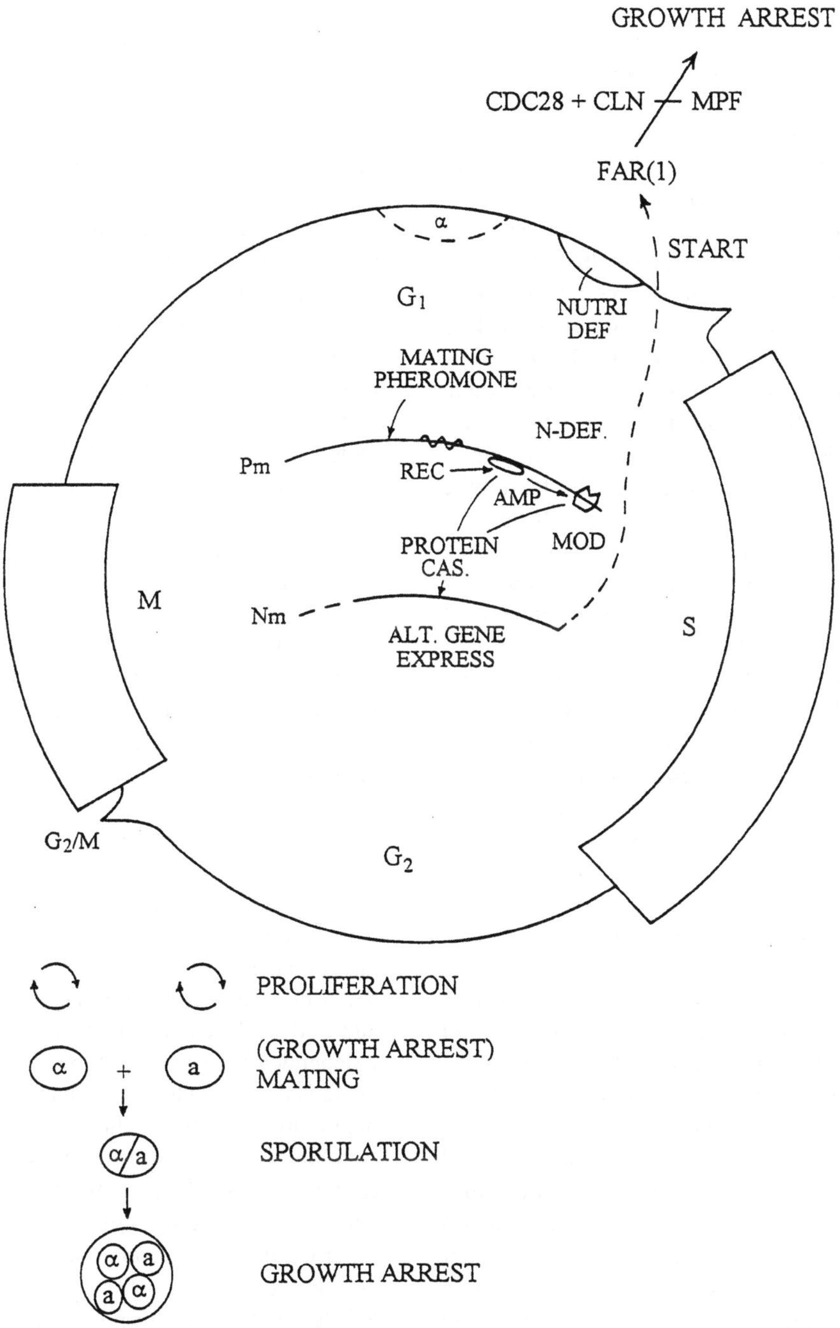

Fig. 1. A typical cell cycle as expressed by budding yeast *Saccharomyces cerevisiae*. Revolution of the cycle requires traverse of two blockpoints, START and G2/M. Attainment of critical size in G1 allows the synthesis of maturation promoting factor MPF), but which can be blocked by the protein product FAR1, which prevents traverse of START. Other arrests are triggered by mating pheromone (α), or deficiency of one or more nutrient ions. Such external signals are received in G1 at the plasma membrane (Pm) and sent to the nuclear membrane (Nm) with a protein amplification cascade of the external signal. Options of the yeast cell during any one cell growth cycle are division, growth arrest, gametogenesis, cell fusion and sporulation. See text for details.

Cell cycles coupled to circadian clock

Designed for another lifestyle, a circadian clock regulates the cell mechanics and dynamics in the Green flagellate *Chlamydomonas.* The clock resides in the endogenous programme of the cell. The cell cycle of this autotroph revolves in phase with the earth's rotation, a feature of the cell's circadian clock. Circadian and cell clocks interact. The start of growth, or G1, anticipates the dawn and begins with liberation of daughter zoospores from the mother-cell wall (Fig. 2; Howell *et al.* 1973). Both features are controlled by the circadian clock (Mihara & Hase, 1971, 1975). The clock also regulates the synthesis of a matrix protein (Cab) necessary for the receptor pigments in the chloroplast organelle (Kay, 1993). Well known is the circadian rhythm in responsiveness to light by the motile cells of *Chlamydomonas* (Bruce & Bruce, 1981).

Embedded in the G1 of the circadian clock entrained cell-growth cycle are transition points for light reactions and the mating/nutritional regulators of growth arrest (Fig. 2). Light at the very beginning activates the division timer. A second light-sensitive timer, positioned 4–6 hours later, frees the cell from further light obligation. Traverse of the START blockpoint may be in darkness, or light (Howell *et al.*, 1973).

Also at the 6-hr location is the deadline for receiving mating pheromone to transform zoospore into gamete. Mating takes place between gametes of opposite mating types to give rise to a thick-walled, diploid zygote, which enters a diapause of arguable duration.

Unlike cells of *Euglena* (Edmunds, 1981), zoospores of *Chlamydomonas* escape the constraints of the circadian clock with the capacity to undergo multiple cycles of mitoses per circadian cycle (Harris, 1989). Cells of sufficient size traverse the START blockpoint at, or about sunset. They then course through S, G and M within a 2-hour interval at summer temperatures. With two such excursions, each zoospore produces 4 daughter cells in one cell division cycle (CDC); four excursions, as in thermophilic strains at 30 °C results in 16 zoospores per circadian cycle. However, if growth is insufficient for division, the cell recycles back to the beginning of G1 for a second round on the following day. Thus, the cell cycle in *Chlamydomonas* is a circadian growth cycle (CGC), with or without one or more cell division cycles.

Circadian clock control of the *Chlamydomonas* cell cycle extends to control of mating and dormancy. Dormancy (plant diapause) in *Chlamydomonas*, as in yeast and waterfleas, is reached via the sexual phase of the life cycle. Gametogenesis, or transformation of the haploid zoospore to a gamete may occur without cell division. The cell is reprogrammed early in G1. A 'transition deadline' is tested by cell-sensitivity to a 'local stimulus', such as nitrogen deprivation. Mating requires two mating types, mt+ and mt−, which may require different parents (heterothallism). Mated cells fuse, secrete a prominent cell wall and enter a dormancy of possibly programmable duration.

A 'two-stimulus' control system of external signals operate to induce the gametogenesis- diapause syndrome in some strains of *Chlamydomonas.* Early investigators (reviewed by Coleman, 1962) found that depletion of fixed nitrogen caused gametogenesis. Other studies, especially those of Ishiura and Iwasa (1973a, b) allude to a two-stimulus control that combines a photoperiod-like reaction with N-depletion. Pair formation follows an oscillation in efficiency with peaks at the beginning and end of subjective night.

Photoperiod reactions include another component of the diapause/dormancy syndrome in *Chlamydomonas.* Lewin (1957) working with the zygotes of *C. moewusii* showed a thickening of cell wall and a deepening of the dormancy in response to duration of the light/dark cycle.

Coupling of circadian and cell clocks

In the *Chlamydomonas* cell cycle, G1 seems entrained to the photophase, and the remainder of the cycle seems uncoupled (reviewed in Harris, 1989). In *Pyrocystis*, a huge (0.9 mm) dinoflagellate, circadian clock and cell cycle clocks are reported to operate quasi-independently (Sweeney, 1982). A 5-day cell division cycle is divided into 5 circadian cycles, with S-phase beginning at the end of the third photophase. Changes in chloroplast characteristics marked the passage of circadian periods. The most tightly coupled relationship between specific phases of cell and circadian clock are reported for a heterotrophic strain of Euglena by Carre & Edmunds (1993, 1991). The major block points, START and G2/M are positioned at locations in the circadian cycle where the phase adjustment system of the clock is most responsive.

Photoperiod reactions

Eukaryote algae show both solar day (circadian) and solar year (circennial) modalities of cell cycling and cell differentiation. They display one of five control

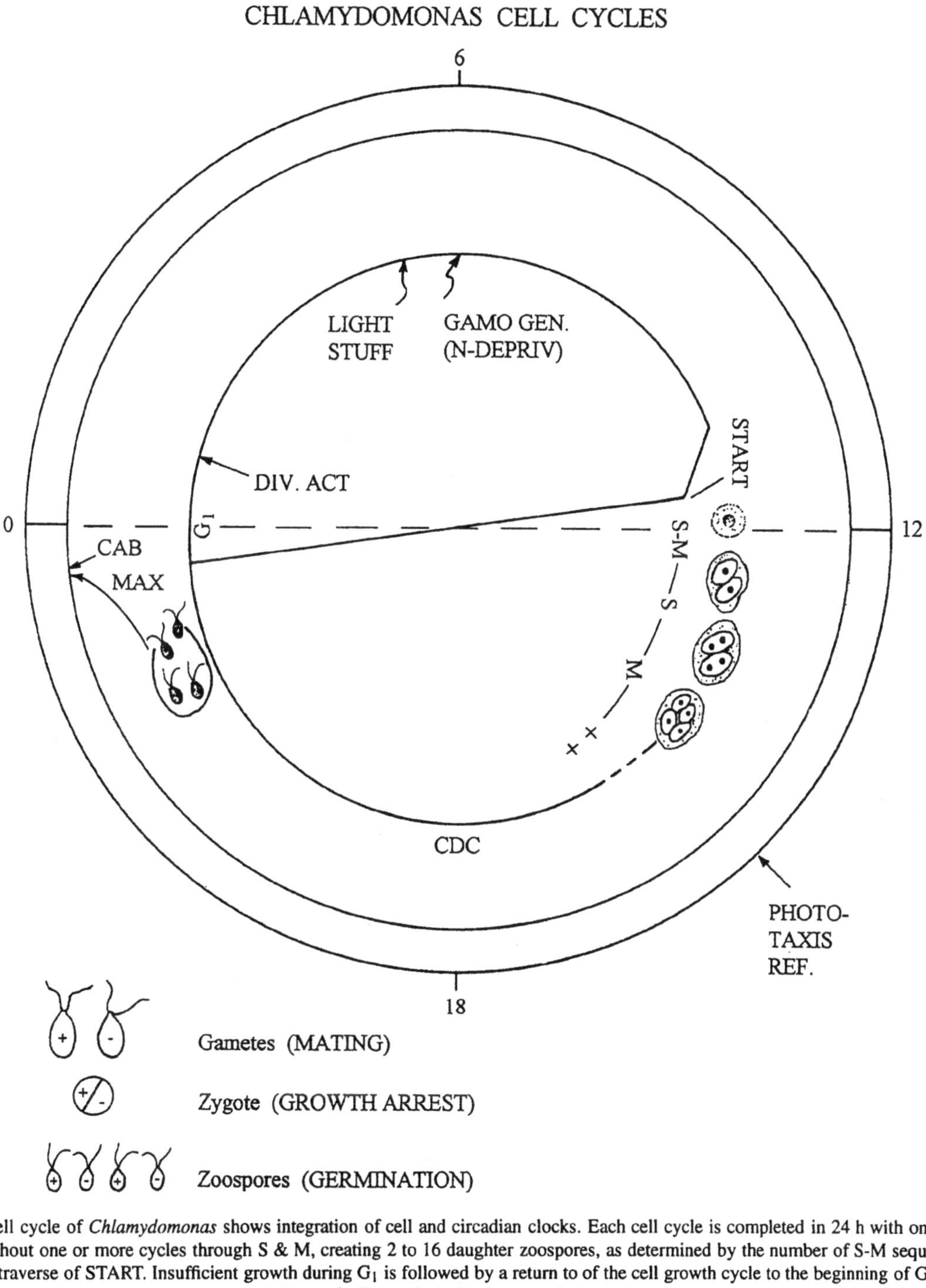

Fig. 2. Cell cycle of *Chlamydomonas* shows integration of cell and circadian clocks. Each cell cycle is completed in 24 h with one G1, and with or without one or more cycles through S & M, creating 2 to 16 daughter zoospores, as determined by the number of S-M sequences that follow the traverse of START. Insufficient growth during G_1 is followed by a return to of the cell growth cycle to the beginning of G_1. Several reference points exist for circadian clock control (e.g., CAB MAX., lysis of maternal cell wall surrounding zoospores, etc.). See text for details.

systems that phase cell cycling to the solar year in multicellular organisms. Luning (1993) and Luning & Kadel (1993) distinguish control by photoperiodic induction from entrainment of an innate circennial cycle such as the annual rhythm of growth in perennial seaweeds. Daylength entrains the annual rhythm of 'growth' velocity! Entrainment was possible with photophase lengths ranging from 10 to 16 h per day, but not with only 8 h of light per 24-h day.

The standard model for photoperiod induction is a 'stationary' daylength of critical duration. A qualitative photoperiod reaction of the subject organisms is characteristically of the threshold type and is displayed in the photoperiod reaction curve (Danilevskii, 1965, Zaslavski, 1988). The reaction characteristically goes from threshold to full response within 1 or 2 h of difference in stationary daylength. Photoperiod induction describes three separable reactions to external signals. The typical reaction is external coincidence (Pittendrigh & Minis, 1964; Bunning, 1936). The photocycle entrains the circadian oscillator, which in turn controls temporal position of a photosensitive reaction. In external coincidence, the light is responsible for entraining phase of the circadian clock and for photo-induction. The latter reaction typifies the photoperiod reaction curve.

A second type of photoperiod reaction arises when a thermocycle may substitute for the light reactions. Referred to as 'internal coincidence' (Saunders, 1973; Saunders & Gilbert, 1990), this category was first described in a parasitoid wasp that may be shielded from light during the inductive stage. Thermoperiod substitutes for light control in the absence of light.

In a third kind of photoperiodism, thermoperiodism dominates the light signal in a temperature oscillating environment. A critical daylength is displayed but only at constant temperatures. Germination of oospores in primary dormancy, such as with *Chara zeylanica*, is activated by a long-day light signal, but not in a short-day thermocycle (see below).

The fourth kind of photoperiod reaction behaves as an 'interval timer'. The timing of the reaction needs to be reset, but not timed by the light phase each day. A salient feature of the interval timer is that it measures night length (Skopik & Bowen, 1976; Skopik *et al.*, 1981). Nevertheless, circadian oscillations in sensitivity to night interruption have been demonstrated (Zaslavski, 1992).

The Daphnia diapause syndrome

Embryonic diapause in the life cycle of the water-flea, *Daphnia* is part of the same syndrome seen in yeasts and Green flagellates. The syndrome starts with the environmental induction of males, continues with the induction of gametogenesis in an (albeit diploid) parthēnogen and concludes with an adjusted duration of the diapause. The control system also contains many similarities to those of protists.

The *Daphnia* life cycle is homothallic (reviewed for *S. cerevisiae* by Hershkowitz, 1988). A mechanism of sex determination, referred to as cassette transposition, appears similar in budding yeast and in *Daphnia*. An alternating set of alleles, one for males another for females, appears at the locus for sex determination in successive generations of yeast cells, and in successive broods in the case of *D. magna* (Hobaek & Larsson, 1990). Induction of the diapause 'syndrome' in *Daphnia* requires external signals similar to those shown to induce diapause in *Chlamydomonas*, with one major advantage.

Signal combinations can be confused in *Chlamydomonas* by the inability to distinguish seasonal patterns within the taxon. In *Daphnia*, the relevant combination of external signals is an essential part of each seasonal pattern of diapause. Two external signals, a global and a local stimulus may operate. Daylength, or photoperiod, of critical duration is necessary to induce diapause in the reference cycle of *D. pulex* Leydig, one of four described by Berg (1931) as the 'autumn monocycle'. When a short-daylength becomes permissive, a second 'density-proportional' signal must be present in culture, as well as in the source (Paul) lake (Stross & Hill, 1965, 1968). Klieven *et al.* (1992) confirmed the effectiveness of two local signals, a 'crowding factor' and a 'food depletion factor' as 'local' signals for a clone of *D. magna*, whose field performance was unstated. The apparent requirement for short daylengths, semi-starvation plus a 'crowding factor' makes the diapause syndrome maladaptive at the species level (Stross unpubl.). That populations of *D. magna* may show variability in the external signals controlling the syndrome is known. Pijanowska (unpubl.) discovered that a predator 'kairomone' may also act in long daylengths as a local stimulus in at least one population.

The prospect that a predator kairomone is responsible for inducing diapause in the fourth cyclic type of *D. pulex*, the 'spring monocycle', has yet to be confirmed.

The 'dicycle', so-called because its representatives undergo two intervals of diapause annually (Berg 1931), enters embryonic diapause in autumn and again in late spring when daylengths are long. It manages to avoid the photoperiod trap when daylengths are long by becoming insensitive to photoperiod at water temperatures comparable to pond temperatures in late May and June (Stross, 1969b). Reproduction in such a population is free to switch back and forth throughout the warm season whenever the 'local stimulus' is sufficient; provided the pond survives evaporation. The dicycle phenotype circumvents real and imagined problems.

Duration of the diapause interval and optimum temperature for diapause development appear to be flexible and programmed by the daylength and temperature at the occasion of egg production (Stross, 1987). The dicycle also avoids the need to develop a signalling system that will bring viable males onto the scene in advance of gametogenesis. In the dicycle, the subitaneous eggs are uniparental, as is standard, and so are the diapausing eggs (i.e., pseudogamogenic). (The photoperiodically induced males either occupy a nonfunctioning position in the life cycle, or a facultative meiosis, unbeknownst, could be triggered by the presence of a male pheromone?)

Photoperiod also controls entrance to embryonic diapause in the dicycle at autumn temperatures. It would appear not to control diapause termination in autumn since summer diapausing embryos hatch in daylengths shorter than the critical amount for the autumn monocycle. A related question is what prevents females emerging in short daylengths of autumn from producing only resting eggs? A mechanism analogous to that in aphids would assure that at least the first brood would be of subitaneous embryos. Although not documented for the dicycle, such a mechanism was observed in the autumn monocycle of two populations (Stross, 1969a, unpubl.).

In two examples of an acyclic type of life cycle, the induction of embryonic diapause behaved much like that in the autumn monocycle. The difference lay in a much lower sensitivity to the density proportional stimulus (Stross, 1987).

Photoperiod induction and annual cycles in Daphnia

Photoperiod control vis-a-vis circennial entrainment is validated by four criteria, all of which are displayed in the autumn monocycle. First the critical photoperiod is well displayed, i.e., the stationary daylength that produces an equal proportion of broods carrying subitaneous and diapausing embryos is consistently realized in assays. The response grades from no evidence of diapause at L14:D10 to 100% reaction at L12:D12. Secondly, the critical photoperiod is latitudinally adjusted, increasing from L13:D11 at 43 °N to L20:D4 at 71 °N when examined at constant temperatures. Thirdly, the clock reaction is temperature compensated at 12 and 19 °C. The Q_{10} ratio is characteristically less than 1.0. Fourthly, the photo-sensitive stage lies with the mother before she is born, as is standard for daylength control of gametogenesis and embryonic diapause (Danilevskii, 1965).

Table 1. Incidence (%) of diapausing broods in a 30-day assay interval of growth and reproduction in an autumnal monocyclic population of *Daphnia pulex* Leydig from Paul Lake, Michigan, as a result of photoperiod induction and continued exposure to a short-day (SD) signal. Data from Hill (unpubl. ms).

Daylength at which			
Parent born	Mother born	Reproduction occurs	Diapausing broods (%)
16:8	16:8	16:8	2
16:8	16:8	11:13	9
16:8	11:13	11:13	78
16:8	11:13	16:8	18
11:13	11:13	11:13	–
11:13	11:13	16:8 (1)	75
11:13	16:8	11:13	8
11:13	16:8	16:8	3

(1) Females transferred after start of reproductive interval.

An experiment by Hill (unpubl.) at 19 °C places the photo- sensitive interval at some time before the birth of the mother. It also shows the need for continual exposure of the mother to an inductive photoperiod. For example, females transferred as pre-adults from long daylengths (LD; L16:D8) to short daylengths (SD; L11:D13) continue to produce broods of subitaneous eggs; only 9% of the broods produced in the next 30 days were of diapausing embryos, compared to 2% of diapausing broods produced by the cohort held in LD (Table 1).

By contrast, a cohort of females born in SD and held there produced 78% diapausing broods, while, the cohort transferred from SD to LD before sexual maturity produced only 18% diapausing broods. Other combinations are shown in Table 1.

Reversal of photoperiod induction in the autumnal monocycle is, judging from two experiments, more rapid in females born to mothers recently transferred from SD. Female young born in the first 6 days of adulthood (day 6) to females transferred from LD to SD produced only 18% diapausing broods in the standard 30-day assay interval when switched back to

LD. By contrast, females born to mothers, held in SD until sometime between days 14 and 22 of adulthood, produced 75% diapausing broods in long daylengths (L16:D8). Females of an arctic population were also slow to reverse following culture in SD (i.e., <L20:D4) for most of their adult lives.

The lack of lags, inductive and reversal, in the reproductive performance was evident in the warm phase of the dicycle. That, plus the absence of a photoperiod response, indicate a complete uncoupling of photoperiod control by the 19 °C temperature.

Diapause & minipause

Adjustments of the dicycle in *D. pulex* extend to the diapause interval, which is completed with a short, mild cooling in early autumn, or in spring with prolonged exposure within, or under ice (Stross, 1987). Davison & Stross (1984) assumed that the diapaused embryos removed from Saratoga Lake, NY, which hosts a spring monocycle, were completing diapause development in spring. However, the 25–45% hatching success achieved with material removed from the lake and tested in spring and summer contrasts with the 90+% success achieved in November and December with embryos that had lain at 4 °C since arrival on the lake bottom (Stross, unpubl.).

Extreme adjustments of the diapause interval are reported for other plankters. Gilbert & Schreiber (1995) reported a 'minipause' cycle in *Synchaeta pectinata*. When food is withheld for only a matter of hours, this rotifer produces an amictic egg that becomes enveloped in egg membranes similar to those of the mictic egg. Development is arrested for only a few days in the culture environment (Gilbert & Schreiber, 1995). The fact that the life cycles of rotifer and waterflea may enter a growth arrest without undergoing gametogenesis is noted.

Eggs of other Crustaceans, such as the Calanoid Copepod *Pontella*, are candidates for minipause. Females release subitaneous eggs, which are surrounded with membrane structures similar to those of diapausing eggs and which contain 'diapause discs' in the ooplasm (Santella & Ianora, 1990).

Minipauses are also described for the Cladocera. Doma (1979) reported a seeming lack of a resting phase in the ephippial eggs of *D. magna*. Stross began research with a culture of Supply House *D. pulex* that resuscitated after all the free swimming forms had died off, presumably because at least one embryo from the ephippial eggs failed to remain in diapause. The point being made is that so many of the accoutrements of the diapause syndrome in the zooplankton are displayed in a range of conditions, at least in the laboratory.

Significance of growth arrest control

A cadre of 'local' signals operate 'universally' with, and in the absence of, photoperiod control in Crustaceans, rotifers, as well as in single-celled eukaryotes. The signals induce sexuality, which is then followed by a non-stalling arrest of the cell-cycle engine, i.e., the diapause syndrome. The system of external signals activates specialized receptors of individuals, whether at the cell membrane, or coordinated by a nerve system. Photoperiod control, when it dominates local signals permits the cell-cycle to 1, anticipate the unfavourable season; 2, depart the active phase while reproduction is favoured; and 3, return when conditions for development of young individuals should be optimal. The essence of photoperiodism is to convert the diapaused phase of the life cycle from survival mode to a mechanism for optimizing a cyclic environment.

Occupying a central function in the cycling of single cells (and multicellular Crustacea), diapause should be regarded as a seme in the sense, used by Margulis (1981). Although the expressions of diapause, such as its locations in the life cycle, are subject to modification in the selective processes that act on demes, the growth arrest itself may not be eliminated from the cell cycle. As seen in cycling of yeast cells, the growth arrest anticipates sexual reproduction.

Thus, an ecology that incorporates the biology of diapause may not be congruent with the purely biosystematic approach, which aims to understand the environment of species evolution. Diapause control and its manifestation in ecology is incorporated in the mechanistic approach.

The ecological implications of diapause in life cycles may be sought in such questions as 1, environmental control of temporal and spatial boundaries in sessile populations; 2, application of knowledge of diapause/dormancy cycles to redesigning ecological communities; and 3, integration of mechanistic and biosystematic approaches in ecology.

Control of spatial boundaries

Life cycle control may explain the formation of spatial boundaries in charophytes, barnacle zones and treelines. Populations of *Nitella flexilis* and *N. opaca* grow on the bottoms of clear lakes with a density that matches a meadow of grass (Stross *et al.*, 1988). A remarkable feature of the *Nitella* meadow is an abrupt termination of plants in deep water, creating a meadow face, which marks the deep boundary. The location of such boundaries may be predicted from knowledge of taxon and light environment, specifically irradiance and spectral quality (Stross *et al.*, 1995).

Light is the probable agent in the formation of the abrupt boundary. Although grazing control of charophytes is well established, there is neither evidence of grazing at the boundary, nor indication that grazers have the capacity to create a sharp depth boundary. Light control is supported by observations (Spence, 1976; Stross, 1979). Of the alternative mechanisms, a photomorphogenic reaction is the more readily supported by the luxuriant growth form of individual plants at the meadow face (Stross *et al.*, 1988).

The photomorphogenic hypothesis of lower boundary control specifically states: a light triggered reaction at some point in the life cycle of the plant becomes critical to revolution of the cell cycle. Plants in the meadow face mark the location in the light gradient where the signal lies at threshold level.

A first step is to identify a photo-sensitive reaction in the life cycle of a charophyte that may be light limiting. At the meadow face the 'light' switch controls vegetative regeneration. Conceivably, a nodal cell at the apex of the overwintering thallus is the starting point for regeneration in spring. Based on an analysis of the light environment at the lower boundary in lakes of Britain and New Zealand, one may expect the light switch to contain red- and blue-sensitive receptors (Stross *et al.*, 1995). Further, one may anticipate 1, the receptors are coupled to a photoperiod switch with a threshold irradiance for induction; and 2, a restricted time window when germination is permitted because of a sensitivity to light that is temporally regulated by the dormancy cycle.

Testing with complete individuals (i.e., thalli with rhizoids) of *N. flexilis* requires special care, and simulation of ultra-oligotrophy is difficult without provision for control. The alternative is to examine germination control in the zygospore in the primary cycle of dormancy, looking for evidence of a photoperiod reaction. Despite intensive effort, early study of an ecologically consistent photoperiod reaction was unsuccessful (Ishikawa & Araki, 1955; Ishikawa & Oohusa, 1956; Black & Wareing, 1955; Vince-Prue, 1975). Stearns & Olson (1958) show an acceptable threshold type response, but only before the seeds of the Eastern Hemlock (*Thuja*) were allowed to after-ripen in the dark.

Fortunately, oospore germination of the chosen clones of *Chara* (*C. zeylanica*) displays both a photoperiod and a thermoperiod controlled reaction (Stross, unpubl.). The observed characteristics of germination included 1, a 6-month interval of after-ripening, following an early-autumn harvest; 2, a light requirement; 3, a separate light reaction in the form of a long-day activation of embryonic development leading to germination with a threshold photophase at some point between 12 and 14 h of light per day; 4, but effectiveness of photoperiod reactions demanded continual exposure of oospores to the light-dark cycle while they are still dormant; and 5, light actions controlling germination were over-ridden by a thermoperiod reaction.

Although fortuitous for the most part, treatment of the dormant oospores was guided by the assumption that the loss of a light and/or photoperiod reaction shown in seed germination by the earlier studies was the result of the after-ripening in continuous darkness (e.g., Black & Wareing, 1955; Stearns & Olson, 1958). Thus, germination may be prevented in oospores of *C. zeylanica*, a subtropical and south temperate latitude species (that inhabits the bottoms in shallow but permanent bodies of freshwater), when the oospores were allowed to overwinter at 43 °N in an unheated greenhouse. Germination was nearly 100%, however, when the oospores were transferred to constant temperatures above threshold, or to oscillating temperatures provided that night temperatures remained at 20 °C or warmer.

That is, the oospores lying exposed in the greenhouse failed to germinate and entered a secondary dormancy, although completely exposed to light, when night temperatures were maintained below 20 °C. The over-riding effect of temperature was shown to be a short- day inductive, thermoperiod reaction, with a threshold between 12 and 11 h of warm phase, and a cool phase below 20 °C for the remaining hours. The photoperiod reaction was best revealed by experiments on a clone held in the greenhouse. The oospores reentered dormancy in spring, which persisted through the summer into autumn, when the oospores again became germinable. They failed to germinate in the greenhouse at autumn daylengths and temperature cycles that dropped below 20 °C at night. However, germination was 100%, or nearly so when oospores were transferred to constant temperatures and the appropriate daylengths. A photoperiod response curve showed nearly complete suppression of germination by short daylengths (10 & 12 h per 24 h) in the ecological range. Long daylengths 14 & 16 h and very short light periods of 3 and 6 hours per 24 h permitted full germination of nearly 100%.

Further study of boundary control, of charophytes in deep water, or of trees on mountain slopes, needs to be aware that photoperiod control of oospore germination was restricted to propagules in the primary dormancy cycle. Buried seeds, or oospores in a secondary cycle may no longer be subject to photoperiod control. Emerging from secondary dormancy, the seed may germinate following activation by a flash of light, or have escaped from light requirements entirely (Black & Wareing, 1955). The shift from photoperiod- to light-flash control has been shown for oospores of the annual, *N. furcata* (Sokol & Stross, 1992, unpubl.)

Students of life cycle control are aware that performance in the field, the phenotype, is the product of external signals acting on expression in one or more genotypes. Although not necessarily demonstrated in Crustacea, the genetic basis of the diapause syndrome is known for other arthropods and for protists, from which extrapolation in a conservative system is permissible. In fact, there is a commonality of external signals that induce the diapause syndrome.

The involvement of circadian and photoperiod clocks in the expression of diapause in the life cycle is conditional as it is in the underlying cell cycles of eukaryotes generally. Temperature sensitivity of clock control seems to be at the centre of the conditional response. The generalizations apply to the seasonal cycles of diapause in the waterflea, *Daphnia*. Only some of cycles (e.g., autumnal monocycle, autumn of the dicycle) overtly employ daylength as an external signal for induction. In oospores and seeds of plants, control of germination by daylength signals may be restricted to the primary cycle of dormancy.

Annual cycles with alternative phases of active and arrested growth are characteristic of eukaryotes. The diapause syndrome when phased to the environmental cycle with biological clocks gives it both a survival and an optimizing quality. Deployment of the clocks is key to alternative seasonal patterns of expression, as it may be to alternative strategies for survival. The signals and the expressions of cell- and life cycles are primary to both understanding and controlling the mechanics that regulate distribution and abundance in populations.

References

Abe, K., S. Fukup & A. Kusanagi, 1984. Estimation of time lengths of cell-cycle phases from asynchronous cultures of the basidiomycetous yeast *Rhodosporidium toruloides*. J. gen. appl. Microbiol. 30: 257–265.

Berg, K., 1931. Studies on the genus, *Daphnia* O. F. Muller: with special reference to the mode of reproduction. Vidensk Meddr. Dansk Nathurhist. Foren. 92: 1–222.

Black, M. & P. F. Wareing, 1955. Growth studies in woody species VII. Photoperiodic control of germination in *Betula pubescens*. Physiol. Pl 8: 300–316.

Bruce, V. G. & N. C. Bruce, 1981. Circadian clock-controlled growth cycle in *Chlamydomonas reinhardtii*. In: H. W. Schweiger (ed.), International Cell Biology 1980–81. Springer, New York: 823–829.

Bunning, E., 1936. Die endogene Tagsrhythmik als Grundlage der photoperiodische Reaktion. Ber. dtsch. bot. Ges. 54: 590–607.

Carre, I. A. & L. N. Edmunds, 1991. cAMP-dependent kinases in the algal flagellate *Euglena gracilis*. J. Biol. Chem. 267: 2135–2137.

Carre, I. A. & L. N. Edmunds, 1993. Oscillator control of cell division in *Euglena*: cyclic AMP oscillations mediate the phasing of the cell division cycle by the circadian clock. J. Cell Sci. 104: 1163–1173.

Coleman, A. W., 1962. Sexuality. In: R. A. Lewin (ed.), Physiology and Biochemistry of Algae. Academic Press, New York, 711–729.

Danilevskii, A. S., 1965. Photoperiodism and Seasonal Development of Insects. Oliver & Boyd London, 283 pp.

Davison, E. A. Jr & R. G. Stross, 1986. A blue light-reversible reaction in an animal system (*Daphnia pulex*). Experientia 42: 620–622.

Doma, S., 1979. Ephippia of *Daphnia magna* Straus – a technique for their mass production and quick revival. Hydrobiologia 67: 183–188.

Edmunds, L. N. Jr., 1981. Clocked cell cycle clocks: ultradian, circadian, and infradian interfaces. In: H. W. Schweiger (ed.), International Cell Biology 1980–81. Springer, New York, 831–845.

Gilbert, J. J. & D. K. Schreiber, 1995. Induction of diapausing amictic eggs in *Synchaeta pectinata*. Hydrobiologia 313/314 (Dev. Hydrobiol. 109): 345–350.

Harris, E. H., 1989. The *Chlamydomonas* Sourcebook. Academic Press, New York, 780 pp.

Hershkowitz, I., 1988. Life cycle of the budding yeast *Saccharomyces cerevisiae*. Microbiol. Reviews, Dec. 1988: 536–553.

Hobaek, A. & P. Larsson, 1990. Sex determination in *Daphnia magna*. Ecology 71: 2255–2268.

Howell, S. H., W. J. Blaschko & C. M. Drew, 1975. Inhibitor effects during the cell cycle in *Chlamydomonas reinhardtii*. J. Cell Biol. 47: 126–135.

Ishiura, M. & K. Iwasa, 1973. Gametogenesis in *Chlamydomonas* I. Effect of light on the induction of sexuality. Pl. Cell Physiol. 14: 911–921.

Ishiura, M. & K. Iwasa, 1973. Gametogenesis in *Chlamydomonas* III. Daily fluctuation of sex-competence. Pl. Cell Physiol. 14: 935–939.

Isikawa, S. & S. Araki, 1955. Effect of light on the germination of forest tree seeds. II. On the photoperiodic tendency. J. Jap. For. Soc. 37: 485–487.

Isikawa, S. & T. Oohusa, 1956. Effects of light upon the germination of spores of ferns. II. Two light periods of *Dryopteris crasirhyzoma* Nakai. Bot Mag. (Tokyo) 69: 132–137.

Kay, S. A., 1993. Shedding light on clock controlled Cab gene transcription in higher plants. Cell Biol. 4: 81–86.

Kleiven, O. T., P. Larsson & A. Hobaek, 1992. Sexual reproduction in *Daphnia magna* requires three stimuli. Oikos 65: 197–206.

Lewin, R. A., 1957. The zygote of *Chlamydomonas moewusii*. Can. J. Bot. 35: 795–804.

Luning, K., 1993. Environmental and internal control of seasonal growth in seaweeds. Hydrobiologia, 260/261 (Dev. Hydrobiol. 85): 1–14.

Luning, K. & P. Kadel, 1993. Daylength range for circennial rhythmicity in *Pterygophora californica* (Alariaceae, Phaeophyta) and synchronization of seasonal growth by daylength cycles in several other Brown algae. Phycologia 32: 379–387.

Margulis, L., 1981. Symbiosis in Cell Evolution. W. H. Freeman New York, 419 pp.

Mihara, S. & E. Hase, 1971. Studies on the vegetative life cycle of *Chlamydomonas reinhardtii* Dangeard in synchronous culture I. Some characteristics of the cell cycle. Pl. Cell Physiol. 12: 225–236.

Mihara, S. & E. Hase, 1975. Studies on the vegetative life cycle of *Chlamydomonas reinhardtii* Dangeard in synchronous culture IV. Effects of 6-methyl purine on the revolution of the cell cycle. Pl. Cell Physiol. 16: 1121–1130.

Murray, A. & T. Hunt, 1993 The Cell Cycle. W. H. Freeman, New York, 251 pp.

Peter, M. & I. Herskowitz, 1994. Direct inhibition of the yeast cyclin-dependent kinase Cdc28-cln by farl. Science, Wash. 265: 1228–1231.

Pittendrigh, C. S. & D. H. Minis, 1964. The entrainment of circadian oscillations by light and their role as photoperiodic clocks. Am. Nat. 98: 261–294.

Santella, L. & A. Ianora, 1990. Subitaneous and diapause eggs in Mediterranean populations of *Pontella mediterranea* (Copepoda: Calanoida): a morphological study. Mar. Biol. 105: 83–90.

Saunders, D. S., 1973. Thermoperiodic control of diapause in an insect: theory of internal coincidence. Science, Wash. 181: 358–360.

Saunders, D. S. & L. I. Gilbert, 1990. Regulation of ovarian diapause in *Drosophila melanogaster* by photoperiod and moderately low temperature. J. Insect. Physiol. 36: 195–200.

Sherlock, G. & J. Rosamond, 1993. Starting to cycle: G1 controls regulating cell division in budding yeast. J. gen. Microbiol. 139: 2531–2541.

Skopik, S. D & M. F. Bowen, 1976. Insect photoperiodism: an hourglass measures photoperiodic time in *Ostrinia nubilalis*. J. Comp. Physiol. 111: 249–259.

Skopik, S. D., M. Takeda & C. Holyoke, 1981. A critical examination of the dual system theory in *Ostrinia nubilalis*. Am. J. Physiol. 10: R322–R329.

Sokol, R. C. & R. G. Stross, 1992. Phytochrome-mediated germination of very sensitive oospores. Pl. Physiol. 199: 1132–1136.

Spence, D. H. N., 1976. Light and plant responses in freshwater. In: C. G. Evans, R. Bainbridge & O. Rackham (eds), Light as an ecological factor. II. Blackwell, London, 93–133.

Stearns, F. & J. Olson, 1958. Interaction of photoperiod and temperature affecting seed germination in *Tsuga canadensis*. Am. J. Bot. 45: 53–58.

Stross, R. G., 1969a. Photoperiod control of diapause in *Daphnia*. II. Induction of winter diapause in the arctic. Biol. Bull. 136: 264–273.

Stross, R. G., 1969b. Photoperiod control of diapause in *Daphnia*. III Two-stimulus control of long-day, short-day induction. Biol. Bull. 137: 359–374.

Stross, R. G., 1979. Densities and boundary regulations of the *Nitella* meadow in Lake George. Aquat. Bot. 6: 285–300.

Stross, R. G., 1987. Photoperiodism and phased growth in *Daphnia* populations: coactions in perspective. In: R. H. Peters & R. de Bernardi (eds), Daphnia. Mem. Ist. Idrobiol. 45: 413–437.

Stross, R. G. & J. C. Hill, 1965. Diapause induction in *Daphnia* requires two stimuli. Science 150: 1462–1464.

Stross, R. G. & J. C. Hill, 1968. Photoperiod control of winter diapause in the freshwater Crustacean, Daphnia. Biol. Bull. 134: 176–188.

Stross, R. G., J. Huvane & R. C. Sokol, 1988. Internal structure of deep-dwelling *Nitella* meadows. Aquat. Bot. 29: 329–345.

Stross, R. G., R. C. Sokol, A-M. Schwarz & C. Howard-Williams, 1995. Lake optics and depth limits for photogenesis and photosynthesis in charophyte meadows. Hydrobiologia 302: 11–19.

Sweeney, B. M., 1982. Interaction of the circadian cycle with the cell cycle in *Pyrocystis fusiformis*. Pl. Physiol. 70: 272–276.

Vince-Prue, D., 1975. Photoperiodism in Plants. McGraw–Hill, New York, 444 pp.

Zaslavski, V. A., 1972. Two-stage photoperiodic reactions as a starting point for the production of a model of the photoperiodic control of arthropod development. Entomol. Revs. 51: 133–145.

Zaslavski, V. A., 1988. Insect Development: Photoperiodic and Temperature Control. Springer, New York, 187 pp.

Zaslavski, V. A., 1992. Light-break experiments with emphasis on the quantitative perception of nightlength in the aphid *Megoura viciae*. J. Ins. Physiol. 38: 717–725.

Hydrobiologia **320**: 119–122, 1996.
V. R. Alekseev & G. Fryer (eds), Diapause in the Crustacea.

The problem of induction and termination of diapause in cyclopoid copepods

V. I. Monchenko
Dept. of Invertebrate Zoology, Institute of Zoology, National Academy of Sciences, B. Khmelnitsky str., 15, Kiev, Ukraine

Key words: Cyclopidae, seasonal activity, diapause

Abstract

The seasonal activity and diapause in Cyclopidae was investigated in natural water-bodies in the vicinity of Kiev and in experiments. It was established that in stenothermal species (cold-loving and warmth-loving) the temperature thresholds of passing into an active state are nearer to the temperature optimum of each species than the temperature thresholds of passing into resting state. But temperature seemingly is not the only factor leading to changes in the activity state of populations. It seems probable that in the appearance of a particular adaptation some definite irritating external factor, regularly repeating each year, and permitting adaptation to it, acquires the role of a switch-factor directing the physiological processes of the organism to move in a new way — entering the resting state or leaving it.

The sharp variations in the number of individuals and biomass of the freshwater plankton in autumn and spring are connected with an abrupt transference of an enormous number of individuals of cyclopoid copepods of several species to the bottom sediments or their return to the plankton. These regular quantitative variations of the composition of the plankton have not hitherto found a full explanation. Still unknown are the physiological processes which accompany the condition of rest or cause it. Even the factors of the environment (or inner causes) which induce the beginning or termination of the condition of rest (or diapause) are still not fully determined. One of the causes of this state of affairs is the diversity of meanings of diapause. In this study diapause is characterized as a torpid, inactive state on the bottom of a water-body in copepodids and rarely adult stages.

The entry of some cyclopoid species into the condition of rest can explain their capacity to endure the drying of the water- body. This capacity is connected with the fact that many such species are widely distributed.

Among about 720 known species, subspecies and forms of the Cyclopidae it is already established that 34 have the capacity to pass into a resting condition. This is proved (1) by obtaining active cyclopoids from moistened bottom sediments collected from a dried up water-body and (2) by detecting resting specimens on the bottom of water-bodies. The species list should almost certainly be enlarged by including species with well expressed differences in seasonal activity, for example *Cyclops abyssorum* s. lat., littoral and interstitial marine cyclopoids the ecology of which is little known, and a number of species found in drying water-bodies, where of course the presence of resting stages is highly probable.

Resting stages are shown mainly by species of the subfamily Cyclopidae, consisting of cyclopoids with a relatively well expressed adaptation to drying water-bodies or with clear cut seasonal activity. On the other hand, resting stages are little known in species of the subfamily Eucyclopinae, which seemingly reflects their generally eurythermal nature. In a series of experiments on obtaining cyclopoids from dry mud, and also in a number of works of other authors (Gurney, 1933; Champeau, 1963, 1966 etc.) it has been shown that in the majority of cases individuals of copepodid stages IV and V returned to an active state. Very rarely they are accompanied by copepodids of stages II–III (Elgmork, 1965 *et al.*) or by sexually mature females (Tore *et al.*, 1991 and others). Resting copepodids of stages I–III occur as a rare exception in *Cyclops strenu-*

us, but it is a rule in *C. scutifer*, all generations of which develop in northern latitudes for 2–3 years. Indications about the existence of resting eggs in *C. strenuus*, *C. furcifer* and *Diacyclops bisetosus* (Wolf, 1905 *et al.*) need verification because the membranes of these described eggs did not differ in thickness from the usual one, and the humidity of the mud was not determined. The same holds true for the data on hiberbation in the egg stage in *Thermocyclops crassus* and *C. abyssorum* (Chapman, 1972 and others).

The question of what factors induce the passing of cyclopoids into a state of rest and return them into an active state has up to now received no universally acceptable answer though several hypotheses have been offered.

Temperature is clearly of great importance. With its seasonal change coincides the state of seasonal activity in copepods. A hindering influence of room temperature on the development of copepodids of stage IV in *Cyclops* sp. (Coker, 1933) has been confirmed by us (Monchenko, 1974, p. 229) when we worked on *Acanthocyclops vernalis* (Fisch.). The larvae developed normally at a temperature 21–22 °C and 8– 9 hours of day-light only up to this stage at which they stayed for a period of 2,5 months and then perished. Only when they were exposed for 50–60 hours at low temperature (4 and 12 °C in the experiment) did the copepodids quickly complete their development at 21–22 °C and began to reproduce. According to another author (Smyly, 1962) the temperature factor conditions the duration of the resting state of the copepodids of *Mesocyclops leuckarti*. This is confirmed by a series of other observations (Zankai, 1987 and others).

This view on the determining importance of temperature in the passing of stenothermal cyclopoids into a passive or active condition was confirmed by our observations on the seasonal dynamics of cyclopoids in the flood-land of the Dnieper near Kiev. For a number of stenothermal species the temperature thresholds of their passing into a state of rest and returning to an active condition were compared with the optimal temperatures for these species (Table 1). It was established that temperature thresholds inducing warm-loving stenothermal species into an active state are much higher than temperature thresholds of their passing into a resting stage. On the contrary, temperature thresholds stimulating cold-loving stenotherms into an active state are considerably lower than the temperature thresholds of their passing into a resting state.

These data seem to be biologically grounded. In fact, in stenothermal species (cold-loving and warm-loving) the temperature thresholds of passing into an active state are nearer to the temperature optimum of each species than the temperature thresholds of passing into a resting state. The population passes to active life as if having some reserve of favourable temperatures. This guarantees the population against possible unfavourable temperature influences in case of change in weather conditions. Other authors admit that the temperature factor has rather often a decisive meaning in the reactivation of cyclopoids. To it is ascribed the role of a 'switch' to a termination of a resting state (Smyly, 1961, etc., Elgmork, 1967 and others).

Other authors suggest that for induction or termination of diapause a decisive role is played by the degree of saturation of water with oxygen (Wierzbicka, 1962, etc.; Kiefer, 1978), change of salinity (Champeau, 1966), of trophic resources (Smyly, 1961), or by photoperiodism (Elgmork, 1959; Einsle, 1967, etc.; Spindler, 1969). Among the named factors light (or photoperiodism) is the most stable and permanent, the most regular in seasonal changes. It is completely independent of other factors, whereas temperature, level of saturation with oxygen, or changes in salinity depend on a number of different causes, beginning with weather and ending with the intensity of development of macrophytes, strength of wind causing mixing of water and so on. It is natural to suggest that adaptation to reactivity towards the light factor is more often fixed in the genetic programme of stenothermal species as a main regulator of seasonal changes in the development of populations than adaptations to the using of other environmental factors.

However, it is unnecessary to give this factor a universal meaning. It is possible to cite not a few examples where cyclopoids belonging to one species are present at the same time in neighbouring water-bodies at different stages of the life cycle (Monchenko, 1974) though the identical influence of the light factor is evident. It is known, that diapausing eggs of *Eurytemota affinis* hatch both in daylight and in darkness (Ban Syuhei, 1992). There is thus no doubt that here the beginning and end of the resting state depends on other factors. We may recall that falling out, or activation, in astatic water-bodies by no means depend on photoperiodism. It should be assumed that in conditions of a diversity of environmental factors and their possible influence it is biologically more justified to expect the presence in organisms of an ability to react to a complex of these factors, or at least to the most important of them.

Table 1. Some temperature indices of the seasonal activity of populations of a number of cyclopoid species.

Species	Temperature thresholds of transition in state		Optimal temperature
	Active	Inactive	
Mesocyclops leuckarti (Claus)	3	1	15-25
Thermocyclops oithonoides (Sars)	6–7	2	15–25
Th. crassus (Fischer)	8–9	4	20–25
Microcyclops rubellus (Lill.)	10–12	5–6	20–25
Megacyclops gigas (Claus)	9	13	1–2
Cyclops vicinus Uljanin	12	21	1–2
C. abyssorum abyssorum Sars	2	11–12	1–2
C. kolensis Lill	2	10–12	1–2
C. insignis Claus	2	7–8	1–2
Diacyclops nanus (Sars)	9	25	2–5

About a complex influence of interacting factors speak the following data. In relatively shallow water-bodies the role of a switch can be played both by temperature and by photoperiodism (Spindler, 1969, etc.; Watson *et al.*, 1971; Hairston *et al.*, 1990 and others). The interaction of such factors as the gaseous regime and photoperiodism has been pointed out (Alekseev, 1987). The importance of an interaction of two factors has been shown also for calanoids (Marcus, 1982, etc.). In deep lakes the decisive factor for changing the activity state is an interaction of temperature and the state of the gaseous regime (Cooley, 1971 and others). Their interaction is especially expressed during thermal stratification of the lake. In this case different factors may play a leading role at different depths (Einsle, 1975), and can explain, for example, the difference in time of falling out, or activation, of the same species in different parts of the lake, or even the presence of a resting state in one part of the lake and its absence in another (Einsle, 1967; Bressac *et al.*, 1983).

Nor should the possibility of a termination of the resting state due to purely physiological causes be disregarded. These may include lasting endogenous mechanisms (biological clocks, activity of neurosecretory and the other endocrine glands), stages of unrolling of a genetic programme (Williams-Howze *et al.*, 1992) without any influence from external factors. In drying-up lakes they may depend on the extent and duration of drying. In the light of various causes leading to changes in the activity state of a population it seems probable that in the appearance of a particular adaptation some definite irritating external factor, regularly repeating each year, and permitting adaptation to it, acquires the role of a switch directing the physiological processes of the organism to move in a new way entering the resting state or leaving it. It must be underlined that the resting state in cyclopoids is not unique in the animal world. It reminds one of diapause in insects (Elgmork *et al.*, 1978; Watson *et al.*, 1971 and others), in oligochaetes (Saussey *et al.*, 1984), *Asellus aquaticus* (Vitagliano *et al.*, 1984), hypobiosis in nematodes (Gibbs, 1986) and many others. The similarity of the resting state in cyclopoids to diapause of insects allows us to extend the analogy into the region of morphophysiological phenomena. It is known that diapause in insects is controlled by the endocrine (neurosecretory) system. At least in calanoids (genus Calanus) analogous structural-functional complexes have already been shown (Carlisle, Pitman, 1961) and are present as a couple of lateral neurosecretory cells at the front edge of the cerebral region. A rich granular structure of these cells in active individuals and the absence of granularity of the protoplasm (i.e. secretory activity) in resting animals leads to the idea of a role of the secretion of these neurosecretory cells in the regulation of diapause and moulting in copepodids.

Seemingly, the resting state is not an obligatory, direct spontaneous reaction to certain unfavourable external influences. In the majority of cases it begins only in immature individuals (about copepodid IV). The organism can only react in such a mode to a complex of external influences at a certain moment of the unrolling of its genetical programme. The resting state is realised as an answering reaction to a most strong influence (in given concrete circumstances) of one or another factor (or two or more important factors) acting as some kind of stress which disturbs the isochromal development in a relatively constant environment. An analogous effect of a change in a most important environmental factor activates, for example, the resting eggs of some rotifer (Blanchot *et al.*, 1982 and others); temperature and salinity activate eggs of *Artemia* (Ivanovsky *et al.*, 1980). A very substantial increase of temperature (from 15–20 °C) determined the start of processes necessary for diapause termination in *Asellus aquaticus* (Tadini-Vitagliano *et al.*, 1982). Though the physiological part of diapause in cyclopoids is far from clear, the biological significance, the general morphological peculiarities, and the inciting forces of the resting state of cyclopoids are in general features beginning to be understandable.

References

Alekseev, V. R., 1987. The role of photoperiodism in diapause phormation in freshwater Cyclopidae (in Russian). Sborn. nauch. trudov NII osern. i rechn. rybn. chos. 233: 95– 102.

Ban Syuhei & Minoda Takashi, 1992. Hatching of diapause eggs of *Eurytemora affinis* (Copepoda: Calanoida) collected from lake-bottom sediments. J. Crust. Biol., 12: 51– 56.

Blanchot, J. & R. Pourriot, 1982. Influence de trois facteurs de l'environnement, lumière, température et salinité, sur l'éclosion des oeufs de durée d'un clone de *Brachionus plicatilis* (O. F. Muller) Rotifère. C.r. Acad. Sci., Sér. 3, 295: 243–246.

Bressac, Y. & A. Champeau, 1983. Le cycle biologique de *Cyclops vicinus* (Copepoda, Cyclopoida) dans la retenue de Serre-Ponson (Hautes-Alpes). Ann. limnol. 19: 187–194.

Carlisle, D. B. & W. J. Pitman, 1961. Neurosecretion and Hormones in Copepoda. Nature (English.). 190: 827–828.

Champeau, A. Note préliminaire sur divers stades de résistance chez les copépodes dans les eaux temporaires de la Camargue. Rapp. et proc.-verb. réun. Comiss. internat. explorat. scient. Mer méditerr. 17: 805- -810.

Champeau, A. Etats de quiescence déterminés chez les copépodes d'eau saumâtre par les variations de chlorinité. C.r. Acad. Sci. Sér. D 261: 1289–1291.

Chapmam, A. M. The annual cycles of the limnetic cyclopoid copepods of Loch Lomond, Scotland. Int. Revue ges. Hydrobiol. 57: 895–911.

Coker, R. E., 1933. Arrêt du dévelopment chez les Copépodes. Bull. Biol. France et Belgique, 67: 276– 287.

Cooley, J. M., 1971. The effect of temperature on the development of resting eggs of *Diaptomus oregonesis* Lill. (Copepoda: Calanoida). Limnol. Oceanogr. 16: 921–926.

Einsle, U., 1967. Die äusseren Bedingungen der Diapause planktisch lebender *Cyclops* Arten. Arch. hydrobiol. 63: 387–403.

Einsle, U., 1975. Revision der Gattung *Cyclops* s. str. Speziell der Abyssorum-Gruppe. Mem. Ist. ital. Idrobiol. 32: 57–219.

Elgmork, K., 1959. Seasonal occurrence of *Cyclops strenuus strenuus* in relation to environment in small water- bodies in Southern Norway. Folia limnol. scand. 196 pp.

Elgmork, K., 1965. Triennial copepod (Crustacea) in the temperate zone. Nature (Lond.), 205: 413.

Elgmork, K., 1967. On the distribution and ecology of *Cyclops scutifer* Sars in New England (Copepoda, Crustacea). Ecology, 48: 867–971.

Elgmork, K. & J. P. Nilssen, 1978. Equivalence of copepod and insect diapause. Verh. Int. Ver. theor. und angew. Limnol., 20: 2511–2517.

Gibbs, H. C., 1986. Hypobiosis in parasitic nematodes – - an update. Adv. Parasitol. 25: 129–174.

Gurney, R., 1933. British freshwater Copepoda. III. *Cyclopoida*. Ray Soc., London, 384 pp.

Hairston, N. G., T. A. Dillon & B. T. De Stasio, 1990. A field test for the cues of diapause in a freshwater copepod. Ecology 71: 2218–2223.

Ivanovsky, Y. A., Y. L. Mitrofanov & J. L. Chaga, 1980. The hatching of nauplii *Artemia salina* by different regimes of salinity and temperature (in Russian). Biol. morja, 4: 25–31.

Kiefer, F., 1978. Freilebenden Copepoda. Binnegewässer, 26, Teil 2: 1–343.

Marcus, A., 1982. Photoperiodic and temperature regulation of diapause in *Labidocera aestiva* (Copepoda: Calanoida). Biol. Bull. 162: 45–52.

Monchenko, V. I., 1974. Cyclopoida Gnathostoma. Cyclopidae (in Ukrainian). Fauna of Ukraine, t. 27, n.3. Nauk. dumka, Kiev: 450 pp.

Naess, T., J. P. Nilssen, 1991. A new pattern of copepod life cycle. Oecologia, 86: 368–371.

Saussey, M. & G. Debout, 1984. Nouvelles données sur le déterminisme de la diapause de *Nicodrilus giardi* (Ribaucourt) (Oligochete, Lumbricien). C. r. Acad. sci. Ser. 3, 299: 35–38.

Smily, W. J. P., 1961. The life-cycle of the freshwater copepod *Cyclops leuckarti* Claus in Esthwaite Water. J. anim. Ecol. 30: 153–169.

Smyly, W. J. P., 1962. Laboratory experiments with stage V copepodids of the freshwater copepod, *Cyclops leuckarti* Claus from Winder-mere and Esthwaite Water. Crustaceana 4: 273–280.

Spindler, K. D., 1969. Experimentelle Untersuchungen zur Dormanz bei *Cyclops vicinus*. Naturwissenschaften 56: 93– 94.

Tadini-Vitagliano, G., F. Valentino & R. Migliore, 1982. Is the mechanism present in *Asellus aquaticus* (L.) an evolutionary stage towards an endogenous calendar? Z. zool. Syst. und Evolutionsforsch. 20: 144–155.

Vitagliano, T. G., E. A. Fano & E. Marchetti, 1984. Diet, photoperiod and reproductive diapause in *Asellus aquaticus* (Crustacea, Isopoda). Boll. zool., 51, Suppl.: 114.

Watson, H. F. Nelson & B. N. Smallman, 1971. The physiology of diapause in *Diacyclops navus* Herrick (Crustacea, Copepoda). Can. J. Zool. 49: 1449–1454.

Wierzbicka, M., 1962. On the resting stage and mode of life of some species of Cyclopoida. Polskie arch. hydrobiol. 10: 215–229.

Williams-Howze, J. & B. C. Coull, 1992. Are Temperature and Photoperiod Necessary Cues for Encystment in the Marine Bentic Harpacticoid Copepod *Heteropsyllus nunni*? Coull Biol. Bull. 182: 109–116.

Wolf, E., 1905. Die Fortpflanzungverhältnisse unserer einheimischen Copepoden. Zool. Jahrb. Abt. f. Syst. 22: 101–280.

Wyngaard, G. A., 1988. Geographical variation in dormancy in a copepod: evidence from population crosses. Hydrobiologia, 167/168 (Dev. Hydrobiol. 47): 367–374.

Zankai, N. P., 1987. Post-embryonic development of cyclopoid copepods in various season at Lake Balaton (Hungary). J. Plankton Res. 9: 1057–1068.

Hydrobiologia **320**: 123–130, 1996.
V. R. Alekseev & G. Fryer (eds), Diapause in the Crustacea.

Essentials of the environmental control of insect seasonality as reference points for comparative studies in other invertebrates

V. A. Zaslavski
Zoological Institute of the Russian Academy of Sciences, 199 034 St Petersburg, Russia

Key words: insects, seasonality, photoperiod, temperature, developmental reactions, modelling

Abstract

A great body of knowledge of photoperiodic and other developmental reactions, governing seasonality in insects, may be helpful for similar investigations which are or will be conducted with other animal groups. The following principles, based on a recent hypothesis, seem the most significant to be tested in comparative studies. (1) On the level of the photoperiodic clock only quantitative perception of photoperiod takes place. (2) The following, presumably neuroendocrine, 'managing' level of the mechanism discriminates between long and short days, determines the type of the response, and integrates photoperiodic and temperature reactions. (3) A working hypothesis, to be tested with both insects and other invertebrates, is that the 'managing' mechanism described in the model of photoperiodic and temperature control of development may have several sensory inputs and, therefore, plays a central role in governing developmental reactions to different environmental cues.

Introduction

The environmental control of seasonality plays an extremely important role as temporal organizer of population and ecosystem functioning. Moreover, there are reasons to speculate that many, if not most, of the population functions of an individual, such as reactions to population density, resource availability, etc., share the same mechanism that executes the reactions to photoperiod and temperature. Needless to say, uncovering the anatomy and physiology of that mechanism is very challenging. However, its anatomical embodiment and physiological essence remain largely mysterious, despite many efforts made during past decades. Due to these difficulties, a predominant interest has been focused on the relevance of diapause and other seasonal phenomena in life cycle strategies and population biology. However, understanding the ecology and evolution of seasonal adaptations depends to a large extent on insights into the underlying physiological and genetic mechanism. Indeed, how may one expect conclusive results from those studies on the genetics of diapause which neglect indications that the mechanism of diapause initiation, maintenance and termination comprises a hierarchy of elements, with specific charactertistics that may well be genetically independent?

As in crustaceans (Alekseev, 1990), the overt manifestations of the environmental control of seasonality in insects are highly diversified (Danilevsky, 1961; Saunders, 1976; Tauber *et al.*, 1986; Danks, 1987; Zaslavski, 1988; Müller, 1992). Nevertheless, I believe that a common and rather simple mechanism behind this diversity is quite probable in insects. An array of essential features which can confidently be ascribed to that mechanism will be offered as points for comparison in corresponding studies in other animal groups. As already noted, the anatomy, cytology, and biochemistry of the mechanism that translates the environmental signals into the internal hormonal commands remain undiscovered. As a consequence, the elaboration of models remains the upper limit of our efforts in this area.

I have advanced a model of a mechanism able to perform the whole diversity of responses, known in insects and Acari, to both natural (24-hour) photoperiods and temperature (Zaslavski, 1972, 1984, 1988, 1995). No substantial inconsistencies have been met with when testing the model against current experimental data. The consistency of the model with experimental data is illustrated in Fig. 1, which shows the diversity of the photoperiodic response curves produced by the model and, in parallel, the corresponding diversity of empirical curves revealed experimentally in different species. It is worth noting that apart from those 'theoretical' curves presented therein, two further kinds were predicted, but without an empirical correlate. Since then, one of these model-produced curves has been demonstrated experimentally (Fig. 2). Apart from the diversity of the photoperiodic curves, the model generates feasible explanations for an array of other phenomena, as outlined briefly below.

The fact that the model has such explanatory capacities suggests that a real unitary mechanism may exist. Therefore, to decide what is essential in the mechanism of insect seasonal control the principal grounds of this model will be presented in this paper.

The diversity of developmental reactions to external factors, and some requirements of the models

The structure and mode of operation of any mechanism can be judged to some extent from the array of output effects that it produces. Models of the photoperiodic determination of diapause consider, with few exceptions, the qualitative response of the long-day type and its simplest interactions with temperature. In reality, however, the diversity of developmental reactions to environmental cues is much larger, and no reasons are seen to presume a separate mechanism for each manifestation.

Consequently, in particular, current models should not neglect the quantitative aspect of the reactions controlling diapause, cyclomorphoses, etc. Further, apart from the long-day type of photoperiodic response the opposite short-day type and a variety of combined types exist and demand explanation. Moreover, both long- and short-day types often were proved to operate in the same individual. All these responses to the daylength exhibit strong interactions with the reaction to temperature. Those interactions, though very diversified, exhibit quite regular trends. Understanding these interactions is essential for the elaboration of models. It is now well established that not only the ordinary response to stationary photoperiod and temperature, but the specific response to their directed changes as well, may be found in a single species (or, more precisely, in the same individual). It would be odd to assume that separate mechanisms are involved. Of significance is the well known fact that not only photoperiod and temperature, but also some other environmental cues, participate and interact in the control of seasonality.

This outline suggests that little progress toward an understanding of the physiology of seasonal control will be achieved through splitting the whole diversity of developmental reactions and searching for a separate explanation for each specific response. A search for the most universal mechanisms seems preferable. It suggests also that this diversity of responses, as already revealed in Insecta, should be taken for comparison in studies of other animals.

Toward universal explanations: photoperiodic reaction

Levels of the mechanism, and the principles of photoperiod perception

Is it really possible to imagine a physiological mechanism capable of executing all of the above mentioned functions? For the sake of convenience let us begin from the reaction to photoperiod.

It is widely recognized that induction of a photoperiodic response is a two-stage process at least, which proceeds at two consecutive levels of the photoperiodic mechanism. The so-called 'photoperiodic clock' (with photoreceptors) is the first level. This 'clock' is commonly thought to qualify photoperiods as either long- or short-day. The second level is commonly called the 'photoperiodic counter'. Its supposed function is merely to accumulate the long- and short-day signals furnished by the clock.

First of all, what is the photoperiodic clock? About ten years ago the great authority on insect photoperiodism, the late A. D. Lees, wrote that despite intensive physiological studies, which have been made during decades, 'the photoperiodic clock mechanisms remain tantalizingly elusive' (Lees, 1986). This remains true, and no doubt will remain true as long as it is widely believed that the photoperiodic clock itself is the site of determination of the qualitative boundary between long and short days, of the shape of the photoperiodic response curve, and of other characteristics observed in experiments.

Another concept (Zaslavski, 1972, 1984, 1988, 1992) holds that it is only the second level of the photoperiodic mechanism that distinguishes between long and short days. Moreover, this second level is consid-

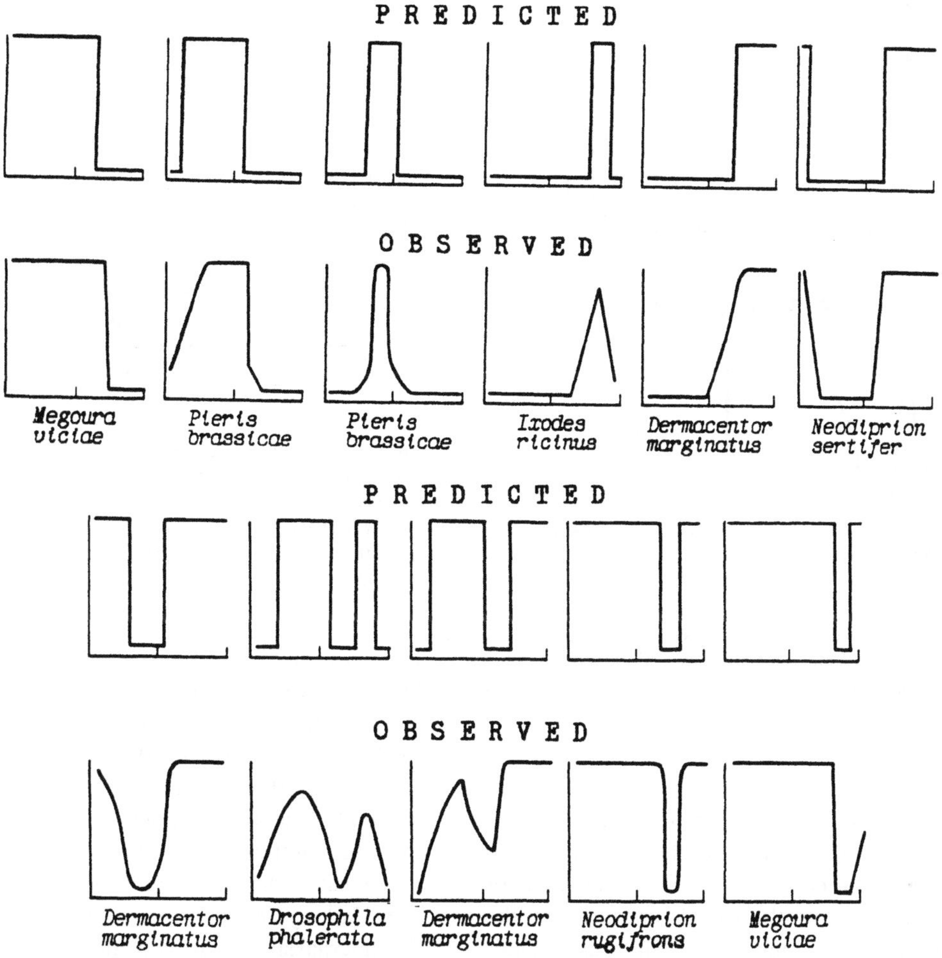

Fig. 1. Diversity of the theoretical photoperiodic response curves derived from the model, and examples of corresponding experimental curves observed in insects and Acari. Explanation of how different photoperiodic curves could be derived from the model is given in Fig. 5 and the corresponding text. Some observed curves belong to different populations of the same species (for details and references see Zaslavski, 1984, 1988). Ordinate — diapause, %; Abscissa — photoperiods from 0 to 24 hours of light per day. From Zaslavski, 1984, modified.

ered fully responsible for the typology of the photoperiodic reactions, and their interactions with temperature and other factors. Therefore, this 'managing' or 'commanding' mechanism is much different from a mere 'counter'.

This concept suggests that the only function of the photoperiodic clock is to accomplish the initial and fundamental step of the photoperiodic reaction, namely the translation of the external temporal information into internal stimuli of strictly specified strength. The clock performs the quantitative measurement of photoperiods, while its proximate effect is a corresponding excitation of the 'commanding' mechanism, which accomplishes the second fundamental step, the processing of that internal quantitative information.

This conclusion was drawn from experiments with a number of insect species. It has been shown that the inductive strengths of different diapause-inducing, as well as different diapause-averting photoperiods, are clearly unequal and vary in quite a regular way (Zaslavski, 1975a, 1984, 1988, 1992; Zaslavski & Fomenko, 1980, 1983; Spieth & Sauer, 1991). The quantitative dependence of the reaction on the daylength cannot emerge after the daylength has been perceived. Therefore, despite the common 'all-or-nothing' appearance, the reaction to photoperiod is primarily quantitative. The variation of the inductive strength of photoperiods has been depicted diagrammatically by the 'primary photoperiodic curves' (Fig. 3). An inspection of the latter indicates that this

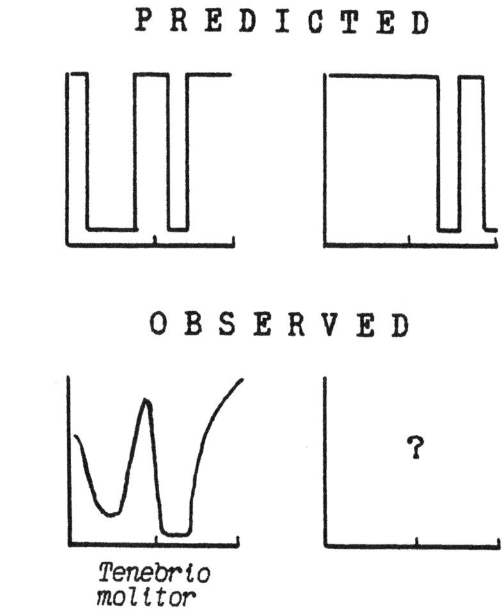

Fig. 2. Two kinds of theoretical photoperiodic response curves which had no experimental correlate when they were derived (Zaslavski, 1984), and ensuing experimental finding of one of them (Tyshchenko & Amadu, 1987, redrawn).

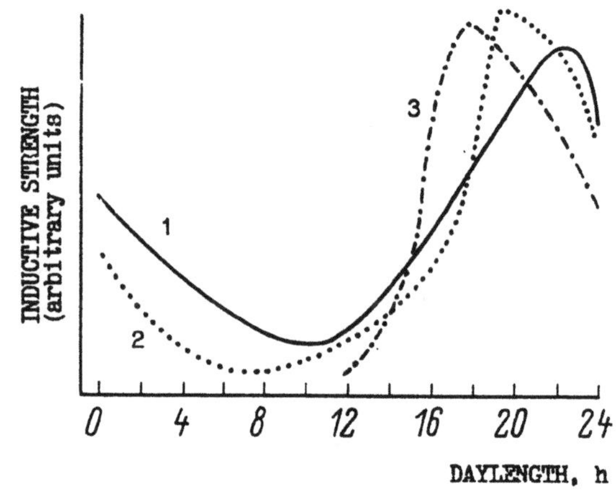

Fig. 3. Diagrammatic delineation of changes of the relative inductive strength (arbitrary units) with which different photoperiods promote diapause-free development in three insect species. 1— *Megoura viciae* Buckt. (aphid), 2 — *Chilocorus bipustulatus* L. (beetle), 3 — *Acronycta rumicis* L. (moth). From Zaslavski, 1984, modified.

dependence is (1) gradual, (2) non-linear, and (3) constitutes a continuum covering both short- and long-day sections of the photoperiodic axis. So, on the level of photoperiod perception, the signals produced by the clock do not mean the 'long day' or 'short day'. Their only characteristic is their strength, or value, which is always positive.

An important point is that the experiment-based 'primary' curves fit well the theoretical curves, which may be derived readily from Tyshchenko's internal coincidence model (Danilevsky *et al.*, 1970), modified by Zaslavski (1988). Contrariwise, models based on the principle of external coincidence cannot produce gradual curves of such a configuaration. Therefore, if the mode of quantitative photoperiod perception, already shown, is actually universal in Insecta, 'internal coincidence' would be the common principle of photoperiodic time measurement.

Moreover, it is not essential that a circadian oscillation underlies a photoperiodic time measurement. This view is supported by findings that (1) the photoperiodic reactions of some species lack any signs of circadian periodicity (Lees, 1973), (2) in others the circadian system was proved to be actually involved in operation of the photoperiodic clock (Vaz Nunes *et al.*, 1990), whereas (3) in several species circadian periodicity was convincingly attributed not to the clock itself, but rather to the 'counter' (Vaz Nunes & Veerman, 1982; Veerman & Vaz Nunes, 1987; Veerman *et al.*, 1988; Kimura & Masaki, 1993). The coincidence/noncoincidence of two internal processes, linked separately with light and dark, appears to be more fundamental than their oscillatory or hourglass nature.

In sum, the quantitative evaluation of external light/dark cycles, and their translation into quantitative internal signals, are considered to be a fundamental component of the insect photoperiodic mechanism.

The mode of processing of internal quantitative information on the second level

Numerous indications exist (Zaslavski, 1975b, 1988) that a single individual may exhibit two 'elementary' photoperiodic reactions, one of the long-day and the other of the short-day type. These facts produced a guideline for modelling the 'commanding' or 'managing' part of the photoperiodic mechanism (Zaslavski, 1972, 1988). It was suggested that quantitative impulses from the photoperiodic clock affect levels of activity of two antagonistic neurosecretory centres. One of these presumed centres is activatory and executes the long-day elementary reaction. The second is inhibitory and executes the response of the opposite, i.e., short-day type. Both centres, in turn, act on an endocrine gland and on each other (Fig. 4), and produce their cumulative effects on these targets. The value of the cumulative effect is established as a balance

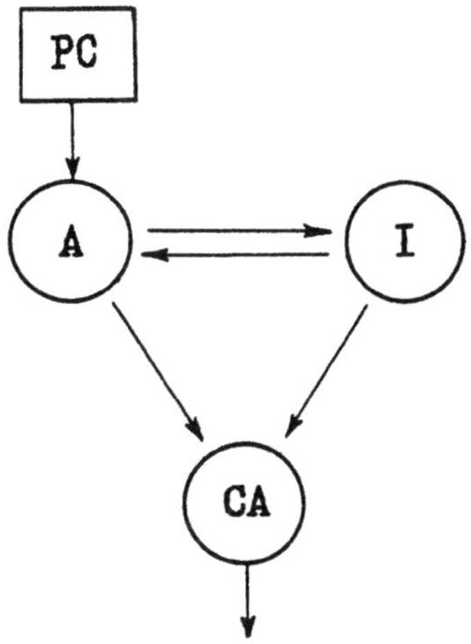

Fig. 4. Scheme of the interrelations within the model of the mechanism accomplishing photoperiodic and temperature reactions in the insect adult stage. PC — photoperiodic clock (with photoreceptors), A — activating centre, I — inhibiting centre, CA — corpora allata (endocrine gland). From Zaslavski, 1984, modified.

between daily augmentation (induction) and simultaneous degradation (deinduction) (Zaslavski, 1995). Neither the critical daylength nor the required day number, but only a threshold value of the cumulative effect is an innate heritable variable which is relevant to the main characteristics of the photoperiodic reaction. Only those photoperiods strong enough to create the cumulative effect exceeding the threshold value exert the long-day effect. Weaker photoperiods act as short days. In this way the commanding mechanism is thought to qualify photoperiods as long- or short-day.

Another fundamental function of this mechanism is to determine the type of the photoperiodic response, that is, whether it belongs to the long-day, the short-day or one of the combined types, whether it is stepwise or non-stepwise under changing regimens, what is the mode of its interaction with temperature, and so on. An analysis of the model shows that the commanding mechanism, briefly outlined here, can execute all of these manifestations, and nearly every combination of characteristics can be selected for. As a consequence, any needed adaptive specialization can be attained, so that the single mechanism is presumed to be multipotential, through free recombination within the limited set of variable features (for details see Zaslavski, 1988). Because the present article cannot expand to full analysis of the model, the operation of the latter will be illustrated by two examples only.

1. Figure 5 illustrates diagrammatically the suggested explanation of the long-day, short-day, and one of the complicated configurations of the photoperiodic response curves. The following conditions define their trajectories. The above-threshold level of the activator and below-threshold level of the inhibitor lead to direct (diapause-free) development, while both the below-threshold level of the activator and above-threshold level of the inhibitor engender diapause.

The following variables, incorporated into the model, are combined in the examples shown in Fig. 5.

(a) The first variable is the presence of the functioning sensory input, that is, the photoperiodic clock (cf. Fig. 4). Without it a species would be photoperiodically neutral; and if only one centre, say, activatory, is linked to the clock, only the elementary reaction, long-day in this case, would operate, whereas the second centre would execute the temperature reaction only.

(b) The second variable condition is the threshold value cumulative effect for the activatory centre and for the inhibitory centre. For the sake of simplicity both are depicted by a single dashed line (Fig. 5).

(c) The third independent variable is the form and amplitude of the primary photoperiodic curve. These properties may be different for the two centres, and may vary to some extent with temperature. However, their configurations are assumed here to be uniform and constant, because their variability has not yet been studied.

(d) The last variable of special importance is the basal level of activity of each centre, defined as the lowest point of the primary curve at a given temperature. The basal level, first, dictates the position of the whole primary curve in relation to the threshold, and, second, exhibits the ordinary positive dependence on ambient temperature. The resulting depressions and elevations of the primary photoperiodic curves constitute, according to the model, a simple and reliable basis to explain the whole variety of interactions between photoperiodic and temperature reactions. For example, the difference between Fig. 5a and Fig. 5b may be due to a rise in temperature, and indeed, inversion of the response from the long-day type under low temperature to the short-day type under high temperature is known (Danilevsky, 1961).

2. A graphic illustration of the model explanation for the 'stepwise', or 'staged' photoperiodic reactions is furnished by Fig. 6. It is based upon the suggested differential kinetics of the cumulative effects of the activatory and inhibitory centres. As seen in Fig. 6, a temporary combination of a high level of the activator

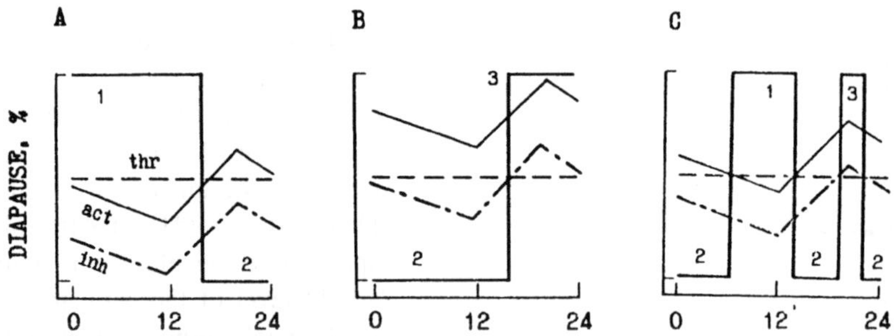

Fig. 5. Model explanation for the photoperiodic response curves of the long-day (A), short-day (B), and one of the combined types (C). act — cumulative effect of the activating centre as dependent on the photoperiod strength (the primary photoperiodic curve); inh — the same of the inhibiting centre; thr — the threshold level for both. 1–3 — parts of the photoperiodic response curve: 1 — diapause caused by the low level of the activator; 2 — non-diapause caused by the high level of the activator accompanied by the low level of the inhibitor; 3 — diapause caused by the high level of the inhibitor. From Zaslavski, 1984, modified.

and a low level of the inhibitor, impossible under any constant daylength, may arise due to the more rapid response of the activator to a shift of photoperiod. As a result, diapause is averted or terminated, though both the initial and final regimens per se induce diapause.

Thus, the specific 'stepwise' reactions to changing photoperiods are thought to proceed through interaction between the two antagonistic 'commanding' centres of the model. An additional variable here is the rapidity of the centre's response after a photoperiodic switch. This variable operates, though largely covertly, in reactions to stationary photoperiods, but is crucial in the stepwise reaction. However, the variable is not simple, but is defined by interaction between the value of the daily impulse, and the rate of deinduction, characteristic of each centre (for details see Zaslavski, 1995).

A relatively simple condition determines the type of the stepwise reaction. If a change of photoperiod initiates more rapid change of the activator than of the inhibitor, then the reaction will be of the short-day/long-day type, where transition from the short to long day causes active development. More rapid change of the inhibitor results in the long-day/short-day reaction, where transition from long to short day is needed to cause active development.

It is not possible here to analyze further recombinations within the model, or give an exhaustive account of the correlation between the results provided by the model and the empirical data. Suffice it to say that the twin processing of the quantitative information furnished by the photoperiodic clock is considered the second fundamental component of the insect photoperiodic mechanism.

Function of the 'commanding' mechanism extended: a working hypothesis

Until now, the reaction to photoperiod has been emphasized, because photoperiod most often governs seasonal development. However, a variety of other environmental cues participate to some extent in the control of seasonality. Besides temperature, these are food, moisture, population density, predators, chemical stimuli, etc. What kinds of interconnections may exist between reactions to those very different factors? According to widespread opinion, the photoperiodic reaction holds a central position, while the other developmental reactions compose a set of loosely connected physiological phenomena and exert only a modifying effect on the photoperiodic response. However, it is doubtful that a number of quite different reactions could perform a common adaptive function without any common physiological mechanism for their co-ordination.

Different reactions are believed to have some degree of unity if their impulses are complementary, because their combined cumulative effect should be formed within a single physiological system. To ascertain if different reactions behave as complementary, it is most convenient to examine their interactions with the photoperiodic reaction because the latter is so diverse. When some reaction affects the level of the photoperiodic response only (e.g., percentage of diapause) without any effect upon configuration of the photoperiodic response curve, then there are no reasons to regard the photoperiodic and the other reaction as complementary and, therefore, as sharing a common physiological basis. Conversely, if the reaction under study causes a regular sequence of transformations of the photoperiodic response curve, it suggests strongly

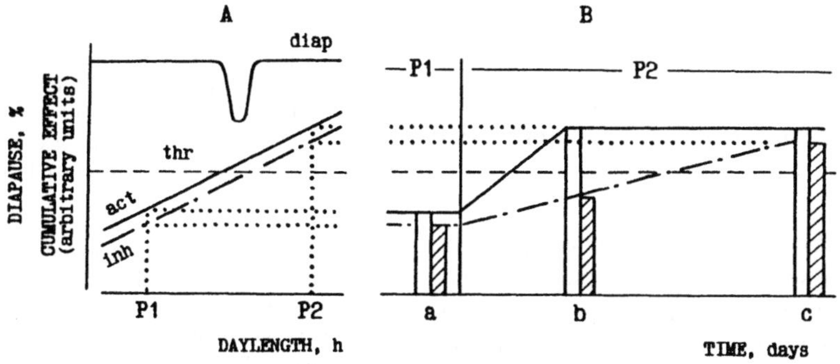

Fig. 6. Scheme showing 'intermediate' photoperiodic reaction at constants photoperiods (diapause under shorter and longer daylengths, and partial non-diapause in the narrow transitional zone) and the stepwise reaction upon an abrupt increase of photoperiod. A — constant photoperiods: cumulative effects (a part of the primary photoperiodic curves) of the activating and inhibiting centres, and a photoperiodic response curve. B — abrupt increase of photoperiod from P1 to P2: dynamics of increase of the cumulative effects of both centres. act — primary curve of the activating centre, inh — the same for the inhibiting centre, thr — threshold level for both centres, diap — photoperiodic response curve. The rate of change of activator cumulative effect is three times as high as that of the inhibitor. The columns a and c give the activator/inhibitor ratio at constant short and long days respectively. Column b shows one of the transitory combinations of the high level of the activator and the low level of the inhibitor occurring upon daylength alteration. This causes temporal activation, which is impossible at constant long-day photoperiod P2. From Zaslavski, 1984, modified.

that this reaction is an outcome of the same system that determines the norm of the photoperiodic reaction.

Convincing evidence (Zaslavski, 1988) favours the view that photoperiodic and temperature reactions share much the same mechanism (see above; Fig. 4). Some data (Charlesworth & Shorrocks, 1980; Tauber *et al.*, 1986; Danks, 1987; Dumortier & Brunnarius, 1989) lead to the same expectation for the alimentary reaction. This conclusion gives rise to a working hypothesis according to which the managing ('commanding') mechanism, presumed in the model, may constitute the physiological basis of the whole variety of developmental reactions. It is suggested that all these reactions share this unique mechanism via their respective sensory inputs. In other words, they are supposed to differ at the sensory level only, and all to be linked together by the common managing mechanism.

Two conclusions stem from such a hypothesis. First, insects are endowed not with a set of unconnected developmental reactions, but with a common physiological system that integrates and executes them. Second, from a purely physiological standpoint, all the reactions mentioned should in principle be equipotential, because they are all potentially able to exert control over the managing mechanism, and to cause the same overt effects (diapause, polyphenism, etc.). This possibility may be important for understanding the evolution of life cycles. Selection would decide the relative role of each reaction in a population, i.e. whether it will be dominating, modifying, or silent.

Hence, according to the views presented here, the managing (commanding) neuroendocrine mechanism plays a deciding role in the physiology of seasonal control, whereas the photoperiodic clock is, possibly, no more than one of several sensory inputs to that mechanism.

Conclusions

This section raises questions for comparative studies of the control of diapause in different animal groups and, especially, in attempts to uncover the operating mechanisms.

- Does phenomenology of the environmental control differ significantly from that revealed in Insecta?
- Is the daylength perceived quantitatively or not? If yes, what is the shape of the 'primary photoperiodic curve' in different species, and how do the curves differ from the 'ordinary' curves of 'diapause/non-diapause'? Answers to these questions suggest the existence or non-existence of the central managing mechanism at the second level of the photoperiodic mechanism.
- Can two opposite reactions to photoperiod and temperature be observed in any species of the group under study? Such reactions imply dual processing of the photoperiodic and other information at the second level of the controlling mechanism.
- Does the quantitative analysis of diapause induction indicate that the effects of the photoperiodic

and other developmental reactions are summated in a common cumulative effect? Such summation would be evidence of the common managing mechanism behind reactions to different environmental cues.

Different invertebrate hydrobionts, and especially crustaceans endowed with strong developmental reactions to a variety of external factors, probably will prove to be very suitable for experimental elucidation of these problems.

Acknowledgments

I thank Drs H. V. Danks and G. Fryer for critical reading of the manuscript and invaluable help to put this paper into proper English. The work was supported by a grant 93-04-06026 from the Russian Foundation for the Basic Researches.

References

Alekseev, V. R., 1990. Diapause in *Crustacea* Nauka, Moscow, 143 pp. (in Russian).

Charlesworth, P. & B. Shorrocks, 1980. The reproductive biology and diapause of the British fungal-breeding *Drosophila*. Ecol. Entomol. 5: 315–326.

Danilevsky, A. S., 1961. Photoperiodism and seasonal development of insects. Leningrad State University, Leningrad, 243 pp. (in Russian; English edition: 1965, Oliver & Boyd, Edingburgh).

Danilevsky, A. S., N. I. Goryshin & V. P. Tyshchenko, 1970. Biological rhythms in terrestrial Arthropods. Annu. Rev. Entomol. 15: 201–214.

Danks, H. V., 1987. Insect dormancy: an ecological perspective. Biological Survey of Canada (Terrestrial Arthropods), Ottawa, 439 pp.

Dumortier, B. & J. Brunnarius, 1989. Diet-dependent switch from circadian to hourglass-like operation of an insect photoperiodic clock. J. Biol. Rhythms 4: 481–490.

Kimura, Y. & S. Masaki, 1993. Hourglass and oscillator expressions of photoperiodic diapause response in the cabbage moth *Mamestra brassicae*. Physiol. Entomol. 18: 240–246.

Lees, A. D., 1973. Photoperiodic time measurement in the aphid *Megoura viciae*. J. Insect Physiol. 19: 2279–2316.

Lees, A. D., 1986. Foreword. In Tauber, M. J., C. A. Tauber & S. Masaki, 1986. Seasonal adaptations in insects. Oxford University Press, New York: VII–VIII.

Müller, H. J., 1992. Dormanz bei Arthropoden. Gustav Fischer, Jena, Stuttgart, New York, 289 pp.

Saunders, D. S., 1976. Insect clocks. Pergamon, Oxford, 279 pp.

Spieth, H. R. & K. R. Sauer, 1991. Quantitative measurement of photoperiods and its significance for the induction of diapause in *Pieris brassicae* (Lepidoptera, Pieridae). J. Insect Physiol. 37: 231–238.

Tauber, M. J., C. A. Tauber & S. Masaki, 1986. Seasonal adaptations in insects. Oxford University Press, New York, 411 pp.

Tyshchenko, V. P. & Sheik Ba Amadu, 1987. Photoperiodic reaction of *Tenebrio molitor* (Coleoptera, Tenebrionidae). Zool. Zh. 66: 51–59 (in Russian).

Vaz Nunes, M., N. A. P. Kenny & D. S. Saunders, 1990. The photoperiodic clock in the blowfly *Calliphora vicina*. J. Insect Physiol. 36: 61–67.

Vaz Nunes, M. & A. Veerman, 1982. Photoperiodic time measurement in the spider mite *Tetranychus urticae*: a novel concept. J. Insect Physiol. 28: 1041–1053.

Veerman, A., M. Beekman & R. L. Veenedaal, 1988. Photoperiodic induction of diapause in the large white butterfly, *Pieris brassicae*: evidence for hourglass time measurement. J. Insect Physiol. 34: 1063–1069.

Veerman, A. & M. Vaz Nunes, 1987. Analysis of the operation of the photoperiodic counter provided evidence for hourglass time measurement in the spider mite *Tetranychus urticae*. J. compar. Physiol. A 160: 421–430.

Zaslavski, V. A., 1972. Two-step photoperiodic reactions as a basis for elaborating a model of the photoperiodic control of development in arthropods. Entomol. Obozr. 51: 217–239 (in Russian).

Zaslavski, V. A., 1975a. Interrelations of quantitative and qualitative reactions in insect photoperiodism. Zool. Zh. 54: 913- -921 (in Russian).

Zaslavski, V. A., 1975b. Extent of occurrence of stepwise photoperiodic reactions among insects and acari. Entomol. Obozr. 54: 291–304.

Zaslavski, V. A., 1984. Photoperiodic and temperature control of insect development. Nauka, Leningrad, 184 pp. (in Russian).

Zaslavski, V. A., 1988. Insect development, photoperiodic and temperature control. Springer, Berlin, New York, 187 pp.

Zaslavski, V. A., 1992. Light-break experiments with emphasis on the quantitative perception of nightlength in the aphid *Megoura viciae*. J. Insect Physiol. 38: 717–725.

Zaslavski, V. A., 1995. Induction and deinduction in insect photoperiodism: experimental evidence and a model. European J. Entomol. 92: 401–413.

Zaslavski, V. A. & R. B. Fomenko, 1980. New data concerning the photoperiodic reaction of *Acronycta rumicis* L. (Lepidoptera, Noctuidae). Entomol. Obozr. 59: 18–25 (in Russian).

Zaslavski, V. A. & R. B. Fomenko, 1983. Quantitative daylength perception in the aphid *Megoura viciae* Buckt. (Homoptera, Aphididae). Entomol. Obozr. 62: 433–441 (in Russian).

Hydrobiologia **320**: 131–140, 1996.
V. R. Alekseev & G. Fryer (eds), Diapause in the Crustacea.

The resting eggs of the Ctenopoda (Crustacea: Branchiopoda): a review

N. M. Korovchinsky & O. S. Boikova
A.N. Severtsov Institute of Animal Evolutionary Morphology and Ecology of Russian Academy of Sciences, Leninsky prospect, 33, 117071 Moscow, Russia

Key words: Sidoidea, resting eggs, morphology, development, dispersal

Abstract

The resting eggs of the Ctenopoda have been studied since the second half of the 19th century. By now they are known in 21 species of the group but only in 4 species *(Sida crystallina, Diaphanosoma brachyurum, Penilia avirostris* and *Holopedium gibberum)* have they been investigated thoroughly. Gross and internal structure of eggs are characterized by considerable diversity. In general their structure has much in common with resting eggs of the Anostraca, Notostraca, "Conchostraca" and Onychopoda as well as with that of some other remotely related groups of freshwater invertebrates (Rotifera, *Hydra*). The taxonomic value of resting eggs is discussed. Probably they play a significant role in determining geographical distribution. Waterfowl are the main agent of their dispersal.

Introduction

The resting stages of the Anomopoda (= Chydoroidea) have been known since the early years of study of the group and may to be considered as comparatively well studied. By comparison the resting eggs of the Ctenopoda (= Sidoidea) (families Sididae and Holopediidae) were only described adequately in the second half of the 19th century (Sars, 1865; Weismann, 1877). Data are sparse and scattered in the literature. Probably only the marine *Penilia avirostris* has been well studied in this respect. Its resting eggs have been described rather thoroughly by Braiko (1965), Della Croce & Bettanin (1969) and Onbe (1972, 1974). For this reason data on this species receive only brief mention in this review.

Resting stages present much of interest from several aspects: (1) ecological (an ability to withstand unfavourable conditions), (2) zoogeographical (significance for dispersal), (3) phylogenetic (morphological similarity to representatives of other groups) and (4) systematic (source of new taxonomic traits). In the context of the "Cladoceran" orders investigation of the resting stages of the evolutionary primitive Ctenopoda have special interest.

Gross structure and clutch size

Initially such researches as Zaddach and Leydig suggested that the Ctenopoda produce ephippia like the Anomopoda (Weismann, *op.cit.*) but in fact they lay only resting eggs with a thick shell.

These eggs may be of spherical *(Holopedium)* or ellipsoid form (the majority of the Sididae) which can vary somewhat even within one population. In *Diaphanosoma volzi* and *D. senegal* they are somewhat flattened (Korovchinsky, 1981, 1993a) (Fig. 8).

Holopedium gibberum, Sida crystallina and *Pseudosida variabilis* have eggs with a smooth surface (Figs 5, 9). In species of *Diaphanosoma* it may be smooth or slightly rough *(D. brachyurum* s.str., *D. chilense, D. fluviatile, D. unguiculatum, D. lacustris, D. mongolianum, D. senegal)* (Figs 1, 2, 6, 7), with small rounded protuberances *(D. birgei)* or with numerous fingerlike prominences *(D. sarsi, D. excisum, D. spinulosum, D. brevireme, D. volzi)* (Figs 3, 4, 10).

In a number of species the resting eggs are covered by a transparent sticky envelope *(S. crystallina, P. variabilis, D. brachyurum, D. chilense, D. fluviatile, D. senegal)* (Figs 6, 7, 9). In the last species it is mostly developed at the ends of the long axis of the egg (Korovchinsky, 1993a) (Fig. 8). As was shown

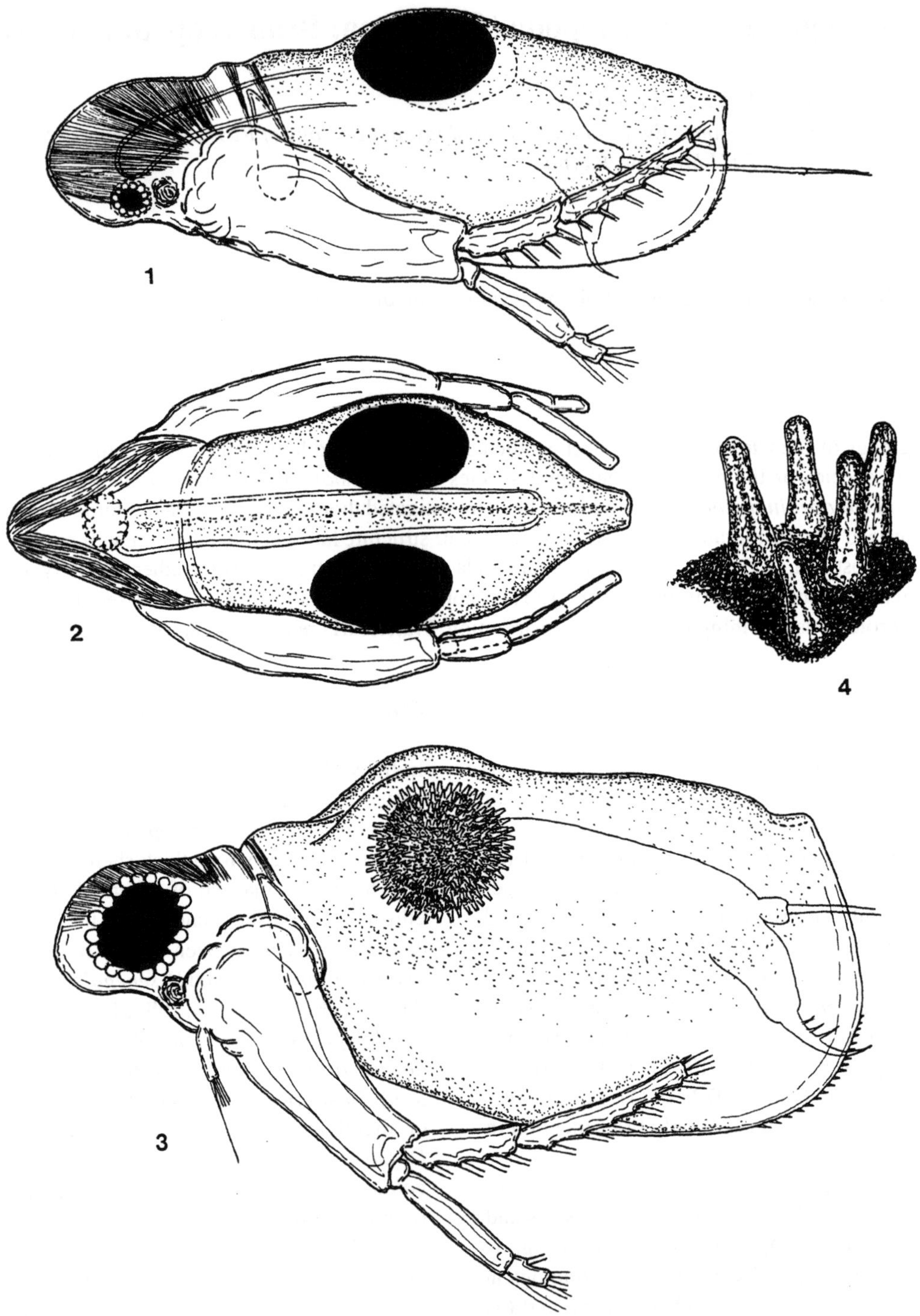

Figs 1–4. Gamogenic females of *Diaphanosoma*: 1. *D. mongolianum* (Kosyno lakes, Moscow, Russia), lateral view; 2. The same, dorsal view; 3. *D. sarsi* (Berlinhafen, Papua New Guinea, Daday's collection), lateral view; 4. prominences of *D. sarsi* resting eggs.

for *D. brachyurum*, the envelope's matter is secreted by the sphermatheca and rapidly swells in the water (Weismann, *op. cit.*).

In comparison with parthenogenetic eggs, resting eggs are larger and their size is related to body size of the females. Data on their dimensions are presented in Table 1. The volume of the resting eggs of *S. crystallina* and *D. brachyurum* is about 11.8–12.5 and 4.29 ± 1.17 ($m^3 \times 10^6$) respectively.

Under the light microscope Ctenopod resting eggs appear dark and of various tints (dark brown or grey-blue). In *S. crystallina* their colour varies from yellow to dark brown (Loidl, 1955). Yet Sars (1865) noted that the carapace of gamogenic females of *D. brachyurum* often becomes more pigmented dorsally. Only this feature along with the dorsal protuberances on the carapace beyond the eggs (Fig. 3) in some species of the genus may be considered as specific characteristics of Sididae gamogenic females.

In *Holopedium*, *Sida* and *Latona* females usually carry many resting eggs in one clutch (up to 22, 5–20 and 8 respectively) though young females may have only 1–2. On the other hand females of *Diaphanosoma*, and probably *Pseudosida*, always carry not more than two resting eggs in the anterior part of the brood pouch (Fig. 2). *S. crystallina* lays several clutches of resting eggs but in northern latitudes (e.g. in the vicinity of St. Peterburg) potential fecundity is not realised (Makrushin, 1976).

Internal structure

Weismann (1877) was the first to make detailed studies on the reproduction of the Ctenopoda and Onychopoda and noted difference in the internal structure of the resting eggs of *Sida* and *Diaphanosoma*. Later, Makrushin also made detailed studies from more extensive material. He stated that resting eggs of *Holopedium gibberum*, *Limnosida frontosa*, *Sida crystallina* and *Penilia avirostris* contain large-grained yolk (grain size 8 μm) and vacuoles, thus resembling in general parthenogenetic (or subitaneous) eggs. On the other hand resting eggs of *Latona setifera*, and especially *Diaphanosoma brachyurum*, differ markedly from these species – they lack vacuoles and contain small-grained yolk of the 'euphyllopod' type (grain size 6 and 3 μm respectively) similar to that of Anostraca, Notostraca and "Conchostraca" (Makrushin, 1978, 1985). Possibly *Latonopsis* eggs have the same structure (Makrushin, 1990).

Ovogenesis and timing

The first detailed description of ovogenesis in *D. brachyurum* was given by Weismann *(op.cit.)*. Like parthenogenetic eggs, resting eggs of Ctenopoda originate from groups of germinal cells. Only one such egg develops in each ovary of *D. brachyurum*. In young or starving individuals the simultaneous development of eggs in both ovaries is not possible. In such case they appear alternately. The process of the egg's release from the ovary and its final formation occupies about half an hour. The outer shell is formed in the brood pouch, and the process of its formation is similar for instance to that of *Polyphemus* (Weismann, *op.cit.*). The germ of fully formed resting eggs is in the gastrula stage.

In temperate regions resting eggs appear in late summer – autumn and their timing depends on species and latitude. In particular in middle European latitudes the beginning of their formation is observed in September–October (Scheffelt, 1908; Heucher, 1916; Schmidt, 1924; Herzig, 1974; Green, 1966; Korovchinsky, 1983) and is usually accompanied by parthenogenetic reproduction. By the end of gamogenesis, which may continue up to the end of December or even January, females carry only resting eggs.

At the same time some unusual cases have been described. The appearance of resting eggs in *D. brachyurum* in Grossen Bullensee (Germany) was recorded as beginning from July 10 (Franken & Franken, 1978). In *D. mongolianum* from Kosyno lakes (Moscow, Russia) they were observed at the end of July (pers.observ.). *H. gibberum* in Northern Germany showed the mass production of such eggs in spring and then in autumn (Lampert & Krause, 1976), in another case, in Northern England, in June–November (Smyly, 1978).

In the subtropics of Bangladesh gamogenetic females of *D. senegal* were recorded from October to December (Korovchinsky, 1993a). *D. celebensis* possibly has the same timing of sexual reproduction in South Vietman and Sri Lanka (Korovchinsky, 1989, 1993b). A gamogenetic population of *D. volzi* in the Philippines (Luzon Island) was registered in the beginning of January (pers. unpubl. observ.). On the other hand gamogenetic eggs of *D. sarsi* from Sri Lanka were found in July when the water bodies become very shallow (Apstein, 1907).

Brehm (1933) suggested the acyclic character of sexual reproduction of *Diaphanosoma* in the tropics. Later investigations by Frey (1982) and Rajapaksa

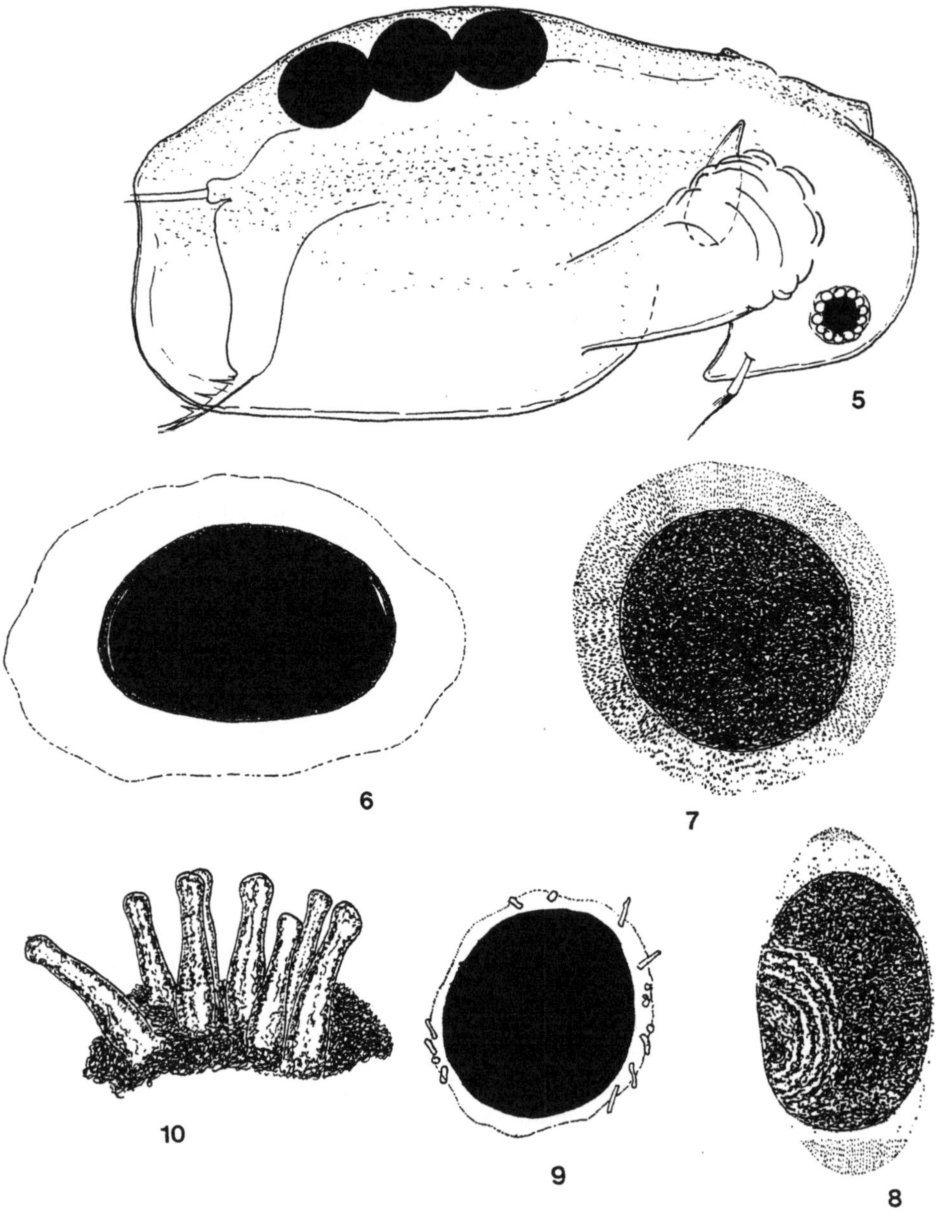

Figs 5–10. Gamogenic female of *Sida crystallina* (Lake Glubokoe, Moscow district, Russia) and Sididae resting eggs: 6. *Diaphanosoma brachyurum* (the same locality); 7. *D. senegal* (Mymensingh, Bangladesh), face view; 8. the same, edge view; 9. *Pseudosida variabilis* (Paraguay, Daday's collection); 10. Prominences of *D. volzi* resting egg (Lake near Evans Head, N.S.W., Australia).

(1986) supported this opinion in respect of various 'cladocerans' and at the same time stressed the heterogeneity of the process within the tropics. In tropical Asia the most intensive gamogenesis of the representatives of the group, including Sididae (in 1.9–8.8% of studied populations), was observed in December–February and June–August, but only a few individuals in these populations were gamogenetic.

In the Southern Hemisphere intensive gamogenesis of the Australian *D. unguiculatum* was recorded in

Table 1. Data on shape and dimension (mm) of Ctenopod resting eggs

Species	Shape	Length (diameter)	Width	Source
Sida crystallina	Spherical or slightly ellipsoid	0.26–0.34	--	Green, 1966
Diaphanosoma brachyurum	Ellipsoid	0.29–0.33	0.17	Weismann, 1877; Herzig, 1984
D. mogolianum	Ellipsoid	0.27	0.17	Pers. observ.
D. senegal	Spherical and slightly flattened	0.20–0.21	--	Pers. observ.
D. sarsi	Spherical or ellipsoid with protuberances	0.18–0.22	0.17–0.18	Apstein, 1907; pers. observ.
D. volzi	Ellipsoid and slightly flattened with protuberances	0.22–0.24	0.18	Pers.observ.
Penilia avirostris	Ellipsoid and strongly flattened	0.21–0.29 av. 0.33	0.14–0.20 av. 0.21	Onbe, 1972 Braiko, 1965
Holopedium gibberum	Spherical	0.20	--	Smyly, 1977

August (Gurney, 1927), *D. volzi* in April (Korovchinsky, 1981), and in Argentinian *D. fluviatile* in November (Paggi, 1978).

Location and number in water bodies

Resting eggs of Ctenopods are released from the brood pouch during the moult, sink to the bottom and attach, if they have a sticky envelope, to gravel, aquatic vegetation or other substrata (Weismann, 1877; Heucher, 1916).

Their number on the bottom may be rather high. In *Holopedium* they were especially numerous (290 cm^{-2}) almost in the centre of the lake but not in the deepest part (Lampert & Krause, 1976). In another case their highest number ($2–5 \times 10^6$ m^{-2}) was recorded at maximal depth (up to 20 m) of three lakes of the English Lake District (Smyly, 1977). Especially interesting is that in Lago Maggiore (Italy) resting eggs of *D. brachyurum* with developing embryos were found at a depth of as much as 250 m, and were especially abundant (3900 m^{-2}) at 150–200 m (Smyly, 1964).

Hatching

All data describing the process concern only European Ctenopods. Their resting eggs must have dormancy periods of different durations and are not capable of autumn reactivation.

In *H. gibberum* from Grasmere (England) hatching from resting eggs begins at the beginning of February at 3 °C near the bottom, is most intensive at the end of March–April, and finishes in May at 7 °C (Smyly, 1978). See also Freidenfelt (1920). Lampert & Krause (1976) recorded the highest degree of hatching in this species at 10–15 °C when 75–80% of eggs had developed. Many resting eggs of *Holopedium* lying on the bottom fail to hatch (only $5–7 \times 10^3$ from $2–5 \times 10^6$ m^{-2}) and it was suggested that most of them are dead. In another case the species was not registered in the plankton for 20 years though there were many resting eggs in bottom sediments (Smyly, 1977). These were almost certainly dead.

D. brachyurum in Neusiedlersee (in fact possibly another species *D. mongolianum*) begins to hatch in the beginning or middle of March at 6–7 °C (Herzig, 1974). Later this author specified that its development began after 60–80 days of dormancy at 2–3 °C but successfully hatched only at temperatures of 4 °C and higher, especially at 10–16 °C. Hatching continues over 46–48 days at 4.5–4.8 °C, 20 days at 10 °C and 6 days at 15 °C (Herzig, 1984).

S. crystallina in Lake Glubokoe (Moscow district, Russia) appears from resting eggs in May at 7 °C and probably continues to hatch during a rather long period (not less than 10 days). This assumption is supported by the permanent presence of individuals of minimal size in the population (Korovchinsky, 1983).

Limnosida and *Diaphanosoma* do not hatch in water bodies of the European Subarctic in years with comparative low water temperature (Vehov, 1987).

The question of the regulation of hatching still remains open. Smyly (1977) suggested that this process may be controlled endogenously because there is no threshold temperature for hatching (in *Holopedium* it continues for 4 months at 3–7 °C).

The last author also described the hatching process and Baldass (1937) the embryogenesis of *Holopedium* resting eggs.

Data on resistance

Since the end of the last century Sididae have been reared from dried mud collected in several regions of the world and stored for a long time, at least some months before steeping (Table 2). Such observations reveal a considerable resistance of dried resting eggs (as well as the ephippia of Chydoridae, Moinidae, Daphniidae and Macrothricidae reared in the same manner).

It is necessary to stress that probably all species listed in the table have an egg structure especially adapted to dehydration, i.e. consisting of small-grained yolk without vacuoles. Resting eggs of *D. brachyurum* can survive in a dry condition (without mud) for 4 days (Makrushin, 1985).

On the contrary species with resting eggs containing large-grained yolk and vacuoles were not reared from dried mud. Resting eggs of *S. crystallina* lost their viability after one day's drying (Makrushin, 1981). Also those of *Limnosida* cannot withstand freezing in contrast to *D. brachyurum* which survives in frozen mountain Subarctic water bodies (Vehov, 1987).

The ability of dried resting eggs of Ctenopods to preserve viability increases in the sequence: *(Holopedium, Limnosida) – Sida – Latona –Diaphanosoma* (Makrushin, 1990).

Discussion

Table 3 summarizes data on investigated Ctenopod resting eggs. In general they are known in about a half the species (21) of the group. In 19 of them they are described morphologically though mainly rather superficially. Only in 6 or 7 species have they been imaged reasonably but without use of SEM which is becoming common in the study of resting stages of other branchiopods (e.g., Glagolev, 1983; Mura & Thiery, 1986; Cesar, 1989). Moreover there are no descriptions even of the external morphology of the resting eggs of such common species as *Limnosida frontosa, Latona setifera* and *Latonopsis australis*. Only 4 species *(S. crystallina, D. brachyurum, P. avirostris, H. gibberum)* have been studied rather thoroughly. On the other hand resting eggs of *Latonopsis australis* and *Sarsilatona serricauda* can be recorded only by virtue of the fact that these species have been reared from dried mud.

In the lack of ephippia Ctenopods differ markedly from anomopods. On the other hand their eggs in some ways resemble those of other branchiopods which lay resting eggs with a thick shell. As in branchiopods that produce only resting eggs Ctenopod eggs display considerable morphological variability, having a smooth, protuberanced or spiny surface. Representatives of one genus may have eggs of different appearance (e.g., *Diaphanosoma* in the Ctenopoda and *Chirocephalus* in the Anostraca). It is also characteristics of the Notostraca to stick their resting eggs on different surfaces (Thiery, 1985; Fryer, 1988). Moreover as noted above some Ctenopods having yolk of "euphyllopod" type, recall the Anostraca. In others *(Holopedium, Sida, Limnosida, Penilia)* inner structure is similar to that of Onychopods which also cannot withstand unfavourable conditions for a long time. All this represents evolutionary convergences and parallelisms existing at different levels within the Brachiopoda. Furthermore, representatives of some other phylogenetically remote groups of freshwater invertebrates (e.g., Rotifera and *Hydra*) have resting eggs of similar structure (Kutikova, 1970; Makrushin, 1985).

The shell of a resting egg serves for mechanical defence of embryos and possibly against dehydration and bacteria, but the precise significance of the elements of the outer structure is not known. Probably their diversity realizes different variations of the shell's attributes. Evidently this morphological diversity should also be comparatively limited due to physical reason and the similarity of environmental factors influencing resting eggs. An outer sticky envelope, besides serving to attach eggs to substrata, may also protect them from dehydration.

The morphological diversity of resting eggs within the genus *Diaphanosoma*, as well as in other genera, may serve for the taxonomy of the group as it does rather successfully in the Anostraca and 'Conchostraca' (Mura & Thiery, 1986; Cesar, 1989; Brendonck

Table 2. Sididae species reared from dried mud

Diaphanosoma excisum	Queensland, Australia	Sars, 1885
Latonopsis australis s.str.	Queensland, Australia	Sars, 1888
Diaphanosoma spinulosum	Vicinity of Sao-Paulo, Brasil	Sars, 1901
D. brevireme	Vicinity of Sao-Paulo, Brasil	Sars, 1901
Sarsilatona serricauda	Vicinity of Sao-Paulo, Brasil	Sars, 1901
Diaphanosoma unguiculatum	Queensland, Australia	Gurney, 1927

Table 3. The degree of investigation of resting eggs of Ctenopoda

Species	Parameters					
	Outer structure	Inner structure	Ovogenesis and timing	Number and distribution in water bodies	Development	Data on resistance
Sida crystallina	+	+	+	–	+	+
Limnosida frontosa	–	+	–	–	–	+
Latona setifera	–	+	–	–	–	–
Latonopsis australis s.str.*	–	–	–	–	–	+
Sarsilatona serricauda*	–	–	–	–	–	+
Diaphanosoma brachyurum	+	+	+	+	+	+
D. chilense	+	–	–	–	–	–
D. fluviatile*	+	–	+	–	–	–
D. excisum*	+	–	–	–	–	+
D. sarsi*	+	–	+	–	–	–
D. senegal*	+	–	+	–	–	–
D. volzi*	+	–	+	–	–	–
D. unguiculatum	+	–	+	–	–	+
D. lacustris	+	–	–	–	–	–
D. mongolianum	+	–	–	–	–	–
D. birgei	+	–	–	–	–	–
D. spinulosum*	+	–	–	–	–	+
D. brevireme*	+	–	–	–	–	+
Pseudosida variabilis*	+	–	–	–	–	–
Penilia avirostris	+	+	+	+	+	+
Holopedium gibberum	+	+	+	+	+	+

* Tropical species are indicated by asterisk

et al., 1990; Thiery & Gask, 1991) and Anomopoda (Glagolev, 1983; Korovchinsky, 1992a). Ctenopod resting eggs need to be studied in detail using SEM. At the same time scarce available data can already help to resolve some taxonomical questions. For instance Behning (1941) mentioned the spiny resting eggs in the Caucasian *D. brachyurum* but probably observed another species because, according to all available descriptions, the eggs of this species are smooth. In another case the morphology of resting eggs gives additional evidence of the conspecificity of '*D. brachyurum*' from Argentina (Paggi, 1978) and *D. birgei*.

Special investigations of resting stages of members of the "cladoceran" assemblage other than the Ctenopoda in nature are not numerous and mainly comprise European species (Murakami, 1961; Herzig, 1985; Mortiz, 1987; Carvalho & Wolf, 1989; Wolf & Carvalho, 1989) however their comparison with data described above for Ctenopods shows many similarities. For all this, comparable information exists only on *H. gibberum, D. brachyurum* and *S. crystallina.* For example gamogenesis may be observed from early spring to late autumn. There is a considerable patchiness in bottom distribution of resting eggs and ephippia; generally the greatest number is found in the deepest part of water bodies. Ephippia of *Daphnia cucullata* present in sediments, but the species not detected in the plankton for more than a year, (Carvalho & Wolf, *op.cit.*) recalls the long failure of *H. gibberum* in one of the lakes studied by Smyly (1977). The last author noted the low intensity of hatching of *Holopedium* comprising only about 0.6% of resting eggs existing in the sediments. His suggestion about the death of most of eggs may not be true because they can remain viable for many years like *Daphnia* ephippia (Wolf & Carvalho, 1989). 12 °C is the most favourable temperature for *Daphnia* hatching as well as 10–16 °C for *H. gibberum* and *D. brachyurum.*

In respect of differences Green (1966) recorded the more intensive character of gamogenesis in a population of *S. crystallina* in comparison with other 'cladocerans'. This species, as well as *D. brachyurum*, completely disappeared in winter. Their spring hatching from resting eggs contributes basically to a new population. It is also interesting that significantly higher densities of *Holopedium* resting eggs in sediments have been found (Lampert & Krause, 1976; Smyly, 1977) than in *Daphnia* (Wolf & Carvalho, 1989) – 16–28 times greater (!). Possibly this is related to the higher fecundity of *Holopedium* gamogenetic females of which carry many resting eggs simultaneously (on av. 3–10, maximum 22) (e.g. Herr, 1917; Lampert & Krause, *op.cit.*).

Recent data (Smyly, 1964) refute initial assumptions (e.g. Burckhardt, 1899) about the ability of Ctenopod resting eggs to develop only at small depths. The same was confirmed for other planktonic crustaceans (Herzig, 1985; Wolf & Carvalho, 1989) though in general the shallower waters are more favourable for hatching. At the same time the problem the activation of resting eggs is still unresolved.

The recent revision of the Ctenopoda (Korovchinsky, 1992b) changed the view of the geographical distribution of many species. However it has become clear that some of them have rather extensive ranges. Obviously they could not disperse over distances of thousands of kilometers without resting eggs, most of which are of especially resistant "euphyllopod" type. At the same time the ranges of species with eggs of the last type (e.g. *D. brachyurum, D. mongolianum*) and those lacking such highly resistant eggs (e.g. *S. crystallina*) are quite comparable. This fact can support the idea of high efficiency of resting egg dispersal in general. The resistance differences are probably important only at invasion into the water body of a particular type (permanent, temporary, etc.).

Various anomopods possess ephippia that are attached to substrata thus, according to Fryer (1972), avoiding their dispersal from proper habitats. By contrast some others, first of all *Daphnia* and *Moina* from temporary waters, have ephippia more specialized for distribution. Judging by the facts, these animals have different strategies in respect of dispersal. In fact however, the situation may be more complicated.

Some ctenopods that have resting eggs with a sticky envelope *(D. brachyurum, D. senegal, D. fluviatile, Pseudosida)* frequently or even preferably, inhabit small and temporary water bodies. Moreover Notostraca, which characteristically frequent temporary water bodies, also stick their eggs on substrata, avoiding their scattering and compensate for a restricted number of eggs per clutch (Thiery, 1985; Fryer, 1988; Thiery & Gask, 1991). It is possible to suggest that the attaching ability of resting eggs may serve both for their effective retention in suitable biotopes as well as for mostly effective dispersal all over them, or in another words these processes are directly connected. The last purpose may be excellently fulfilled by hydrophilous carriers, first of all by waterfowl, constantly moving between water bodies. Eggs may be dispersed sticking to their legs and bills, within the plumage, and with ingested food, which frequently includes aquatic vegetation.

It has been demonstrated that many waterfowl can carry resting stages and even individuals of freshwater crustaceans on body and in the gut (e.g. Proctor, 1964; Proctor & Malone, 1965; Proctor *et al.*, 1967; Vilcek, 1978). The first of the mentioned papers contains data on findings of viable resting eggs of *Sida crystallina* and *Sarsilatona serricauda* (syn. *Latonopsis fasciculata*) passed unharmed through the gut of ducks, then hatched and reared to adults. Other animals also can to be dispersal agents (Maguire, 1963) but probably only birds have the capacity to determine the large-scale distribution of individual species and regional faunas of freshwater ctenopods and anomopods (Korovchinsky, 1982).

Acknowledgements

We are deeply indebted to Professor G. Fryer (Windermere, England) for the correction of the English text and valuable comments.

References

Apstein, C., 1907. Das Plankton im Colombo-See auf Ceylon. Zool. Jb. Syst. Okol. Geogr. Tiere 25: 201–244.

Baldass, F. von, 1937. Entwicklung von *Holopedium gibberum*. Zool. Jb. Anat. Ontog. Tiere 63: 399–454.

Behning, A. L., 1941. Kladotsera Kavkaza, Tbilisi, 284 pp. (Cladocera of the Caucasus).

Braiko, V. D., 1965. Biologia zimnyh jaits *Penilia avirostris*. Doklady AN SSSR, sektsia biol. nauk 164: 681–683. (Biology of *Penilia avirostris* winter eggs).

Brehm, V., 1933. Die Cladoceren der Deutschen Limnologischen Sunda-Expedition. Arch. Hydrobiol., Suppl. Bd. 11: 631–771.

Brendonck, L., A. Thiery & A. Coomans, 1990. Taxonomy and biogeography of the Galapagos Branchiopod fauna (Anostraca, Notostraca, Spinicaudata). J. Crust. Biol. 10: 676–694.

Burckhardt, G., 1899. Faunistische und systematische Studien über das Zooplankton der grosseren Seen der Schweiz und ihrer Grenzgebiete. Rev. suisse Zool. 7: 353–713.

Carvalho, G. R. & H. G. Wolf, 1989. Resting eggs of lake-*Daphnia*. I. Distribution, abundance and hatching of eggs collected from various depth in lake sediments. Freshwat. Biol. 22: 459–470.

Cesar, I. I., 1989. Comparative study on the resting eggs of several Anostracans (Crustacea). Key for the determination of the species based upon the egg structure and diameter. Stud. Neotrop. fauna Environ. 24: 169–181.

Della Groce, N. & S. Bettanin, 1969. Formazione delle uova durevoli in *Penilia avirostris* Dana. Cah. Biol. mar. 10: 95–102.

Franken, W. & M. Franken, 1978. Limnologische Untersuchungen am Grossen Bullensee, einem sauren Heidesee Norddeutschlands, II. Zooplankton. Arch. Hydrobiol. Suppl. 54: 80–122.

Freidenfelt, T., 1920. Zur Kenntnis der Biologie von *Holopedium gibberum* Zaddach. Arch. Hydrobiol. 12: 725–749.

Frey, D. G., 1982. Contrasting strategies of gamogenesis in northern and southern populations of Cladocera. Ecology 63: 223–241.

Fryer, G., 1972. Observations on the ephippia of certain macrothricid cladocerans. Zool. J. linn. Soc. 51: 79–96.

Fryer, G., 1988. Studies on the functional morphology and biology of the Notostraca (Crustacea: Branchiopoda). Phil. Trans. r. Soc., Lond., B., 320: 27–124.

Glagolev, S. M., 1983. Struktura poverhnosty efippiuma Daphniidae po dannym skanirujushey elektronnoy mikroskopii. Zool. J. 62: 1422–1425. (Structure of surface of Daphniidae ephippium as shown with SEM).

Green, J., 1966. Seasonal variation in egg production by Cladocera. J. anim. Ecol. 35: 77–104.

Gurney, R., 1927. Some Australian freshwater Entomostraca reared from dried mud. Proc. zool. Soc. Lond.: 59–71.

Herzig, A., 1974. Some population characteristics of planktonic crustaceans in Neusiedler See. Oecologia 15: 127–141.

Herzig, A., 1984. Temperature and life cycle strategies of *Diaphanosoma brachyurum*: An experimental study on development, growth, and survival. Arch. Hydrobiol. 101: 143–178.

Herzig, A., 1985. Resting eggs – a significant stage in the life cycle of the crustaceans *Leptodora kindti* and *Bythotrephes longimanus*. Verh. int. Ver. Limnol. 22: 3088–3098.

Heucher, H., 1916. Das Zooplankton des Zurichsees mit besonderer Berucksichtigung der Variabilitat einiger Planktoncladoceren. Arch. Hydrobiol. 11: 1–81.

Korovchinsky, N. M., 1981. Taxonomic and faunistic revision of Australian *Diaphanosoma* (Cladocera: Sididae). Aust. J. mar. Freshwat. Res. 32: 813–831.

Korovchinsky, N. M., 1982. Sravnitelnaya morfologia, ekologia i systematika vetvistousyh rakoobraznyh nadsemeystva Ctenopoda fauny mira. Ph.D. Thesis, Moscow, 324 pp. (Comparative morphology, ecology and systematics of superfamily Ctenopoda of the world fauna).

Korovchinsky, N. M., 1983. O biocenoticheskyh vzaimootnoshenijah v zaroslah kubyshki *(Nuphar luteum)* i zykle populacii *Sida crystallina*. In N. N. Smirnov (ed.), Biocenozy mezotrophogo ozera Glubokogo. Nauka Publ., Moscow: 104–117. (On the biocenotic relationships in brushwoods of *Nuphar luteum* and population cycle of *Sida crystallina*).

Korovchinsky, N. M., 1989. Redescription of *Diaphanosoma celebensis* Stingelin, 1900 (Crustacea, Cladocera). Hydrobiologia 184: 7–22.

Korovchinsky, N. M., 1992a. Sovremennoye sostojanie i problemy sistematiki vetvistousyh rakoobraznyh. In N. N. Smirnov (ed.), Sovremennye problemy izuchenija vetvistousyh rakoobraznyh. Gidrometeoizdat Publ., St.-Petersburg: 4–45. (Modern state and problems of Cladoceran taxonomy).

Korovchinsky, N. M., 1992b. Sididae & Holopediidae. Guides to the identification of the microinvertebrates of the continental waters of the World 3, SPB Acad. Publ., The Hague, 82 pp.

Korovchinsky, N. M., 1993a. Description of the male, analysis of morphological variability and geographical distribution of *Diaphanosoma senegal* Gauthier, 1951. Hydrobiologia 259: 9–16.

Korovchinsky, N. M., 1993b. New records of *Diaphanosoma celebensis* Stingelin, 1900 (Crustacea: Daphniiformes: Sididae) in tropical Asia with remarks on the morphological variability and biology of the species. Rev. Hydrobiol. trop. 26: 119–125.

Kutikova, L. A., 1970. Kolovratki fauny SSSR. Nauka Publ., Leningrad, 701 pp. (Rotifers of the USSR fauna).

Lampert, W. & I. Krause, 1976. Zür Biologie der Cladocere *Holopedium gibberum* Zaddach im Windgefallweiher (Schwarzwald). Arch. Hydrobiol. Suppl. 48: 262–286.

Loidl, R., 1955. Ein Wasserfloh legth Eier. Microkosmos 44: 145–148.

Maguire, B., 1963. The passive dispersal of small aquatic organisms and their colonization of isolated bodies of waters. Ecol. Monogr. 33: 161–185.

Makrushin, A. V., 1976. Nekotorye osobennosti vosproizvoditelnoy systemy Cladocera (Crustacea). Zool. J. 55: 1143–1148. (Some peculiarities of Cladoceran reproductive system).

Makrushin, A. V., 1978. Angidrobioz i struktura zheltka jaitz Cladocera. Zool. J. 57: 364–374. (Anhydrobiosis and Structure of Cladocera egg's yolk).

Makrushin, A. V., 1981. Ustoichivost k vysushivaniu latentnyh jaitz *Sida crystallina* i *Moina macrocopa* (Cladocera, Crustacea). Zool. J. 60: 933–935. (Resistance of resting eggs of *Sida crystallina* and *Moina macrocopa*).

Makurshin, A. V., 1985. Angidrobioz pervichnovodnyh bespozvonochnyh: sohranenye zhyznesposobnosty v vysushennom sostojanii. Nauka Publ., Leningrad, 101 pp. (Anhydrobiosis of primary water invertebrates).

Makrushin, A. V., 1990. O prichine i sledstvijh embrionalnoy diapauzy u pervichnovodnyh bespozvonochnyh i ascidia. J. Obschey Biol. 51: 476–482. (On the cause and consequences of embryonal diapause in primarywater invertebrates and ascidians).

Moritz, C., 1987. A note on the hatching and viability of *Ceriodaphnia* ephippia collected from lake sediments. Hydrobiologia 145 (Dev. Hydrobiol. 35): 309–314.

Müller, O. F., 1785. Entomostraca seu Insects Testacea, quae in Daniae et Norvegiae reperit, descripsit et iconibus illustravit. Lipsiae et Havniae, 135 pp.

Mura, G. & A. Thiery, 1986. Taxonomical significance of scanning electron microscopic morphology of the Euphyllopods' resting eggs from Morocco. Part I. Anostraca. Vie Milieu 36: 125–131.

Murakami, I., 1961. Studies on the winter eggs of the water flea, *Moina macrocopa* Straus. J. Fac. Fish. Anim. Husb. Hiroshima Univ. 3: 323–346.

Onbe, T., 1972. Occurrence of the resting eggs of a marine cladoceran, *Penilia avirostris* Dana on a sea bottom. Bull. Jap. Soc. Sci. Fish. 38: 305.

Onbe, T., 1974. Studies on the ecology of marine cladocerans. J. Fac. Fish. Anim. Husb. Hiroshima Univ. 13: 83–179.

Paggi, J. C., 1978. Revision de la especies argentinas del genero *Diaphanosoma* Fisher (Crustacea, Cladocera). Acta zool. lilloana 33: 43–65.

Proctor, V. W., 1964. Viability of Crustacean eggs recovered from ducks. Ecology 45: 656–658.

Proctor, V. W. & C. R. Malone, 1965. Further evidence of the passive dispersal of small aquatic organisms via the intestinal tract of birds. Ecology 46: 728–729.

Proctor, V. W., C. R. Malone & V. L. De Vlaming, 1967. Dispersal of aquatic organisms, viability of disseminules recovered from the intestinal tract of captive Kildeer. Ecology 48: 672–676.

Rajapaksa, R., 1986. A contribution to the taxonomy, biogeography and gamogenesis of freshwater Cladocera, with special reference to the tropical region. Ph.D. Thesis, Waterloo (Canada), 220 pp.

Sars, G. O., 1865. Norges Ferskvandskrebsdyr Forste afnit Branchiopoda, I. Cladocera Ctenopoda (Sididae & Holopediidae). Efter det Acad. Collegium, Christiania, 79 pp.

Sars, G. O., 1885. On some Australian Cladocera raised from dried mud. Forhandl. Vidensk. Selsk. Christiania: 1–46.

Sars, G. O., 1888. Additional notes on Australian Cladocera raised from dried mud. Forhandl. Vidensk. Selsk. Christiania 7: 1–74.

Sars, G. O., 1901. Contribution to the knowledge of the freshwater Entomostraca of South America, as shown by artificial hatching from dried material. Arch. Math. Naturv. 23: 1–102.

Scheffelt, E., 1908. Die Copepoden und Cladoceren des sudlichen Schwarzwaldes. Arch. Hydrobiol. 4: 104–198.

Schmidt, C., 1924. Die Phyllopoden und Copepoden des Obinger und Seeoner-Sees. Arch. Hydrobiol. 15: 179–208.

Smyly, W. J. P., 1964. An investigation of some benthic entomostraca of three lakes in northern Italy. Mem. Ist. ital. Idrobiol. 17: 33–56.

Smyly, W. J. P., 1977. A note on the resting egg of *Holopedium gibberum* Zaddach (Crustacea: Cladocera). Microscopy (Lond.) 33: 170–171.

Smyly, W. J. P., 1978. The crustacean zooplankton of Grasmere before and after a change in sewage effluent treatment. Int. Rev. ges. Hydrobiol. 63: 389–403.

Thiery, A., 1985. Ponte et ultrastructure de l'oeuf de *Triops granarius* Lucas (Crustacea, Notostraca): adaptations à l'assèchement de l'habitat. Verh. int. Ver. Limnol. 22: 3024–3028.

Thiery, A. & C. Gasc, 1991. Resting eggs of Anostraca, Notostraca and Spinicaudata (Crustacea, Branchiopoda) occurring in France: identification and taxonomical value. Hydrobiologia 212 (Dev. Hydrobiol. 64): 245–259.

Vehov, N.V., 1987. Rasprostranenie, biotopicheskoe raspredelenie i zhiznennye cikly teplovodnyh vetvistousyh v subarkticheskyh vodojomah Evropy. Gidrobiologitchesky J. 23: 13–18. (Distribution, biotopical dispersal and life cycles of warmwater Cladocera in subarctic water bodies of Europe).

Vilcek, F., 1978. Vplyv vtakov na druhove zlozenie perloociek (Cladocera) v priehrade Zemplinska Sirava. Biologia (Bratislava) 33: 651–656.

Weismann, A., 1877. Beitrage zur Naturgeschichte der Daphnoiden. II. Die Eibildung bei den Daphnoiden. Z. wiss. Zool. 28: 95–175.

Wolf, H. G. & G.R. Carvalho, 1989. Resting eggs of lake-*Daphnia*. II. In situ observations on the hatching of eggs and their contribution to population and community structure. Freshwat. Biol. 22: 471–478.

Hydrobiologia **320**: 141–152, 1996.
V. R. Alekseev & G. Fryer (eds), Diapause in the Crustacea.

Ecological and evolutionary significance of resting eggs in marine copepods: past, present, and future studies

Nancy H. Marcus
Department of Oceanography, Florida State University, Tallahassee, FL 32306, USA

Key words: Copepod, dormancy, diapause, resting eggs

Abstract

The occurrence of a resting egg phase in the life cycle of marine and freshwater planktonic copepods is well documented and receiving increasing attention by investigators. The species generally occur in coastal marine waters, freshwater ponds and lakes in areas that undergo strong seasonal fluctuations, though examples have been reported for tropical and sub-tropical areas not subject to such extreme fluctuations. Typically such species disappear from the water column for portions of the year, but remain in the region as benthic resting eggs. Studies to date have focused on the conditions that promote the occurrence of resting eggs, the factors that affect their survival and hatching from sediments, the existence of egg banks in sediments, and the impact of resting eggs on plankton community structure. Benthic resting eggs of copepods include diapause eggs as well as subitaneous (non-diapause) eggs that are quiescent due to conditions in the sediments. As with other groups of organisms the resting egg phase is viewed as being critical for the perpetuation of species year after year, especially those that disappear from the water column for portions of the year. Some data indicate that eggs can survive for many years in sediments which would expand their influence to evolutionary time scales. This paper summarizes our understanding of embryonic dormancy in marine copepods.

Introduction

A recent taxonomic classification of the subclass Copepoda (Huys & Boxshall, 1991) distinguished 10 orders: Platycopioida, Calanoida, Mormonilloida, Harpacticoida, Poecilostomatoida, Siphonostomatoida, Monstrilloida, Misophrioida, Cyclopoida, and Gelyelloida. The ancestral copepod for the group is believed to have come from the shallow-water marine environment. Most copepods only reproduce sexually and eggs are carried by the female in sacs or spawned freely into the water column. The usual life cycle includes six naupliar stages, five copepodid stages, and the final adult stage. Generation times range from days to weeks to years.

To date dormancy has been reported to occur in only three orders of copepods: the Calanoida, Harpacticoida, and Cyclopoida. A wide variety of organisms, insects in particular possess resting or dormant stages. Dormancy has been defined as a 'state of suppressed development' (Danks, 1987), but is best represented as a spectrum of suppressed development, the extremes being quiescence and diapause. Danks defined quiescence as 'an immediate direct response to a limiting factor' and diapause as a 'more profound interruption that routes the metabolic programme of the organism away from direct developmental pathways and into a much more clearly organized break in development'. In the case of quiescence, development resumes immediately upon the resumption of favourable conditions. For example if temperature or oxygen concentration falls below a threshold, development is suppressed and remains so until conditions are no longer limiting. In the case of diapause, the state of suppressed development is not the result of a response to one's immediate conditions. Rather the diapause state is cued by some factor in the environment which normally precedes the deterioration of environmental conditions. Individuals in the diapause state do not resume development even if conditions are favourable for development to proceed, rather diapause must first be broken. The duration of diapause may be affected by environmental conditions e.g. temperature.

In those copepods for which dormancy has been noted the dormant phase varies from species to species. Examples have been reported of embryonic dormancy in marine (see Grice & Marcus, 1981) and freshwater (e.g. Brewer, 1964) calanoids, delayed naupliar development in a marine harpacticoid and a marine calanoid (see Grice & Marcus, 1981), copepodid dormancy in marine calanoids (see Grice & Marcus, 1981; Davis, 1984) and freshwater cyclopoids (see Elgmork, 1980), and adult encystment in a marine harpacticoid (Coull & Grant, 1981). There are no reports of a dormant stage in marine cyclopoids. Reviews of dormancy in copepods have focused on evolutionary aspects of diapause in freshwater copepods (Elgmork, 1980), and embryonic dormancy in marine copepods (Grice & Marcus, 1981; Uye, 1985). A recent review of diapause in Crustacea by Alekseev (1990) considered copepods. This paper examines our understanding of embryonic dormancy in marine copepods, focusing on research conducted since the earlier reviews of the subject.

Background

Zillioux & Gonzalez (1972) first referred to eggs of planktonic copepods found in the sea bed as 'resting'. Subsequently, two types of resting (dormant) eggs were distinguished, subitaneous (non-diapause) and diapause (Grice & Marcus, 1981). Subitaneous eggs were defined as eggs that hatch within a few days at ambient temperatures, but can be induced to become quiescent in response to adverse conditions e.g. temperature, salinity, or oxygen concentrations. Diapause eggs were defined as those eggs that do not hatch within a few days even if conditions are favourable. Instead diapause eggs enter a refractory phase which may last several months, during which development and/or metabolic processes apparently cease or drop to undetectable levels. Upon completion of the refractory phase, eggs become competent to resume development and hatch if conditions are permissible. It was noted that some copepods produce only subitaneous eggs, but some produce both subitaneous and diapause eggs (Grice & Marcus, 1981).

Although studies of other organisms have indicated there is tremendous variation in the expression of dormancy and it is apparent that the two terms diapause and quiescence are not sufficient to describe all of the various forms of developmental suppression occurring in nature (see Danks, 1987) studies of marine copepods have generally defined resting eggs as either quiescent subitaneous or diapause. However, results of work underway in my laboratory suggest that intermediate states of egg dormancy also typify the life cycle of some copepod species (Chen & Marcus, unpub.). *Pontella meadi*, *Labidocera scotti*, and *L. mirabilis* were collected from the northeastern Gulf of Mexico and incubated in dishes at ambient temperatures. Eggs produced overnight were collected and incubated in filtered seawater at temperatures comparable to those in the field at the time of collection. These eggs hatched over a period ranging from a few to 81 days. Eggs with hatching times greater than a few days and less than a couple of months do not correspond to the definition of subitaneous eggs, nor do they correspond to the definition of diapause eggs. They appear to represent some intermediate class. The production of eggs with such different development rates might be a form of bet-hedging. Rather than putting 'all of their eggs in one basket (i.e. cohort)' that might fail, development and population growth is spread out over a longer time frame.

Based upon field collections of sea-bottom sediments or eggs spawned by females in the laboratory, 38 marine copepod species are known to have a resting or dormant egg phase (Table 1). They represent five families of the order Calanoida: Temoridae, Centropagidae, Pontellidae, Acartidae, and Tortanidae. Lindley (1992) suggested that resting eggs pre-adapted species to colonize inland waters more readily than those species that lack resistant eggs. All, except *Eurytemora* spp. spawn their eggs freely into the water.

Of the 38 species, 14 have been shown to produce diapause eggs: *Labidocera aestiva. L. miribillis, Pontella meadi, P. mediterranea, Anomolocera patersoni, Tortanus forcipatus, T. derjuginii, Eurytemora affinis, Centropages hamatus, C. tenuiremis (= Yamadai), Acartia californiensis, A. tonsa, A. clausi*, and *A. latisetosa*. Diapause eggs were distinguished from subitaneous eggs by morphological differences and/or differences in hatching and survival when incubated in the laboratory.

Despite the fact that resting eggs are recognized as being important ecologically, we still have a very poor understanding of the mechanism(s) underlying the dormancies of most copepods. Research on embryonic dormancy has focused on the relationship of resting eggs to population dynamics. No studies have examined the biochemical and physiological basis of qui-

Table 1. Marine calanoid copepod species for which resting eggs have been found in sediments or spawned in the laboratory. Entries in bold are new since Uye (1985).

Family/Species	Geographic area	Reference
Acartiidiae		
Acartia clausi	Inland Sea of Japan	Kasahara *et al.*, 1974
	Onagawa Bay, Japan	Uye ,1980
	Mission Bay, California	Uye & Fleminger, 1976
	Jakle's Lagoon, Washington	Landry, 1975
	Buzzards Bay, Massachusetts	Marcus, 1984
	Northern California	**Marcus, 1990**
	Norway	**Naess, 1991**
Acartia hudsonica	**Narragansett Bay, Rhode Island**	**Sullivan & McManus, 1986**
Acartia tonsa	Northeast United States	Zillioux & Gonzalez, 1972
	La Jolla, California	Uye & Fleminger, 1976
	Northern California	**Marcus, 1990**
	Northern Gulf of Mexico	**Marcus, 1991**
Acartia californiensis	Mission Bay, California	Uye & Fleminger, 1976
	Yaquina Bay, Oregon	Johnson, 1980
Acartia erthryea	Inland Sea of Japan	Kasahara *et al.*, 1974
Acartia pacifica	Inland Sea of Japan	Uye, unpub.
	Jiaozhou Bay, China	**Zhong & Xiao, 1992**
	Xiamen, China	**Chen & Li, unpub.**
Acartia pulmosa	Inland Sea of Japan	Uye, 1985
Acartia tsuensis	Inland Sea of Japan	Uye, 1985
Acartia steuri	Onagawa Bay, Japan	Uye, 1980
Acartia bifilosa	**Baltic Sea**	**Viitasalo, 1992**
	Jiaozhou Bay, China	**Zhong & Xiao, 1992**
Acartia latisetosa	**Adriatic coast, Italy**	**Belmonte, 1992**
		Fanelli *et al.*, 1992
Acartia spinacauda	**Xiamen, China**	**Chen & Li, unpub.**
Centropagidae		
Centropages ponticus	Black Sea	Sazhina, 1968
Centropages hamatus	White Sea	Pertzova, 1974
	Buzzards Bay, Massachusetts	Marcus, 1984
	George's Bank, North Atlantic	**Marcus, unpub.**
	North Sea, English Channel	**Lindley, 1986**
	Northern Gulf of Mexico	**Marcus, 1989**
Centropages abdominalis	Inland Sea of Japan	Kasahara *et al.*, 1974
Centropages yamadai	Inland Sea of Japan	Kasahara *et al.*, 1974
(*tenuiremis*)	**Xiamen, China**	**Chen & Li, unpub.**
Sinocalanus tenellus	Fukuyama, Japan	Uye, 1985
Centropages typicus	**North Sea, English Channel**	**Lindley, 1990**

Continued on p. 144

escence or diapause. This contrasts dramatically with research on *Artemia* which has focused primarily on physiological, biochemical, and genetical aspects of cyst dormancy. The situation with copepods is unfortunate because knowledge of what species are genetically and physiologically capable of doing under certain environmental conditions is essential for predicting their growth and dynamics in the field.

While explanations for the occurrence of diapause in insects, seeds and other organisms including cope-

Table 1. Continued.

Family/Species	Geographic area	Reference
Centropages furcatus (**velificatus**)	**Northern Gulf of Mexico**	**Marcus, 1989**
Pontellidae		
Anomalocera patersoni	**Gulf of Naples**	**Ianora & Santella, 1991**
Anomalocera ornata	**Northern Gulf of Mexico**	**Chen & Marcus, unpub.**
Pontella mediterranea	Black Sea	Sazhina, 1968
	Mediterranean Sea	Grice & Gibson, 1981
Pontella meadi	Woods Hole, Massachusetts	Grice & Gibson, 1977
	Northern Gulf of Mexico	**Chen & Marcus, unpub.**
Labidocera bipinnata	Inland Sea of Japan	Uye *et al.*, 1979
Labidocera trispinosa	La Jolla, California	Uye, unpub.
Labidocera aestiva	Woods Hole, Massachusetts	Grice & Lawson, 1976
	Northern Gulf of Mexico	**Chen & Marcus, unpub.**
Labidocera wollastoni	**North Sea, English Channel**	**Lindley, 1986**
Labidocera scotti	**Northern Gulf of Mexico**	**Marcus, 1989**
Epilabidocera longipedata	Yaquina Bay, Oregon	Johnson, 1980
	Northern California	**Marcus, 1990**
Calanopia thompsoni	Inland Sea of Japan	Kasahara *et al.*, 1974
Calanopia americana	**Northern Gulf of Mexico**	**Chen & Marcus, unpub.**
Temoridae		
Eurytemora affinis	Yaquina Bay, Oregon	Johnson, 1980
	Southern Hokkaido, Japan	**Ban, 1992b**
	Norway	**Naess, 1991a**
	Rhode Island	**Marcus** *et al.*, **1994**
	Baltic Sea	**Viitasalo, 1992**
Eurytemora pacifica	Onagawa Bay, Japan	Uye, 1985
Eurytemora americana	Buzzards Bay, Massachusetts	Marcus, 1984b
	Rhode Island	**Marcus** *et al.*, **1994**
Temora longicornis	**George's Bank, North Atlantic**	**Marcus, unpub.**
	North Sea, English Channel	**Lindley, 1986**
	Baltic Sea	**Viitasalo, 1992**
Tortanidae		
Tortanus forcipatus	Inland Sea of Japan	Kasahara *et al.*, 1974
	Xiamen, China	**Chen & Li, 1991**
Tortanus discaudatus	**Rhode Island**	**Marcus** *et al.*, **1994**
	Northern California	**Marcus, 1991**
Tortanus derjugunii	**Xiamen, China**	**Chen & Li, 1991**

pods have often focused on the cold-hardiness of diapause stages and their capacity to survive adverse periods, Danilevsky (1965) and Tauber & Tauber (1981) emphasized that diapause is of fundamental importance in the synchronization of insect life cycles such that development, growth, and reproduction were optimized. In the case of planktonic copepods the diapause response might therefore promote temporal partitioning of the environment and effectively lead to reproductive isolation and speciation as suggested by Marcus (1984a). In addition resting eggs of copepods may play a role in dispersal (Boileau & Hebert, 1991), retention in a region (Uye, 1980; Marcus, in press), the prevention of overcrowding (Uye, 1980), and slowing down the rate of evolutionary change (Hairston & De Stasio, 1988).

Distribution and abundance

Most studies of the distribution and abundance of resting eggs in the field have been conducted in north temperate waters between 30 and 60 °N (Table 1). Few studies (Marcus, 1989, 1991; Chen & Li, 1991) have been conducted in sub-tropical waters and there are none from tropical regions. One study of a temperate brackish lake in New South Wales, Australia noted the hatching of copepod nauplii from 'cysts' (66 to 80 μm in diameter) previously thought to be dinoflagellate in origin (McMinn *et al.*, 1992).

Most of the habitats examined have been shallow (<20 m water depth). Only three studies have examined the occurrence of eggs in deeper water. In a study of waters around the British Isles, Lindley (1990) found eggs to a depth of 94 m though most were found between depths of 20 and 80 m. Marcus (1990, in press) found eggs at water depths ranging from 60 to 120 m off the coast of Northern California, and noted the occurrence of some eggs at a depth of 165 m of the edge of Georges Bank (Marcus, unpub.).

Concentrations of eggs in the sediments are variable. The highest numbers (10^6 m^{-2}) have generally been found in muddy sediments characteristic of depositional environments e.g., estuaries and lagoons. Marcus & Fuller (1986) suggested that this distributional pattern reflected the physical charcteristics and settling velocities of eggs which they showed were similar to silt sized particles. Positive correlations between egg abundance and the proportion of silt sized particles in the sea bed were reported by Marcus & Fuller (1989) and Lindley (1990). Lindley (1990) concluded that eggs would be most abundant in areas of the open sea that were 20 to 80 m deep and where bottom stress due to tidal currents was $\leq$10 dynes cm^{-2}. In areas where bottom stress is greater e.g., shallow waters, bottom sediments are suspended more often resulting in egg hatching. Off the coast of California eggs occur in greater numbers inshore at 60 m than offshore at 90 and 120 m despite the fact that muddy sediments are accumulated offshore (Marcus, in press). Taking into account development time and sinking rates of the eggs I suggested that if the eggs were non-diapause they had a greater probability of reaching the sea floor prior to hatching at 60 m rather than 90 or 120 m. Alternatively, the impact of predation on the eggs in the sediments may have been greater at the slightly deeper sites. Sediment resuspension affects the vertical distribution of eggs in the sea bed. When sediments and eggs are resuspended, eggs may be reburied or remain at the water sediment interface depending upon their settling velocity relative to that of the surrounding sediment particles (Marcus & Taulbee, 1992). If the particles are large eggs sink more slowly and become layered on top: if the particles are small eggs sink with the sediments and are re-buried.

Morphology

Few studies have considered structural and morphological differences between subitaneous and diapause eggs. Based upon SEM Grice & Lawson (1976) reported that the 'shell' of *Labidocera aestiva* 'summer' (subitaneous) eggs was thinner than that of 'winter' (diapause) eggs. Diapause eggs of *Labidocera aestiva* (Marcus, unpub.) and *L. bipinnata* (Uye *et al.*, 1979) are tougher than subitaneous eggs and resist being ruptured when poked with a needle. The same is true in *Tortanus derjugunii* (Chen & Li, 1992). The first detailed TEM work of marine copepod diapause eggs was that of Santella & Ianora (1990) who reported that the egg envelope of smooth subitaneous eggs of *Pontella mediterranea* was very thin whereas that of spiny subitaneous or spiny diapause eggs was complex and multi-layered. The layers of the spiny subitaneous eggs were approximately half the thickness of the diapause eggs. The authors suggested that the spiny subitaneous eggs represent a transition phase as females switch from producing subitaneous to diapause eggs. Because the oocytes that gave rise to the two types were similar in structure they concluded that the layers were assembled after spawning. Later, Santella & Ianora (1992) demonstrated that the multi-layered envelope characterizing the diapause eggs arises from the exocytosis of four different types of cortical granules, following egg activation. In a subsequent study of *Anomalocera patersoni* Ianora & Santella (1991) reported that diapause and subitaneous eggs were similar in size, smooth, and dark in appearance, but like *P. mediterranea* the egg envelope was thin in subitaneous eggs and multi-layered in the diapause eggs. Similar differences are found in the egg envelope of *Acartia latisetosa* (Fanelli *et al.*, 1992).

The differences between subitaneous and diapause eggs of marine copepods are similar to those of freshwater copepods (Champeau, 1970; Hairston & Olds, 1984; Ban & Minoda, 1991). Differences in thickness and number of layers have been noted. The multilayered, tough outer covering of diapause eggs is believed to protect the embryo from destruction by predators (Ianora & Santella, 1991) since diapause eggs show greater survival after gut passage than subitaneous eggs (Hairston & Munns, 1984; Marcus, 1984). Ianora & Santella (1991) also speculated that the thicker coat might play a role in the maintenance of diapause, and Chen & Li (1992) observed that *Tortanus derjugunii* produced eggs with cases that differed in thickness. Those with thick cases survived incubation in the laboratory longer than those with thin cases. For freshwater copepods inhabiting temporary ponds it was suggested that the thicker coating protects the embryo from desiccation (Hairston & Olds, 1984). Resting eggs of freshwater copepods can survive desiccation as evidenced by their re-appearance following the inundation of ponds that have been dry for prolonged periods. On the other hand, Naess (1991b) reported that no nauplii hatched from marine sediments that were dried or frozen. Except for copepods inhabiting tidepools it is unlikely that marine species will experience desiccation. Although, marine copepod eggs may not experience desiccation, they may be subjected to salinity fluctuations. The permeability of copepod eggs has been little studied (e.g., Miller & Marcus, 1994) and no study has examined diapause eggs. Based on the hatching of eggs of *Acartia steuri* and *A. clausi* at a range of salinities, Uye (1980) concluded that eggs of the latter were less sensitive to changes in osmotic pressure. Lindley (1992) noted that eggs of species of *Acartia*, *Centropages*, and *Temora* did not respond conspicuously to osmotic stress whereas those of *Calanus* and *Pseudocalanus* became malformed in low salinity water. No experimental details were provided.

Except for the studies of Santella & Ianora (1990) and Ianora & Santella (1992) the stage at which diapause eggs cease development is not known. Subitaneous and diapause eggs of *Labidocera aestiva* when first spawned are uniform in colour to the extreme edge of the envelope, but after approximately one day a clear perimeter is evident (Marcus, 1980, unpub.). This is indicative of development and suggests that eggs enter diapause at a multi-celled stage. Based upon TEM sections of eggs of *P. mediterranea*, Santella & Ianora (1990)initially concluded that development in diapause eggs is arrested prior to initiation of first cleavage. However, Ianora & Santella (1991) later concluded that diapause eggs of *P. mediterranea* undergo cleavage because the disk like structures they had originally been unable to identify were in fact nuclei. Examination of the developing embryos of *Anomalocera patersoni* (Ianora & Santella, 1991) revealed that diapause embryos ceased developing at the 32-cell stage. Subsequent examination at two weeks, and periodically thereafter up to four months indicated that development of embryos incubated in dishes at room temperature did not progress beyond that stage. Only after six months were embryos observed at the blastula stage suggesting that diapause was broken sometime between four and six months. Hatching occurred after seven months of incubation.

Biochemistry and physiology

Biochemical and physiological studies of subitaneous and diapause eggs are noticeably lacking. Based upon differences in the settling velocities and densities of subitaneous and diapause eggs, Marcus & Fuller (1986) suggested that the chemical composition of the eggs was different. Analyses of *Calanus* sp. have revealed that wax esters are the main storage compound for the diapausing CV stages of some species (e.g., *Calanus finmarchicus*) (see Corner & O'Hara, 1986). However, since diapausing eggs typically occur in anoxic sediments wax esters might not be useful as a storage compound because oxygen is required for their metabolism. It might be more productive to compare the biochemical composition of copepod resting eggs to other dormant crustacean eggs e.g. *Artemia* cysts that occur in anoxic environments. Such studies would provide much needed insight into the mechanism of diapause in copepods.

Induction of diapause

Much early work on embryonic dormancy focused on the conditions of temperature, salinity, and oxygen that inhibited hatching. Dormancy was of the quiescent type (an immediate response to external conditions) because upon the resumption of favourable conditions the eggs would hatch if exposure had not been long enough to reduce viability. The factors that trigger the induction of diapause egg production have been revealed for only two marine copepods, *Labidocera aestiva* and *Acartia clausi*. For both species, photoperiod (modified by temperature) acts as a cue signalling the future deterioration of environmental conditions. Short daylengths and cold temperatures are effective for *L. aestiva* (Marcus, 1980, 1982b) a summer/fall species, whereas long daylengths and warm temperatures are effective for *A. clausi* (Uye, 1985) a winter/spring species. Latitudinal variation in the response threshold of these species (i.e. the particular conditions of photoperiod and temperature that induce diapause) were also reported (Marcus, 1984a; Uye 1985). The delayed hatching of *A. hudsonica* eggs at 20 °C from Narragansett, Rhode Island may not have been an example of diapause since there was no evidence of a refractory phase (Sullivan & McManus, 1986). Long daylengths also induce diapause egg production in the freshwater, *Diaptomus sanguineus* (Hairston & Olds, 1987) and short daylengths and cool temperatures induce it in freshwater *Eurytemora affinis* (Ban, 1992a). Interestingly some diapause eggs were observed in crowded cultures of *E. affinis* under long daylengths that normally resulted in subitaneous egg production indicating that crowding or inadequate food levels might play a role in the induction of diapause (Ban, 1992a). Ban also noted that *E. affinis* produced diapause eggs not only in late autumn, but also in late spring and suggested that crowding promoted diapause egg production in the late spring when the species was at its maximum abundance. Belmonte (1992) obtained spiny and smooth eggs from *Acartia latisetosa* reared in the laboratory under constant temperature and 24 hr light. Since the spiny eggs did not hatch within one day as the smooth eggs did the author concluded that the eggs were diapause eggs and suggested that they were not produced for overwintering, but as a result of crowding. The factor(s) mediating diapause egg production due to crowding is(are) not known.

The studies which implicated photoperiod and temperature as the major factors responsible for the induction of embryonic diapause in copepods were not designed to assess which stage was sensitive to the cues. Marcus (1982a) suggested that *Labidocera aestiva* produced diapause eggs in response to a cumulative number of short daylengths and that once the threshold was passed, the animals switched to making diapause eggs. She also determined that the type of egg produced by *Labidocera aestiva* females could be reversed by exposing females to alternative photoperiod and temperature conditions. However, Ban (1992a) found that conditions perceived by the naupliar stages determined the type of egg produced by *Eurytemora affinis* and that the programming of egg type was irreversible. Such differences are not surprising given the tremendous variation expressed by insects.

Diapause is generally viewed as a mechanism to ensure survival during periods of environmental adversity. While photoperiod and temperature act as independent cues signalling the future deterioration of environmental conditions, few studies (Hairston & Munns, 1984; Hairston & Olds, 1984; Hairston *et al.*, 1985; Hairston & Walton, 1986) have sought to reveal the ultimate cause of diapause. Hairston and his colleagues examined the timing of diapause in the freshwater, *Diaptomus sanguineus* in relation to the onset of fish predation and pond desiccation. They concluded that fish predation in large permanent ponds and desiccation in small temporary ponds were the ultimate stresses regulating the onset of diapause. The switch to diapause egg production typically preceded the onset of fish predation by at least one generation, and the timing of the switch became earlier with greater variation in the onset of fish predation from year to year. The results of a fortuitous natural selection experiment in the field revealed that the switch to diapause egg production became later in the absence of predation. While various other factors e.g., adverse temperature and salinity, poor food conditions have been suggested as other ultimate reasons for diapause no study has actually demonstrated whether or not animals can survive during the 'adverse' periods.

Maintenance/Termination

Diapause eggs are capable of extended (months to years) survival, but become competent to resume development much sooner. Eggs of *Labidocera aestiva* required approximately one to two months of chilling at 5 °C before warming at 25 °C resulted in synchronous hatching (Marcus, 1980) and Ban & Minoda (1991) also observed that chilling followed by warming resulted in synchronous hatching in *Eurytemora affinis*. On the other hand, Marcus (1989) found that a period of warming followed by chilling resulted in the termination of diapause in *Centropages hamatus* and synchronous hatching. Similar results were obtained with diapause eggs of *Tortanus derjugunii* (Chen & Li, 1991). Subsequent experiments (Marcus, 1987) on diapause eggs of *L. aestiva* from Woods Hole produced at different times during the fall revealed that the later eggs were produced in the season the fewer days of chilling were necessary to result in synchronous hatching when warmed. Ban & Minoda (1991) also noted that eggs of *E. affinis* produced later in the season had a shorter refractory period. This pattern of variation in the duration of diapause means that all eggs will be ready to resume development at the same time and hatching will be synchronous which should promote reproductive success.

Survival, hatching and recruitment

While early studies of resting egg abundance in the sea bed focused primarily on the upper 1–2 cm of the sediments, subsequent studies (Marcus, 1984, submitted; Marcus & Fuller, 1989; Marcus *et al.*, 1994; Viitasalo, 1992; Ban, 1992b) revealed that viable eggs occur much deeper in the sediments. In fact the concentration of eggs at depth was often greater than in surficial layers. Viitasalo (1992) proposed that this vertical distribution reflected a 3-tier system in which hatching occurred at the surface, no hatching occurred but survival was sustained at moderate depths, and mortality increased with depth due to adverse chemical conditions. Elucidation of the significance of resting eggs to recruitment and the population dynamics of species requires information on the conditions necessary for hatching to occur and how long eggs survive in the sediments.

Although eggs occur below 2 cm, they do not hatch while buried (e.g., Kasahara *et al.*, 1975; Uye & Fleminger, 1976). Mechanisms that lead to the uncovering of buried eggs are therefore important. For shallow waters resuspension of bottom sediments due to tidal and wind driven currents promotes the resuspension of eggs enabling hatching to occur. The activities of 'conveyor-belt feeding' polychaetes also translocates eggs of *Labidocera aestiva* to the water sediment interface where they can hatch (Marcus & Schmidt-Gengenbach, 1986) which suggests that in deep waters not subject to physical resuspension, bioturbation promotes the recruitment of copepod nauplii from the sea bed by bringing eggs to the mud surface. Variation in the abundance of eggs in the sediments might therefore reflect the patchy distribution of benthic animals.

For all species represented by resting eggs in the sea bed it is important to know how long eggs can survive in the sediments. Survival from year to year is ecologically important because hatching perpetuates the species year after year. On the other hand if eggs can survive many years in the sea bed then the slow accumulation of eggs will sustain a pool of genetic information (i.e. an egg bank) that may slow down the rate of evolutionary change (Hairston & De Stasio, 1988).

One approach to determining the survival capacity of eggs is to conduct experiments that examine the affect of various factors on survival. Early studies (e.g., Kasahara *et al.*, 1975; Uye *et al.*, 1979) shed some light on the duration of survival, but the focus of the work was on factors that inhibited hatching (see Grice & Marcus (1981) for a summary of these effects).

To date the longest survival of eggs collected from sediments and subsequently incubated, is 80 weeks for *Centropages* spp. and *Temora longicornis* (Lindley, 1990) and *Eurytemora affinis* (Ban, 1992b). The age of the eggs was not known, nor was it clear whether the eggs were subitaneous or diapause. Thus, incubation times should represent minimum survival times.

Survival is greatly affected by temperature. Uye (1980) showed that non-diapause eggs of *Acartia steuri* and *A. clausi* survived 100 and 165 days respectively at 5 °C, but only 75 and 70 days at 20 °C. *A. clausi* was better able to survive cold temperatures. Diapause eggs of *L. aestiva* incubated in jars of seawater at 5 °C for periods as long as 300 days in the laboratory continued to hatch at high levels when transferred to warmer temperatures (Marcus, 1987).

Several early studies considered the effects of hypoxia/anoxia on egg hatching and survival since pore waters of bottom sediments are typically anoxic below a few mms. The results (reviewed by Grice & Marcus, 1981) showed that generally eggs did not hatch at concentrations of oxygen less than 0.15 ml l^{-1}. More recently Lutz *et al.* (1992) also found that hatching of subitaneous eggs of *Labidocera aestiva*, *Acartia tonsa*, and *Centropages hamatus* was greatly reduced at concentrations less than 0.15 ml l^{-1} oxygen, but noted that hatching increased if eggs were incubated longer (e.g. for *C. hamatus* 11 days vs 5 days). Egg survival was also variable at low oxygen concentrations (i.e. <0.15 ml l^{-1} oxygen) and it was suggested that the variation might be due to age differences. Subsequently Lutz *et al.* (1994) reported that at oxygen concentrations where initial hatching did not occur, eggs of *A. tonsa* that were within a few hours of hatching survived such exposure better than newly spawned eggs, as evidenced by higher hatches when the eggs were returned to normoxic seawater. When subitaneous eggs of *A. tonsa*, *C. hamatus*, and *L. aestiva* were subjected to anoxia for 32 days (Marcus & Lutz, 1994) only newly spawned eggs of *C. hamatus* and *L. aestiva* survived the 32 day exposure. Eggs of *A. tonsa* were the most sensitive; 50% of the eggs were dead after 4 to 8 days. Again differences in survival were noted for eggs that were at different stages of embryonic development when exposed to the anoxia.. Curiously the results were opposite to those reported by Lutz *et al.* (1994) i.e., newly spawned eggs survived better than those that were fully-developed. Only two other studies examined the effect of anoxia on the long-term survival of subitaneous copepod eggs. Eggs of *Eurytemora affinis* survived 15 days when incubated at 15 °C (Ban & Minoda, 1992) and those of *Acartia clausi* survived 25 days at 17.5 °C (Uye & Fleminger, 1976). Even fewer studies have examined the effect of anoxia on the survival of diapause eggs. Ban & Minoda (1992) isolated eggs of *E. affinis* from bottom sediments and classified them as diapause. When they exposed these eggs to anoxia for 51 days, survival decreased with time, but some survived the 51 day exposure. These results support the hypothesis that diapause eggs are better able to tolerate adverse environmental conditions than subitaneous eggs. Additional support comes from work in progress (Lutz & Marcus, unpub.). Subitaneous and diapause eggs of *C. hamatus* were exposed to anoxia for periods up to six months and then exposed to normoxic conditions. In two separate experiments, the hatching of subitaneous eggs declined to zero after being incubated under anoxia at ambient field temperatures for eight to twelve weeks. On the other hand the hatching of diapause eggs is >80% after 6 months of incubation under anoxia and ambient field temperatures.

The sediments where eggs occur are often sulphitic. Since sulphide is extremely toxic to most organisms, egg survival may be adversely affected. Uye *et al.* (1984) reported low egg numbers for *Acartia erythreae*, *A. pacifica*, and *A. clausi* at stations in Fukuyama Harbor, Japan and suggested that the heavily polluted sediments led to increased egg mortality. It was not clear if the eggs were non-diapause or diapause eggs. Subitaneous eggs of *A. tonsa*, *C. hamatus*, and *L. aestiva* are vulnerable to sulphide and mortality is greater for eggs exposed to anoxia and sulphide than anoxia alone (Marcus, unpub). On the other hand diapause eggs of *C. hamatus* appear to be unaffected by sulphide.

An alternative approach to determining the survival capacity of eggs is in situ monitoring of naupliar recruitment. Hatching of eggs *in situ* has been successfully monitored for freshwater species only (Hairston & De Stasio, 1988; De Stasio, 1989). Using emergence traps inserted into the bottom of a pond, the appearance of nauplii over a period of 3 years was observed and related to the general population growth of the zooplankton community. Similar efforts in the marine system have been attempted, but have been unsuccessful due to turbulence and clogging of the traps (Marcus, pers. obs.). Hairston & De Stasio provided evidence for the existence of a long-term benthic egg bank similar to the seed banks of terrestrial plants. The existence of an egg bank was revealed recently for a marine system. In an anoxic basin in the Pettaquamscutt estuary, RI viable copepod eggs were found at a depth of 22 cm (Marcus *et al.*, 1994). Lead-210 dating of the sediments indicated that the eggs were at least 40 years old.

Summary and recommendations for future work

The occurrence of resting eggs in the sea bed is more widespread than originally believed but much remains to be learned about their geographic distribution, particularly in sub-tropical, tropical, and boreal regions for which there is scant evidence, and about deeper areas of the continental shelf. We need to know why eggs accumulate in certain areas. While the general view is that resting eggs are confined to sediments it would be interesting to look for them in waters overlying deeper portions of the ocean. Conceivably some species produce eggs that are neutrally buoyant at intermediate depths. To support the continued study of world-wide egg distributions it is recommended that an SEM/TEM catalogue of egg morphotypes be developed to facilitate the identification of specimens. In addition, a method that enables the rapid classification of eggs as subitaneous or diapause is essential.

Differences in survival of subitaneous and diapause eggs are similar to those in other organisms. Diapause confers the capacity for long-term survival whereas quiescence enables individuals to survive shorter and less predictable periods of adverse conditions. Given the wide range of development times in the eggs of some species (Chen & Marcus, unpub.) it would be interesting to compare the biochemistry, physiology, and survival capacity of slow and fast developing eggs. Slowly developing eggs might be expected to be more tolerant of conditions in the sea bed as such eggs are likely to spend more time there.

Resting eggs are a critical component of the life cycle of species that disappear entirely from the plankton for portions of the year, but their significance in species that do not disappear is not so obvious. For these species it is particularly important to quantify the contribution of resting eggs to recruitment. Resurgence of the population might still depend on recruitment from resting eggs. Experiments that examine rates of egg burial and resuspension are needed here.

Studies on egg banks in different systems are needed. To know how long eggs can survive in the sediments is essential for predicting responses to environmental change. Efforts should include biochemical, physiological, and genetical analyses. Comparisons between freshwater and marine systems should be undertaken. Studies of eggs from different depths within the sea bed, particularly from unmixed sediments of anoxic basins may yield insights into the evolution of species and the structure of zooplankton communities of the past.

Acknowledgments

I thank F. Chen, L. Stalder, two anonymous reviewers and G. Fryer for their comments on the manuscript.

References

Alekseev, V., 1990. Diapauza rakoobraznykh: Ekologicheskiye i fiziologicheskiye aspekty (Diapause in Crustaceans: Ecological and Physiological Aspects). Nauka Press, Moscow, 143 pp.

Ban, S., 1992a. Effects of photoperiod, temperature, and population density on induction of diapause egg production in *Eurytemora affinis* (Copepoda:calanoida) in Lake Ohnuma, Hokkaido, Japan J. Crust. Biol. 12: 361–367.

Ban, S., 1992b. Seasonal distribution, abundance, and viability of diapause eggs of *Eurytemora affinis* (Copepoda:Calanoida) in the sediment of Lake Ohnuma, Hokkaido. Bull. Plank. Soc. Japan 39: 41–48.

Ban, S. & T. Minoda, 1991. The effect of temperature on the development and hatching of diapause and subitaneous eggs in *Eurytemora affinis* (Copepoda:Calanoida) in Lake Ohnuma, Hokkaido, Japan. Bull. Plank. Soc. Japan, Spec. Vol.: 299–308.

Ban, S. & T. Minoda, 1992. Hatching of diapause eggs of *Eurytemora affinis* (Copepoda:Calanoida) collected from lake-bottom sediments. J. Crust. Biol. 12: 51–56.

Belmonte, G., 1992. Diapause egg production in *Acartia (Paracartia) latisetosa* (Crustacea,Copepoda, Calanoida). Boll. Zool. 59: 363–366.

Boileau, M. G. & P. D. N. Hebert, 1991. Genetic consequences of passive dispersal in pond-dwelling copepods. Evolution 45: 721–733.

Brewer, R. H., 1964. The phenology of *Diaptomus stagnalis* (Copepods:Calanoida): the development and hatching of the egg stage. Physiol. Zool. 37: 1–20.

Champeau, A., 1970. Etude de la vie latent chez les calanoides (Copepodes) charactéristiques des eaux temporaires de Basse-Provence. Ann. Fac. Sci. Marseille 44: 155–189.

Chen, F. & S.-J. Li, 1991. On seasonal distribution and diapause eggs in *Tortanus* from Xiamen waters. Acta Oceanologica Sinica 13: 721–727. (Chinese).

Chen, F. & S.-J. Li, 1992. Seasonal variation of two physiological types of *Tortanus* eggs from Xiamen waters, and their temperature effect. J. appl. Ecol. 3: 62–68. (In Chinese with English summary).

Corner, E. & S. O'Hara, 1986. The Biological Chemistry of Copepods. Clarendon Press, Oxford.

Coull, B. & J. Grant, 1981. Encystment discovered in a marine copepod. Science 212: 342.

Danilevsky, A., 1965. Photoperiodism and Seasonal Development of Insects. 1st English Edition. Oliver & Boyd, Edinburgh.

Danks, H. V., 1987. Insect Dormancy: An Ecological Perspective. Tyrell Press, Gloucester, 439 pp.

Davis, C. C., 1984. Planktonic Copepoda (including Monstrilloida). pp. 67–91. In, Marine Plankton Life Cycle Strategies. K. A. Steidinger and L. M. Walker (eds), CRC Press, Boca Raton.

De Stasio Jr., B. T., 1989. The seed bank of a freshwater crustacean: copepodology for the plant ecologist. Ecology 70: 1377–1389.

Elgmork, K., 1980. Evolutionary aspects of diapause in freshwater copepods. pp. 411–417. In, Evolution and Ecology of Zooplankton Communities. W. C. Kerfoot (ed.). University Press of New England, Hanover.

Fanelli, G., A. Ianora & L. Santella, 1992. Produzione de uova di diapausa nel ciclo biologico del copepode *Acartia latisetosa* (Kriczaguin). Oebelia suppl, XVII: 295–302.

Grice, G. D. & V. Gibson, 1977. Resting eggs in *Pontella meadi* (Copepoda:Calanoida). J. Fish. Res. Bd Can. 34: 410–412.

Grice, G. D. & V. Gibson, 1981. Hatching of eggs of Pontella mediterranea Claus (Copepoda:Calanoida). Vie Milieu 31: 49–51.

Grice, G. D. & T. J. Lawson, 1976. Resting eggs in the marine calanoid copepod, *Labidocera aestiva* Wheeler. Crustaceana 30: 9–12.

Grice, G. D. & N. H. Marcus, 1981. Dormant eggs of marine copepods. Oceanogr. Mar. Biol. annu. Rev. 19: 125–140.

Hairston Jr., N. G. & B. T. De Stasio Jr., 1988. Rate of evolution slowed by a dormant propagule pool. Nature. 336: 239–242.

Hairston Jr., N. G. & W. R. Munns Jr., 1984. The timing of copepod diapause as an evolutionary stable strategy. American Naturalist 123: 733–751.

Hairston Jr., N. G. & E. J. Olds, 1984. Population differences in the timing of diapause:adaptation in a spatially heterogenous environment. Oecologia (Berlin) 61: 42–48.

Hairston Jr., N. G. & E. J. Olds, 1987. Population differences in the timing of diapause: a test of hypotheses. Oecologia 71: 339–344.

Hairston Jr., N. G., E. J. Olds & W. R. Munns Jr., 1985. Bet-hedging and environmentally cued diapause strategies of diaptomid copepods. International Vereinigung fur Theoretische und angewandte Limnologie, Verhandlungen 22: 3170–3177.

Hairston Jr., N. G. & W. E. Walton, 1986. Rapid evolution of a life history trait. Proc. natn. Acad. Sci. USA 83: 4831–4833.

Huys, R. & G. A. Boxshall, 1991. Copepod Evolution. The Ray Society, London, 468 pp.

Ianora, A. & L. Santella, 1991. Diapause embryos in the neustonic copepod *Anomalocera patersoni*. Mar. Biol. 108: 387–395.

Johnson, J. K., 1980. Effects of temperature and salinity on production and hatching of dormant eggs of *Acartia californiensis* (Copepoda) in an Oregon estuary. Fish. Bull. 77: 567–584.

Kasahara, S., S. Uye & T. Onbe, 1974. Calanoid copepod eggs in sea-bottom muds. Mar. Biol. 26: 167–171.

Kasahara, S., S. Uye & T. Onbe, 1975. Calanoid copepod eggs in sea-bottom muds. II. Seasonal cycles of abundance in the populations of several species of copepods and their eggs in the Inland Sea of Japan. Mar. Biol. 31: 25–29.

Landry, M. R., 1975. Dark inhibition of egg hatching of the marine copepod *Acartia clausi* Giesbracht. J. exp. mar. Biol. Ecol. 20: 43–47.

Lindley, J. A., 1986. Dormant eggs of calanoid copepods in sea-bed sediments of the English Channel and southern North Sea. J. Plankton Res. 8: 399–400.

Lindley, J. A., 1990. Distribution of overwintering calanoid copepod eggs in sea-bed sediments around southern Britain. Mar. Biol. 104: 209–217.

Lindley, J. A., 1992. Resistant eggs of the Centropagoidea (Copepoda:Calanoida): a possible preadaptation to colonization of inland waters. J. Crust. Biology 12: 368–371.

Lindley, J. A. & H. G. Hunt, 1989. The distribution of *Labidocera wollastoni* and *Centropages hamatus* in the North Atlantic Ocean and the North Sea in relation to the role of resting eggs in the sediment. In: Ryland, J. S., Tyler, P. A. (eds), Reproduction, genetics, and distributions of marine organisms. Olsen and Olsen, Fredensborg: 407–413.

Lutz, R. V., N. H. Marcus & J. P. Chanton, 1992. Effects of low oxygen concentrations on the hatching and viability of eggs of marine calanoid copepods. Mar. Biol. 114: 241–247.

Lutz, R. V., N. H. Marcus & J. P. Chanton, 1994. Hatching and viability of copepod eggs at two stages of embryological development: anoxic/hypoxic effect. Mar. Biol. 199–204.

Marcus, N. H., 1980. Photoperiodic control of diapause in the marine calanoid copepod *Labidocera aestiva*. Biol. Bull. 159: 311–318.

Marcus, N. H., 1982a. The reversibility of subitaneous and diapause egg production by individual females of *Labidocera aestiva* (Copepoda:Calanoida). Biol. Bull. 162: 39–44.

Marcus, N. H., 1982b. Photoperiodic and temperature regulation of diapause of *Labidocera aestiva* (Copepoda:Calanoida). Biol. Bull. 162: 45–52.

Marcus, N. H., 1984a. Variation in the diapause response of *Labidocera aestiva* (Copepoda:Calanoida) from different latitudes and its importance in the evolutionary process. Biol. Bull. 166: 127–139.

Marcus, N. H., 1984b. Recruitment of copepod nauplii into the plankton: importance of diapause eggs and benthic processes. Mar. Ecol. Prog. Ser. 15: 47–54.

Marcus, N. H., 1987. Differences in the duration of egg diapause of *Labidocera aestiva* (Copepoda:Calanoida) from the Woods Hole, Massachusetts, region. Biol. Bull. 173: 169–177.

Marcus, N. H., 1989. Abundance in bottom sediments and hatching requirements of eggs of *Centropages hamatus* (Copepoda:Calanoida) from the Alligator Harbor region, Florida. Biol. Bull. 176: 142–146.

Marcus, N. H., 1990. Calanoid copepod, cladoceran, and rotifer eggs in seabottom sediments of northern California coastal waters:identification, occurrence, and hatching. Mar. Biol. 105: 413–418.

Marcus, N. H., 1991. Planktonic copepods in a sub-tropical estuary:seasonal patterns in the abundance of adults, copepodites, nauplii, and eggs in the sea bed. Biol. Bull. 181: 269–274.

Marcus, N. H. In press. Population dynamics of copepods in northern California coastal waters: benthic-pelagic coupling. Mar. Biol.

Marcus, N. H. & C. M. Fuller, 1986. Subitaneous and diapause eggs of *Labidocera aestiva* (Copepoda:Calanoida): differences in fall velocity and density. J. exp. mar. Biol. Ecol. 99: 247–256.

Marcus, N. H. & C. M. Fuller, 1989. Distribution and abundance of eggs of *Labidocera aestiva* (Copepoda:Calanoida) in the bottom sediments of Buzzards Bay, Massachusetts USA. Mar. Biol. 100: 319–326.

Marcus, N. H. & R. V. Lutz, 1994. Effects of anoxia on the viability of subitaneous eggs of planktonic copepods. Mar. Biol. 121: 83–87.

Marcus, N. H., R. V. Lutz, W. Burnett & P. Cable, 1994. Age, viability, and vertical distribution of zooplankton resting eggs from an anoxic basin: Evidence of an egg bank. Limnol. Oceanogr. 39: 154–158.

Marcus, N. H. & J. Schmidt-Gengenbach, 1986. Recruitment of individuals into the plankton: the importance of bioturbation. Limnol. Oceanogr. 31: 206–210.

Marcus, N. H. & K. Taulbee, 1992. Potential effects of a resuspension event on the vertical distribution of copepod eggs in the sea bed: a laboratory simulation. Mar. Biol. 114: 249–251.

McMinn, A., C. Bolch & G. Hallegraeff, 1992. Cobricosphaeridium Harland and Sarjeant: Dinoflagellate cyst or copepod egg. Micropaleontology 38: 315–316.

Miller, D. D. & N. H. Marcus, 1994. The effects of salinity and temperature on the density and sinking velocity of eggs of the calanoid copepod *Acartia tonsa* Dana. J. exp. mar. Biol. Ecol. 179: 235–252.

Naess, T., 1991a. Marine calanoid resting eggs in Norway: abundance and distribution of two copepod species in the sediment of an enclosed basin. Mar. Biol. 110: 261–266.

Naess, T., 1991b. Tolerance of marine calanoid resting eggs: effects of freezing, dessication and Rotenone exposure-afield and laboratory study. Mar. Biol. 111: 455–459.

Pertzova, N. M., 1974. Life cycle and ecology of a thermophilous copepod *Centropages hamatus* in the White Sea. Zool. Zh. 53: 1013–1022 (in Russian).

Sazhina, L. I., 1968. On hibernating eggs of marine Calanoida. Zool. Zh. 47: 1554–1556 (in Russian).

Santella, L. & A. Ianora, 1990. Subitaneous and diapause eggs in Mediterranean populations of *Pontella mediterranea* (Copepoda:Calanoida): a morphological study. Mar. Biol. 105: 83–90.

Santella, L. & A. Ianora, 1992. Fertilization envelope in diapause eggs of *Pontella mediterranea* (Crustacean, Copepoda). Molecular Reproduction and Development 33: 463–469.

Sullivan, B. K. & L. T. McManus, 1986. Factors controlling seasonal succession of the copepods *Acartia hudsonica* and *A. tonsa* in Narragansett Bay, Rhode Island: temperature and resting egg production. Mar. Ecol. Prog. Ser. 28: 121–128.

Tauber, C. & M. Tauber, 1981. Insect seasonal life cycles: Genetics and Evolution. Annu. Rev. Ecol. Syst. 12: 281–308.

Uye, S., 1980. Development of neretic copepods *Acartia clausi* and *A. steuri*. I. Some environmental factors affecting egg development and the nature of resting eggs. Bull. Plank. Soc. Japan 27: 1–9.

Uye, S., 1985. Resting egg production as a life history strategy of marine planktonic copepods. Bull. Mar. Sci. 37: 440–449.

Uye, S. & A. Fleminger, 1976. Effect of various environmental factors on egg development of several species of *Acartia* in Southern California. Mar. Biol. 38: 253–262.

Uye, S., S. Kasahara & T. Onbe, 1979. Calanoid copepod eggs in sea-bottom muds. IV. Effects of some environmental factors on the hatching of resting eggs. Mar. Biol. 51: 151–156.

Uye, S., M. Yoshiya, K. Ueda & S. Kasahara, 1985. The effect of organic seabottom pollution on survivability of resting eggs of neretic calanoids. Crustaceana 46: 390–405.

Viitasalo, M., 1992. Calanoid resting eggs in the Baltic Sea: implications for the population dynamics of *Acartia bifilosa* (Copepoda). Mar. Biol. 114: 397–405.

Zhong, X-F. & Y.-C. Xiao, 1992. Resting eggs of *Acartia bifilosa* Giesbrecht and *A. pacifica* Steuer in Jiaozhou Bay. Marine Sci-

ences (Qingdao). No. 5: 55–59. (In Chinese with English summary).
Zillioux, E. J. & J. G. Gonzalez, 1972. Egg dormancy in a neritic calanoid copepod and its implications to overwintering in boreal waters. In, B. Battaglia (ed.), Fifth European marine biology Symposium. Piccin Editore, Padova: 217–230.

Hydrobiologia **320**: 153–159, 1996.
V. R. Alekseev & G. Fryer (eds), Diapause in the Crustacea.

Copepod eggs survive a decade in the sediments of the Baltic Sea

Tarja Katajisto
Finnish Institute of Marine Research, P.O. Box 33, FIN-00931 Helsinki, Finland;
Tvärminne Zoological Station, FIN-10900 Hanko, Finland

Key words: calanoid copepods, resting eggs, long-term viability, the Baltic Sea

Abstract

Many planktonic calanoid copepod species have been proved to spend a part of their life cycle as benthic resting eggs. In addition to avoiding seasonally occurring unfavourable conditions resting stages may also be used as a long-term survival strategy. The aim of this study was to find out for how long eggs of calanoid copepods retain their viability in the sediments of the Baltic Sea. The occurrence of viable copepod eggs in sea bottom sediment was studied in Pojovik Bay, SW coast of Finland. Eggs were found throughout a 25 cm deep core but deeper than 20 cm they were very scarce. Eggs were incubated at 12 °C in order to check the viability, and their age was estimated by determining the sediment accumulation rate with ^{137}Cs-method. Viable eggs were estimated to be 10–13 yrs old, some possibly even 18–19 yrs. Most eggs in the top 8 cm were viable, their age being up to 7–8 yrs. Nauplii that hatched from the eggs belonged to *Acartia bifilosa* and *Eurytemora affinis*, *A. bifilosa* dominating the hatchers of the top sedment layers and *E. affinis* the deeper layers. Preliminary evidence is presented that *E. affinis* produces true diapause eggs in the Baltic Sea.

Introduction

Many planktonic copepod species spend part of their life cycle as benthic resting eggs in both marine (Grice & Marcus, 1981; Uye, 1985) and freshwater environments (Hutchinson, 1967). Resistant eggs have been applied to several life cycle strategies: (1) avoidance of seasonally occurring unfavourable conditions when the population totally disappears from the plankton (e.g. Grice & Gibson, 1975; Kasahara, Uye & Onbé, 1975; Hairston & Munns, 1984; Marcus, 1989), (2) maintenance of a more constant perennial planktonic population by time-released hatching from sediments (Uye, 1980, 1985), (3) colonization of shallow coastal waters where freely released eggs probably fall to the bottom prior to hatching (Lindley, 1992). In addition to these 'short-term' strategies resting stages may also be used as a (4) long-term survival strategy against unpredictable catastrophes; resistant eggs in the sediment thus act as an egg bank, analogous to the seed banks of plants (Hairston & De Stasio, 1988; De Stasio, 1989). A seed pool can greatly reduce the need of a population to respond genetically to all conditions met in cyclic or randomly varying environments (Templeton & Levin, 1979).

Most studies on the survival of calanoid eggs cover time scales needed for surviving an unfavourable season. These include laboratory experiments on diapausing (e.g. Grice & Gibson, 1975; Ban & Minoda, 1992) and subitaneous (e.g. Uye, 1980) eggs, as well as field evidence of populations recovering from benthic eggs after being absent from the plankton for a part of the year (e.g. Kasahara, Uye & Onbé, 1975; Næss, 1991). To act as a seed bank the eggs must retain their viability in the sediment for extended time-periods. Direct evidence of two to three years survival in sediments have been obtained in field and laboratory studies (Hairston & De Stasio, 1988; De Stasio, 1989; Ban & Minoda, 1992), but indirect methods have to be applied to find out if the eggs remain viable for much longer time periods. The long-term survival of copepod eggs has been studied by determining the age of the sediment at depths where viable eggs are found. Marcus *et al.* (1994) showed that calanoid eggs retain their viability for over 40 years in marine sediments; recent results by Hairston *et al.* (1995) indicate that eggs found in more than 200 years old lake sediments may hatch. In the Baltic Sea only a few studies have been made on calanoid resting eggs (Viitasalo, 1992; Viitasalo & Katajisto, 1994). The aim of this study was to determine for how long calanoid resting eggs retain their viability in the sediments of an estuarine bay in the northern Baltic Sea.

Material and methods

The study site, Sällvik deep, is situated in the deepest part (approx. 42 m) of a long narrow sea inlet, Pojo Bay, off the SW coast of Finland, northern Baltic Sea. It is separated from the more open archipelago by a shallow sill (6–7 m). A layer of fresh water mostly originating from the River Mustionjoki at the head of the bay flows seawards on top of more brackish deep water. The resulting halocline, at about 10 m depth, effectively prevents mixing of these layers. Renewal of the deep water is mostly dependent on occasional inflows of dense water over the sill, which most often takes place during winter and spring. The temperature near the bottom is below 1 °C in winter and rises usually to only 4–5 °C in summer; the salinity ranges from 4 to 5.5‰. In late summer and autumn the oxygen concentration of the deep water is very low (Niemi, 1978). Probably due to this, a benthic macrofauna is missing (J. Rissanen, University of Helsinki, pers. comm.). In the lack of bioturbation no mixing of different aged sediment occurs which is also indicated by the lamination of the sediment (pers. obs.). The brown oxygenated layer extends down to *c.* 1–2 cm depth, and beneath that, blackish layers alternate with greyish ones. Age determination of the eggs found in different depth layers of sediment seemed possible by determining the sedimentation rate.

The zooplankton community at Sällvik is a mixture of limnetic and brackish water species. The large calanoid *Limnocalanus macrurus* G.O. Sars occurs almost throughout the year; other more or less frequently occurring calanoids are *Eurytemora affinis* (Poppe) and *Acartia bifilosa* Giesbrecht (Levander, 1915).

Sampling was conducted on 12 November 1992. Two sediment cores (diameter 94 mm, depth 24 to 25 cm) were taken with a 'Limnos' sediment corer, one for egg analysis and one for sediment age determination. The cores were divided into subsamples at 1 cm intervals using a special cutting device of the corer (described in Kansanen *et al.*, 1991). Water to be used in the hatching experiments was taken from 35–40 m depth at the sampling station and was later filtered through a Whatman GF/C filter. At the time of sampling the deep water temperature was 5.7 °C.

The sediment samples for egg analysis were placed in plastic bags and kept in a cool-box until they were stored in a temperature control room at 3 °C. After a month they were transferred to Helsinki in a cool-box and stored at 3–5 °C. The incubations were started 33–39 days after the sampling. The eggs were extracted from the sediments by sugar flotation method introduced by Onbé (1978) and modified by Marcus (1989). Samples were first suspended in artificial seawater (salinity 4–5‰) which was prepared by adding sea salt to distilled water. They were then sonicated for 30 s with a Soniprep 150 cell distruptor and washed with the artificial seawater in a 50 μm sieve. The material remaining on the sieve was washed into a centrifugation tube with sugar solution (1000 g sucrose in 1000 ml distilled water) and centrifuged at 3000 rpm (1268 *g*) for 3 min. Neither sonication (Marcus, 1984; De Stasio, 1989) nor centrifugation in sugar solution (Onbé, 1978) is reported to affect seriously the viability of calanoid eggs. The supernatant was transferred back to the 50 μm sieve and washed carefully with GF/C-filtered seawater from Sällvik. The material on the sieve was transferred into a petri-dish and examined with a dissecting microscope. Eggs were collected with a pasteur pipette and put in wells of a microtitration plate and incubated at 12 °C up to 14 days. Hatched nauplii and unhatched eggs were counted at intervals of a few days during the incubation period. Also the petri-dishes were placed at 12 °C and checked for hatched nauplii and representatives of other zooplankton taxa.

The sediment age was estimated by determining the distribution of radionuclide ^{137}Cs in the sediment core. Subsamples (1 cm slices) from the core were transferred to petri-dishes and stored frozen until analyzed. ^{137}Cs was determined gamma-spectrometrically from freeze dried and homogenized samples by using a multi-channel analyser equipped with a high-purity germanium detector (HPGe, GeM25195).

There is a possibility that some water and eggs from upper sediment layers may slide along the sides of the sampler. Therefore an additional sample was taken in July 1993 to ensure that nauplii hatching from the deepest layers really originated from those layers. A 'Limnos' sampler was used and the subsampling was conducted from three depth layers, 10–13 cm, 14–18 cm and 19–22 cm. Some sediment was carefully taken with a spoon from the middle of the sampler to ensure that no sediment from the sides of the core came into the samples. The samples were sonicated, sieved and centrifuged as described above, and eggs were placed in petri-dishes at 12 °C and room temperature. The petri-dishes were checked for hatched nauplii after 5, 7 and 14 days.

Results

Calanoid eggs were found throughout the 25 cm core. Eggs were not counted but relative abundance estimates were made. In the top 8–9 cm eggs were numerous and most of them were viable (Table 1). Egg abundances of the surface sediment were monitored in a seasonal study during 1992; in November the abundance of calanoid eggs in the top 1–2 cm was 3.7×10^6 m^{-2} (Katajisto, unpubl. data). Fewer eggs occurred in deeper layers and their hatching performance was also low. Deeper than 20 cm eggs were very scarce and appeared to be decaying. None of them hatched (Table 1).

Nauplii hatched from eggs taken from over 10 cm depth, some even from 17.5–19.5 cm (Table 1). However, the total number of nauplii recovered from sediment depths between 10 and 20 cm was only 21 in the incubation experiments, plus some that hatched in the petri-dishes (9.5–11.5 cm samples). As there was a possibility that deeper layers were contaminated with viable eggs from upper layers these results must be viewed with caution. In the additional samples taken in 1993 many nauplii hatched from the 10–13 cm depth interval, which confirms the occurrence of viable eggs in these layers. In contrast, no nauplii hatched from 14–22 cm depth in 1993. Contamination can therefore not be totally excluded as an explanation for the presence of viable eggs in these layers in 1992.

In the ^{137}Cs profile a sudden fall in the concentrations was detected at 8 cm depth, indicating the fallout after the Chernobyl accident in April 1986 just above it (Fig. 1). This suggests a sediment accumulation rate of about 1 cm yr^{-1} below the loose surface sediment. The first two layers (0–2 cm) corresponded probably to the sediment deposition during the year 1992. During another study, the station Sällvik was visited every two weeks in 1992 and thickening of the dark brown surface layer could be observed during this time (pers. obs.). In addition, the dry weight of 1–2 cm thick surface sediment samples taken in November (Katajisto,

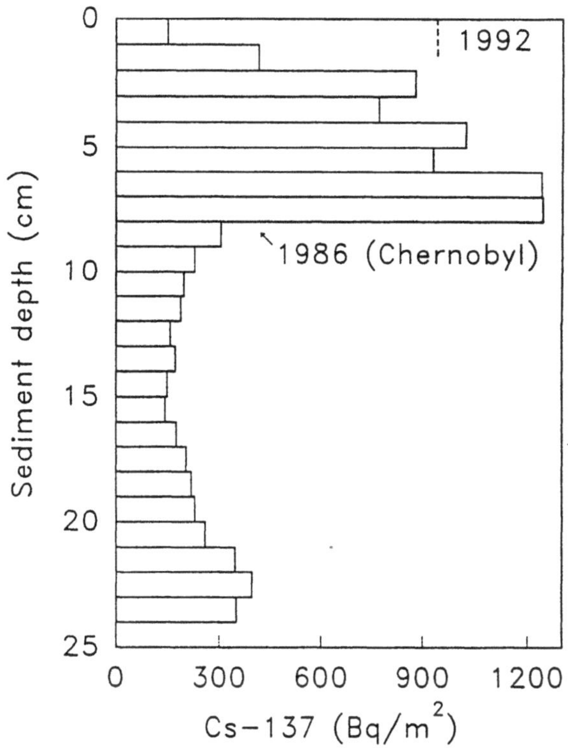

Fig. 1. Cesium-137 concentration in different sediment depth layers at Sällvik 12.11.1992.

unpubl. data) was equal to the cumulative sedimentation of dry matter from June to November measured with sediment traps at 30 m depth (A.-S. Heiskanen, unpubl. data).

The results suggest that the eggs that hatched from the 10–13 cm depth interval were at least 9 to 13 yrs old. If the viable eggs from 17.5–19.5 cm depth did not originate from the upper parts of the core they were at least 17–19 yrs old (Table 1).

Nauplii of *Acartia* sp. and *Eurytemora affinis* were found to hatch from the eggs. The *Acartia* nauplii were probably *A. bifilosa*, which is the only species of the genus in the Pojo Bay (Levander, 1915; M. Koski, unpubl. data). The first naupliar stages of *E. affinis* are difficult to distinguish from those of *Limnocalanus macrurus*, but in some later incubations of Sällvik benthic eggs only *E. affinis* and *A. bifilosa* nauplii have been successfully reared into identifiable later stages (pers. obs.). It can thus be assumed that the *E. affinis/L. macrurus* nauplii were mostly *E. affinis*. Both *A. bifilosa* and *E. affinis* nauplii were found hatching throughout the 'viable part' of the core, but *A. bifilosa* nauplii dominated the hatchers from surface samples (0–1.5 cm) and *E. affinis* the deeper layers (Table 1).

Table 1. Calanoid egg abundance and hatching of eggs exracted from different sediment layers. Abundances are estimated in a relative scale: + + + abundant, ++ plentiful, + present. Hatching is indicated after two incubation intervals. Some of the incubation wells dried out which makes N different for the initial and final incubations and accounts for the missing values (–). Aca, Eur = the number of *Acartia* and *Eurytemora* nauplii that were identified to species among hatchlings. Sediment age was estimated from the Cs-137 profile.

					Initial hatching				Final hatching			
Sediment depth (cm)	Sediment age (estim.) (yrs)	Egg abund.	Aca	Eur	Incubation time (days)	N	Hatched nauplii	%	Total incub. time (days)	N	Hatched nauplii	%
0 - 0.5	0	+++	34	5	6	159	39	25	14	159	45	28
0.5– 1.5	0– 1	+++	140	13	6	318	153	48	–	–	–	–
1.5– 2.5	1– 2	++	8	21	4	67	34	51	11	67	45	67
2.5– 3.5	2– 3	++	15	47	6	98	67	68	–	–	–	–
3.5– 4.5	3– 4	+++	5	57	5	104	88	85	–	–	–	–
4.5– 5.5	4– 5	++	–	31	6	70	42	60	13	49	37	76
5.5– 6.5	5– 6	+++	4	29	6	198	145	73	13	152	133	88
6.5– 7.5	6– 7	++	18	38	6	120	63	53	12	120	103	86
7.5– 8.5	7– 8	+++	–	49	6	241*	85	35	13	255*	176	69
8.5– 9.5	8– 9	++	–	4	6	40	4	10	14	25	10	40
9.5–10.5	9–10	++	–	9	4	57	3	5	11	57	10	18
10.5–11.5	10–11	+	–	5	5	50	4	8	10	50	5	10
11.5–12.5	11–12	++	–	1	6	53	0	0	12	29	1	3
12.5–13.5	12–13	+	1	–	5	21	1	5	–	–	–	–
13.5–14.5	13–14	+			6	27	0	0	14	14	0	0
14.5–15.5	14–15	+	2	4	6	37	6	16	–	–	–	–
15.5–16.5	15–16	+	1	–	6	3	1	33	10	3	1	33
16.5–17.5	16–17	+	3	–	4	22	0	0	9	22	3	14
17.5–18.5	17–18	+	2	1	6	20	3	15	–	–	–	–
18.5–19.5	18–19	+	1	–	6	9	1	11	9	9	1	11
19.5–20.5	19–20	(+)			6	13	0	0	10	13	0	0
20.5–21.5	20–21	(+)			6	2	0	0	10	2	0	0
21.5–22.5	21–22	(+)			6	4	0	0	10	4	0	0
22.5–23.5	22–23	(+)			6	1	0	0	10	1	0	0
23.5–24.5	23–24	(+)			6	1	0	0	10	1	0	0

* A different set of wells was counted after the initial and final incubation periods.

This study was conducted to determine the longevity of calanoid eggs in sediments, but some observations on other zooplankton taxa were also made. Cladoceran ephippia occurred throughout the core; *Daphnia* spp. and *Bosmina* spp. hatched in samples down to 16 cm depth. In the 1993 sampling and incubations, rotifers (mostly *Synchaeta* spp.) hatched in all samples, even from the 19–22 cm depth interval.

Discussion

The results presented here are among the longest records in calanoid egg survival, together with data of Marcus *et al.* (1994) and Hairson *et al.* (1995). These studies indicate that calanoid eggs can survive in anoxic sediments for years, which far exceeds the previous laboratory estimates. Viable eggs were recovered from sediments that had been stored in airtight bags in laboratory for 15 months (Kasahara, Onbé & Kamigaki, 1975). In controlled laboratory experiments subitaneous and diapause calanoid eggs have survived exposure to low oxygen conditions for up to 32 (Marcus & Lutz, 1994) and 51 days (Ban & Minoda, 1992), respectively. The role of oxygen deficiency in the long-term survival of resting eggs is unclear. Although egg viability has been suppressed after exposure to low oxygen conditions (e.g. Uye *et al.*, 1979; Ban & Minoda, 1992; Lutz *et al.*, 1992, 1994; Marcus & Lutz, 1994), it has also been suggested that anoxic or low oxygen conditions might prolong dormancy by slowing down metabolism (Andrew, 1993) and inhibiting hatching (Ban & Minoda, 1992). Eggs survive longer in anoxic conditions at low temperature (e.g. Ban & Minoda, 1992), thus the low temperatures prevailing for most of the year in the deep water of the study area may have prolonged the survival of the eggs in the sediment.

Resting eggs of copepods can be defined as 'diapause' or 'quiescent' eggs (Grice & Marcus, 1981). Diapause refers to a genetically controlled state of arrested development while quiescence is a state of retarded development, induced and maintained by adverse environmental conditions. In diapause, development can only resume after a certain refractory period even in favourable environmental conditions. Most of the eggs incubated from the surface layer (0–0.5 cm) did not hatch in 14 days. However, the eggs proved to be highly viable down to 8 cm depth (Table 1). It seems reasonable to suggest that the recently deposited surface sediment contained viable diapause eggs that were not ready to resume development. In contrast, the eggs in the deeper layers had already been laid in the previous years, and were ready to hatch. Diapause eggs may also require an exposure to low temperature (chilling) before they can hatch (e.g. Sullivan & McManus, 1986). The deep water temperature at Sällvik is at its highest in the autumn and drops to 2 °C or less in winter (Niemi, 1978). Thus a possible requirement for chilling would prevent the eggs from hatching in the autumn and recruiting to the plankton at the onset of winter. During winter, deposition of new eggs into the sediment must be small due to cold temperatures and low abundance of adults in the plankton (M. Koski, unpubl. data). Hence most eggs in surface sediments in spring have probably already been laid in the previous autumn and they should be ready to hatch in suitable conditions. This indeed happened when incubations were conducted with eggs from surface sediment (0–1 cm) taken in June 1994. The total hatching percentage was 93% and after 5 d indubation at 13 °C (Katajisto, unpubl. data).

The unhatched eggs could not be identified to species but several facts suggest that *Eurytemora affinis* produces diapause eggs in the study area. Only a few *E. affinis* hatched from the surface layers (0–1.5 cm) although it was the dominant hatcher in deeper layers (Table 1). In contrast, in the June 1994 incubation 86% of the hatched nauplii were *E. affinis*. *E. affinis* has been shown to produce true diapause eggs in Japan (Ban, 1992). It carries its eggs in an egg sac. Subitaneous eggs remain in sacs until hatching, but diapause eggs are released and fall to the bottom (Ban, 1992). Thus most *E. affinis* eggs that are found in the sediments should be diapause eggs.

As eggs were both abundant and mostly viable down to 8 cm depth it can be concluded that a long-lived calanoid egg pool exists in the sediments of the studied area. Hairston & De Stasio (1988) showed that in a small lake a new copepod generation could be established from the sediment egg pool after total failure of reproduction during two consecutive years. However, in larger water bodies migration may also be used to repopulate areas after local catastrophes. More probably the longevity of the eggs in marine environment acts as a buffer against fluctuations in population density caused by varying reproductive success in different years (cf. Warner & Chesson, 1985).

The longevity of the eggs is not the only factor governing the significance of the benthic egg bank for the planktonic population. Copepod eggs to not hatch while buried in sediment (Uye, 1980; Uye *et al.*, 1979; Kasahara, Onbé & Kamigaki, 1975), thus the fate of eggs deposited on the sediment surface is dependent on factors which affect burial and vertical movement of the eggs in sediments. The rate of sediment accumulation defines how fast eggs are buried to the depth at which probability of getting back to the sediment surface to hatch is minimal. Physical forces affect the resuspension and mixing of sediments, but especially bioturbation by benthic animals has been shown to be essential for the hatching of benthic eggs (Marcus, 1984; Marcus & Schmidt-Gengenbach, 1986). Bioturbation makes sediment more susceptible to resuspension (Rhoads & Young, 1970) and mixes the sediment deeper than physical forces (e.g. Stanford, 1992).

At the study site, Sällvik deep, the sediment accumulation rate is high, so that a few years after deposition eggs get buried under a thick sediment cover. The eggs probably also stay buried due to the lack of sediment mixing by benthic macrofauna. The deepest parts of basins tend to accumulate most eggs (De Stasio, 1989; Ban, 1992; Næss, 1991) as eggs resuspend from the bottom and are transported and deposited again (Marcus, 1984; Lindley, 1990). Thus the 'egg bank' at the study site can result from the accumulation of eggs from a larger area, rapid burial, and low probabilities of hatching soon after burial. However, the longevity of eggs at this particular site is an indication that this trait occurs also in a larger area.

It is difficult to evaluate the importance and functioning of the benthic egg bank in marine environments. The existence of an egg bank was shown at study sites where actual recruitment to the plankton is restricted to the youngest eggs (this study) or does not occur (Marcus *et al.*, 1994). In mixed sediments different aged eggs may contribute to the recruitment to the population, but age-determination of eggs may be impossible. The role and prevalence of egg banks in these mixed sediments clearly call for further studies.

Acknowledgements

I thank T. Sjölund, A.-S. Heiskanen, M. Viitasalo and J. Ekebom for assistance in the field, K. Pellikka for assistance in the laboratory and S. Saesmaa for helping in identification of the nauplii. Tvärminne Zoological Station and the Finnish Institute of Marine Research provided facilities for the sampling and laboratory work. T. Jaakkola from the Dept. of Radiochemistry, University of Helsinki, provided the ^{137}Cs analysis and H. Vahtera prepared the samples for it. The critical comments of M. Viitasalo, T. Onbé and one anonymous reviewer helped to improve the manuscript. Special thanks are due to A.-S. Heiskanen for companionship on board as well as for making the ^{137}Cs analysis possible. The study was financially supported by the Academy of Finland, Maj and Tor Nessling Foundation and Walter and Andrée de Nottbeck Foundation.

References

Andrew, T. E., 1993. The influence of oxygen concentration on the respiration rate of the resting eggs of Leptodora kindti (Focke, 1844) and Bythotrephes longimanus Leydig 1860. Arch. Hydrobiol. 128: 409–414.

Ban, S., 1992. Seasonal distribution, abundance and viability of diapause eggs of Eurytemora affinis (Copepoda: Calanoida) in the sediment of lake Ohnuma, Hokkaido. Bull. Plankton Soc. Japan 39: 41–48.

Ban, S. & T. Minoda, 1992. Hatching of diapause eggs of Eurytemora affinis (Copepoda: Calanoida) collected from lake-bottom sediments. J. Crustacean Biol. 12: 51–56.

De Stasio, B. T., Jr., 1989. The seed bank of a freshwater crustacean: copepodology for the plant ecologist. Ecology 70: 1377–1389.

Grice, G. D. & V. R. Gibson, 1975. Occurrence, viability and significance of resting eggs of the calanoid copepod Labidocera aestiva. Mar. Biol. 31: 335–337.

Grice, G. D. & N. H. Marcus, 1981. Dormant eggs of marine copepods. Oceanogr. Mar. Biol. annu. Rev. 19: 125–140.

Hairston, N. G., Jr. & B. T. De Stasio, Jr., 1988. Rate of evolution slowed by a dormant propagule pool. Nature 336: 239–242.

Hairston, N. G., Jr. & W. R. Munns, Jr., 1984. The timing of copepod diapause as an evolutionary stable strategy. Am. Nat. 123: 733–751.

Hairston, N. G., Jr., R. A. Van Brunt, C. M. Kearns & D. R. Engstrom, 1995. Age and survivorship of diapausing eggs in a sediment egg bank. Ecology 76: 1706–1711.

Hutchinson, G. E., 1967. A treatise in limnology. 2. Introduction to lake biology and the limnoplankton. John Wiley & Sons, New York, 1115 pp.

Kansanen, P. H., T. Jaakkola, S. Kulmala & R. Suutarinen, 1991. Sedimentation and distribution of gamma-emitting radionuclides in bottom sediments of southern Lake Päijänne, Finland, after the Chernobyl accident. Hydrobiologia 222: 121–140.

Kasahara, S., T. Onbé & M. Kamigaki, 1975. Calanoid copepod eggs in sea-bottom muds. III. Effects of temperature, salinity, and other factors on the hatching of resting eggs of Tortanus forcipatus. Mar. Biol. 31: 31–35.

Kasahara, S., S. Uye & T. Onbé, 1975. Calanoid copepod eggs in sea-bottom muds. II. Seasonal cycles of abundance in the populations of several species of copepods and their eggs in the inland sea of Japan. Mar. Biol. 31: 25–29.

Levander, K. M., 1915. Zur Kenntnis der Bodenfauna und des Planktons der Pojowiek. Fennia 35: 1–39.

Lindley, J. A., 1990. Distribution of overwintering calanoid copepod eggs in sea-bed sediments around southern Britain. Mar. Biol. 104: 209–217.

Lindley, J. A., 1992. Resistant eggs of the Centropagoidea (Copepoda: Calanoida): a possible preadaptation to colonization of inland waters. J. Crustacean Biol. 12: 368–371.

Lutz, R. V., N. H. Marcus & J. P. Chanton, 1992. Effects of low oxygen concentrations on the hatching and viability of eggs of marine calanoid copepods. Mar. Biol. 114: 241–247.

Lutz, R. V., N. H. Marcus & J. P. Chanton, 1994. Hatching and viability of copepod eggs at two stages os emryological development: anoxic/hypoxic effect. Mar. Biol. 119: 199–204.

Marcus, N. H., 1984. Recruitment of copepod nauplii into the plankton: importance of diapause eggs and benthic processes. Mar. Ecol. Prog. Ser. 15: 47–54.

Marcus, N. H., 1989. Abundance in bottom sediments and hatching requirements of eggs of Centropages hamatus (Copepoda: Calanoida) from the Alligator Harbor region, Florida. Biol. Bull. 176: 142–146.

Marcus, N. H. & R. Lutz, 1994. Effects of anoxia on the viability of subitaneous eggs of planktonic copepods. Mar. Biol. 121: 83–87.

Marcus, N. H. & J. Schmidt-Gengenbach, 1986. Recruitment of individuals into the plankton: the importance of bioturbation. Limnol. Oceanogr. 31: 206–210.

Marcus, N. H., R. Lutz, W. Burnett & P. Cable, 1994. Age, viability, and vertical distribution of zooplankton resting eggs from an anoxic basin: evidence of an egg bank. Limnol. Oceanogr. 39: 154–158.

Næss, T., 1991. Marine calanoid resting eggs in Norway: abundance and distribution of two copepod species in the sediment of an enclosed marine basin. Mar. Biol. 110: 261–266.

Niemi, Å., 1978. Ecology of phytoplankton in the Tvärminne area, SW coast of Finland. III. Environmental conditions and primary production in Pojoviken in the 1970s. Acta Bot. Fennica 106: 1–28.

Onbé, T., 1978. Sugar flotation method for sorting the resting eggs of marine cladocerans and copepods from sea-bottom sediment. Bull. Jap. Soc. Sci. Fish. 44: 1411.

Rhoads, D. C. & D. K. Young, 1970. The influence of deposit-feeding organisms on sediment stability and community trophic structure. J. Mar. Res. 28: 150–178.

Sanford, L. P., 1992. New sedimentation, resuspension, and burial. Limnol. Oceanogr. 37: 1164–1178.

Sullivan, B. K. & L. T. McManus, 1986. Factors controlling seasonal succession of the copepods Acartia hudsonica and A. tonsa in Narragansett Bay, Rhode Island: temperature and resting egg production. Mar. Ecol. Prog. Ser. 28: 121–128.

Templeton, A. R. & D. A. Levin, 1979. Evolutionary consequences of seed pools. Am. Nat. 114: 232–249.

Uye, S., 1980. Development of neritic copepods Acartia clausi and A. steueri. I. Some environmental factors affecting egg development and the nature of resting eggs. Bull. Plankton Soc. Japan 27: 1–9.

Uye, S., 1985. Resting egg production as a life history strategy of marine planktonic copepods. Bull. mar. Sci. 37: 440–449.

Uye, S., S. Kasahara & T. Onbé, 1979. Calanoid copepod eggs in sea-bottom muds. IV. Effects of some environmental factors on the hatching of resting eggs. Mar. Biol. 51: 151–156.

Viitasalo, M., 1992. Calanoid resting eggs in the Baltic Sea: implications for the population dynamics of Acartia bifilosa (Copepoda). Mar. Biol. 114: 397–405.

Viitasalo, M. & T. Katajisto, 1994. Mesozooplankton resting eggs in the Baltic Sea: identification and vertical distribution in laminated and mixed sediments. Mar. Biol. 120: 455–465.

Warner, R. R. & P. L. Chesson, 1985. Coexistence mediated by recruitment fluctuations: a field guide to the storage effect. Am. Nat. 125: 769–787.

Hydrobiologia **320**: 161–168, 1996.
V. R. Alekseev & G. Fryer (eds), Diapause in the Crustacea.

Benthic resting eggs of calanoid copepods in Norwegian enclosures used in mariculture: abundance, species composition and hatching

Tore Næss
Institute of Marine Research, Austevoll Aquaculture Research Station, N-5392 Storebø, Norway

Key words: benthic copepod resting eggs, marine ponds, Norway

Abstract

Abundance, species composition and viability of benthic resting eggs of calanoid copepods were investigated in the sediments of seven Norwegian enclosed pond systems used in mariculture as well as in the sediments of the nearby open sea. The ponds represented different types of morphometry and zooplankton communities and were located from approximately 60 to 70 °N. Differences in environmental conditions in deep water were also typical with anoxia and the presence of hydrogen sulphide at most of the localities. The lowest and highest egg densities in the enclosures were $7.5 \pm 6.4 \times 10^4$ and $5.5 \pm 2.5 \times 10^6$ m^{-2}, respectively, higher than in the sediments of the nearby open sea. Highest densities were found at larger or median depths. Altogether, egg densities higher than 10^3 m^{-2} were found of only four calanoid species; *Eurytemora affinis, Acartia teclae, Acartia clausi, Temora longicornis.* Hatching success was variable (0–92%), and no significant correlation to the presence of anoxia and hydrogen sulphide was found. This high hatching success was achieved with eggs exposed to hydrogen sulphide for several months. The recruitment potential to the plankton of nauplii was estimated from 4.8 to 1682 nauplii l^{-1} for the different localities. The results showed that resting egg production is a common life history characteristic of some neritic calanoid species in Norwegian waters. In addition to copepod eggs, low densities ($< 10^3$ m^{-2}) of eggs from rotifers and 'cladocerans' were found in most of the localities.

Introduction

During the last 15 years the use of enclosed pond systems in the production of marine cold water fish has developed and is now a common method in Norway. Such ponds are operated to maintain a high production of zooplankton used as feed for larval fish. Recent studies have documented densities of several millions per square meter of calanoid copepod resting eggs in the sediments of such a system (Næss, 1991a, b). In similar systems, a sudden peak in copepod nauplii populations during early spring has been observed as a yearly event, which strongly suggests that hatching of eggs from the sediment occurs. Copepod resting egg production is now well documented as a part of the life cycle of several genera of neritic marine calanoid copepods of the northern hemisphere (reviewed by Grice & Marcus, 1981; Uye, 1985). This study is a further evaluation of the occurrence of the phenomenon in Norwegian waters, especially in enclosed marine ecosystems used in mariculture. Uye *et al.* (1984) have documented a negative correlation between viability of copepod eggs and the degree of organic pollution i.e. presence of anoxia and hydrogen sulphide. Since the present localities differed in oxygen conditions and presence of hydrogen sulphide in the deeper parts, this factor has been given special attention. Recent studies have verified the importance of oxygen conditions on hatching and viability of calanoid copepod eggs (Lutz *et al.*, 1992, 1994).

Material and methods

Localities

The investigation took place in November 1992, at a time of the year when the seasonal egg production has

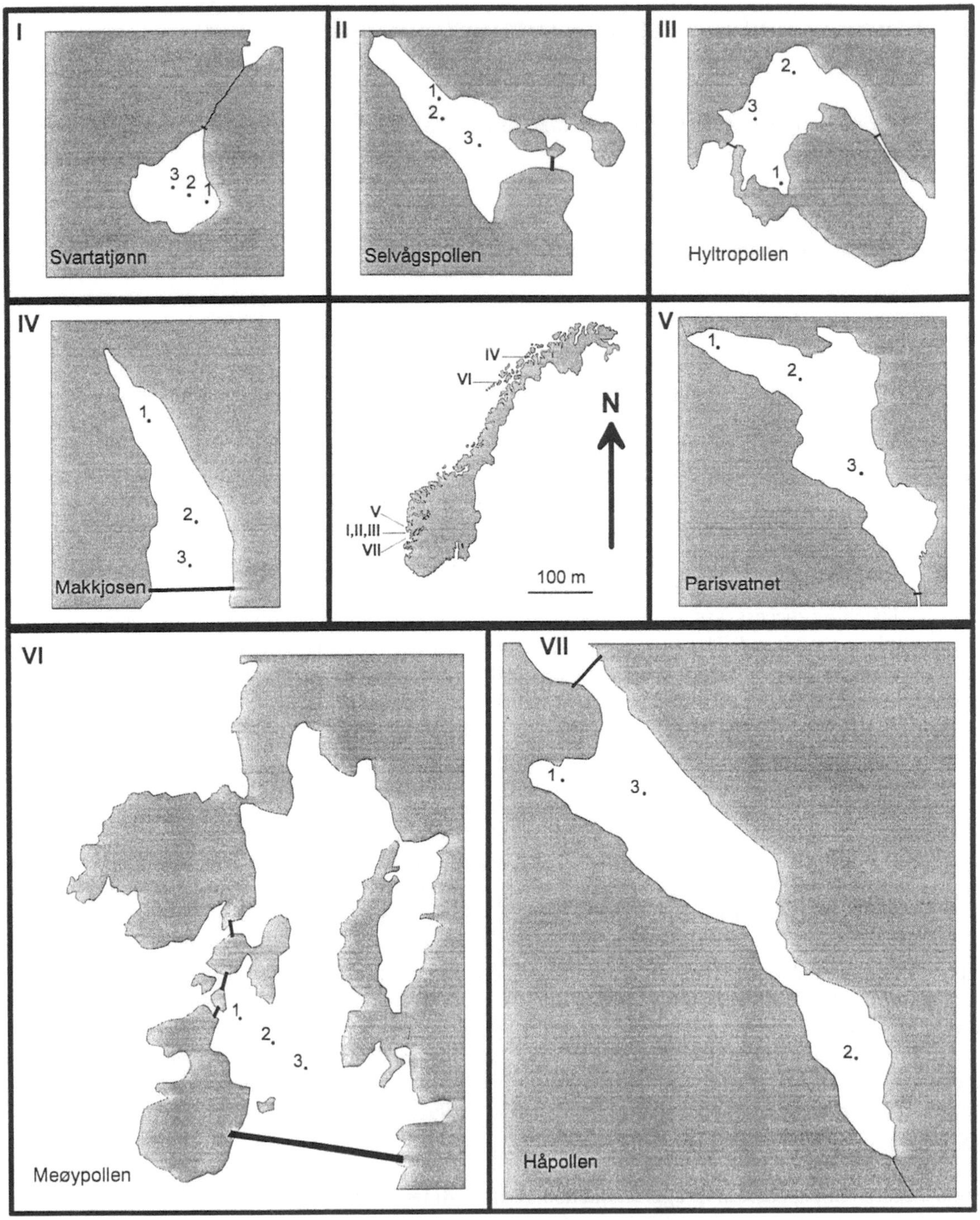

Fig. 1. Localization and morphology of studied pond systems. Sampling stations are indicated.

Table 1. Characteristics of the studied localities.

Locality	Geographical localization	Volume (m^3)	Surface area (m^2)	Max. depth (m)	Drained	Rotenone treated	First occ. mariculture	Dominating copepod species
Svartatjønn	60.06 °N 0.516 °E	22 000	9 700	3.5	Yes	Yes	1984	*Acartia teclae* *Eurytemora affinis* *Centropages hamatus*
Strønopollen	60.09 °N 05.22 °E	53 500	17 000	6	Yes	No	1930*	*Temora longicornis* *Acartia teclae*
Hyltropollen	60.04 °N 05.15 °E	60 000	20 000	6	Variable	Yes	1980	*Calanus finmarchicus* *Acartia clausi*
Makkjosen	69.51 °N 18.52 °E	37 500	20 500	3	Partly	Yes	1987	*Calanus finmarchicus* *Eurytemora affinis*
Parisvannet	60.38 °N 44.9 °E	270 000	50 000	10	No	Yes	1987	*Temora longicornis* *Centropages hamatus*
Meøypollen	68.75 °N 12.47 °E	250 000	103 000	4.5	Partly	Yes	1991	*Calanus finmarchicus* *Oithona similis*+
Håpollen**	59.51 °N 05.15 °E	500 000	73 000	17	No	No	1988	*Temora longicornis* *Oithona similis*+

* For oyster production.
** Closed with concrete in 1992.
\+ Cyclopoida.

declined. Sampling was conducted in seven marine pond systems located in Western and Northern Norway, from approximately 60 to 70 °N. They represented different types of morphometry and zooplankton species composition as well as operational procedures (draining, sea water supply, rotenone treatment (a toxin), turbulence and fertilization) due to their use in mariculture (Fig. 1, Table 1). Water exchange with the open sea was operated either through filter sieves in the concrete dam, which allowed surface water exchange, or deep water supply through pumping. Five localities were used exclusively as zooplankton production units, while in Meøypollen and Parisvannet, the water column was used directly as feeding habitat for released cod larvae (*Gadus morhua*) during spring and summer. The two ponds located in Northern Norway, Makkjosen and Meøypollen, were characterized by a sandy, oxygenated sediment in all parts, in contrast to the presence of an anoxic, hydrogen sulphide-contaminated, black mud in deeper parts of the others. The depths of the sampling stations are presented in Table 2. Hyltropollen and Håpollen consisted morphologically of two or more basins and sampling station 2 represented the deepest area of a second internal basin. The other localities consisted of one basin with continuous deeper parts. Before the man-made closing with concrete, Makkjosen and Meøypollen were geomorphologically more closely connected to open water areas than the others which were more or less natural enclosures (Fig. 1). Svartatjønn was a former freshwater lake, converted into a marine ecosystem in 1984 (cf. Næss, 1991b). The localities are henceforth referred to by the two first letters of their names.

Methods

Sediment samples were taken with a Kajak-type core sampler (Kajak, 1971), inner diameter 46 mm. Three replicate samples were taken from three separate sampling stations in each pond, representing different depth areas. In addition, 3 replicate samples were taken of the sediment of the nearby open sea to evaluate egg densities at similar depths in more natural environments of that region. The sediment cores were pushed upwards in the sampler using a piston and the sediment layers, 0–1 and 1–3 cm, were individually placed in 250 ml plastic boxes and kept at < 4 °C and in darkness until analysis more than one month later to ensure the termination of the refractory phase (Grice & Marcus, 1981). In the laboratory, the samples were

Table 2. Sampling stations. Water depth and occurrence of H_2S in the sediment. ox = oxygenated sediment, + = low concentration of H_2S, ++ = high concentration of H_2S.

Sampling station	SV		ST		HY		MA		PA		ME		HÅ	
	Depth	H_2S	Depth	H_2S	Depth	H_2S	Depth	H_2S	Depth	H_2S	Depth	H_2S	Depth	H_2S
Stn. 1	1	+	1.5	ox	2	+	1	ox	1	ox	1	ox	3–4	ox
Stn. 2	2–2.5	+	3.5	ox	4	++	2	ox	5–6	ox	2	ox	7–8	++
Stn. 3	3.5	++	6	++	6	++	3	ox	10	ox	4	ox	15	++
Stn. Sea	5	ox	3	ox	5	ox	–	–	5–6	ox	5	ox	–	–

washed through 350, 150 and 60 μm plankton screens. Typically, the material on the 150 and 60 μm screens were added to tubes containing a saturated solution (1/1) of sucrose and distilled water and centrifuged at 3000 rpm for 5 min (cf. Onbé, 1978). However, in the preparation of the first replicate or in all samples from ponds inhabited by *Eurytemora affinis*, all the size fractions were analyzed because of the nature of egg sac formation in this species and the observation of whole or partial egg sacs in pilot samples. After centrifugation, the material remaining in suspension was washed through a 60 μm sieve and transferred to a counting apparatus for identification and counting under a dissecting microscope.

To evaluate egg viability, eggs of dominant species from selected stations were individually added to sterile multiwell dishes (Nunc, Denmark), which contained 2 ml of 0.2 μm filtered 25 ppt autoclaved seawater, and placed in 24 h light at 10 ± 1 °C for daily hatching analysis. The hatching percentages given are synonymous with eggs hatching during the first ten days. This procedure has earlier shown to be beneficial for hatching analysis of such eggs (Næss & Bergh, 1994).

The recruitment potential of resting eggs to the plankton was calculated by dividing the pond systems in three separate depth layers, limited by the isolines of the mean depth between two sampling stations, and the egg data from the included sampling station was used as mean egg densities of that layer. Integration over the three layers (1) gave the total number of eggs in the locality. Assuming a synchronous hatching of all eggs, this number is synonymous with the potential amount of nauplii. Corresponding nauplii densities are calculated according to (2):

$$E_{tot} = \sum_{i=1}^{3} A_i * D_i , \qquad (1)$$

$$D_{naupl} = E_{tot}/V_{tot} . \qquad (2)$$

E_{tot} = number of eggs in a pond system;
i = number of sampling depths;
A_i = area of depth layer i;
D_i = egg density of layer i;
D_{naupl} = nauplii density assuming homogeneous distribution;
V_{tot} = water volume of locality.

Oxygen conditions in HY and SV were analysed using the Winkler method. In the other localities portable oxygen meters were used. The occurrence of H_2S was typically identified by smelling the sediment cores. However, in SV and HY, the redox potential, Eh, was also measured using a pH-meter with a platinum electrode according to the recommendations given in Fenchel & Hemmingsen (1974).

Results

Oxygen and hydrogen sulphide

At sampling, oxygenated water was found close above the sediment surface of all the pond systems at Stn. 1 and of all but one, HY, over the sediment at Stn. 2 (Table 3). The redox potential of the sediment pore water showed weak reducing conditions (0 to −100 mV) at Stn. 1 and 2 of SV, and high reducing conditions (< −150 mV) at Stn. 3 in SV and HY in agreement with the hydrogen sulphide smell test (Table 3). Occurrence of hydrogen sulphide was also common in three other pond systems, but not in the two northern ponds, MA and ME (Table 2). Oxygenated sediment surface was found in all samples from outside waters (Stn. Sea).

Table 3. Oxygen concentration close above the sediment surface and redox potential of the pore water at the sampling stations of SV and HY.

Locality		Stn. 1		Stn. 2		Stn. 3	
		O_2 (ml l^{-1})	red-ox (mV)	O_2 (ml l^{-1})	red-ox (mV)	O_2 (ml l^{-1})	red-ox (mV)
SV	1 cm	4.3	-27 ± 19	0.7	-40 ± 21	0.0	-170 ± 10
	2–3 cm		-60 ± 9		-70 ± 7		-162 ± 2
HY	1 cm	10.2		0.0		0.0	-153 ± 28
	2–3 cm						-200 ± 4

Benthic resting eggs

The separate sediment layers, 0–1 and 1–3 cm, were individually analyzed. However, due to large internal differences with no systematic correlation between the number of eggs found in the different layers, egg densities (Table 4, 5) are given as total numbers in the upper 3 cm.

Benthic copepod resting eggs were found at all samplings in all the pond systems, but at very variable densities (Table 4). Typically, big differences in the number of eggs between replicates were found, indicating a nonhomogeneous distribution of eggs in the sediment. Highest egg densities were generally found in the deeper parts, and decreased with decreasing depth. The highest observed egg densities at Stn. 3 were found in SV, $5.5 \pm 2.5 \times 10^6$ eggs m^{-2} and the lowest in MA, $7.5 \pm 6.4 \times 10^4$ eggs m^{-2}. Altogether, egg densities higher than 10^4 m^{-2} were found for four calanoid species; *Eurytemora affinis*, *Acartia teclae*, *Acartia clausi* and *Temora longicornis* (Table 5). Eggs from *E. affinis* were found to be numerous in four of the seven locations, while the highest recorded egg densities originated from *Acartia teclae* ($4.4 \pm 2.2 \times 10^6$ m^{-2}) and *Temora longicornis* ($2.9 \pm 2.7 \times 10^6$ m^{-2}). In addition to the species listed in Table 4, low densities ($< 10^3$ m^{-2}) of eggs from other calanoids and from cladocerans and rotifers were found in most samples. Large differences in egg abundance also appeared in the estimated recruitment potential of nauplii to the plankton (Table 4). In three of the pond systems, a synchronous hatching of all eggs corresponded to nauplii densities more than 100 per litre, highest in SV with 1682 nauplii per litre.

In the sediments of the nearby open sea (Stn. Sea), copepod resting eggs were found at three of six analyzed localities, at densities of 10^3–10^4 eggs m^{-2}. At two of the stations with no observed eggs at all (outside SV and MA), high densities of benthic harpacticoid copepods were found associated with the sediment surface.

Hatching success

The hatching success of the eggs showed large variations (Table 6). Highest hatching success was achieved with the eggs of *Temora longicornis* from Stn. 3 in HÅ (91.7%), lowest for *Acartia clausi* from Stn. 3 in HY (0.0%). Large differences in hatching percentages were also found within the same species from different localities, e.g. for *A. clausi* in HY and PA. For *Eurytemora affinis*, the most widespread species, the hatching success varied between 30.4 and 79.2%.

Discussion

This study showed that a benthic resting egg stage as a part of the life cycle is common for some neritic calanoid copepods in Norwegian waters, thus generalizing the earlier findings from the basin, SV (Næss, 1991b). Maximum egg densities found in this study, i.e. millions per square meter, are of the same order of magnitude as reported from shallow areas of other parts of the northern hemisphere (Kasahara *et al.*, 1975; Marcus, 1984; Uye *et al.*, 1984).

No significant correlation between the number of benthic eggs and the morphometry of the systems (max. depth, volume, area) could be detected in the present material. However, the two ponds directly used as feeding habitat for cod larvae both had relatively low egg densities. Care should be taken, however, to use data from just one year, as egg densities of one species can vary by more than an order of magnitude from one year to another (author's unpublished data). One surprising finding was the relatively low egg densities

Table 4. Densities of benthic copepod eggs (numbers m^{-2}) in the upper 3 cm of the sediment, estimated total number of eggs in localities, corresponding potential nauplii densities and area of depth layers (m^2) used in the estimation.

Locality		Stn. 1	Stn. 2	Stn. 3	Stn. Sea	Total number of eggs	Nauplii per litre
SV		$1.1 \pm 0.7 \times 10^6$	$2.4 \pm 0.8 \times 10^6$	$5.5 \pm 2.5 \times 10^6$	0	3.7×10^{10}	1682
	Area	2.610	1410	5680			
ST		$1.0 \pm 0.3 \times 10^4$	$7.2 \pm 5.1 \times 10^3$	$1.2 \pm 0.7 \times 10^5$	$9.2 \pm 0.9 \times 10^3$	5.1×10^8	9.5
	Area	8.262	5.542	3.196			
HY		$1.9 \pm 1.2 \times 10^5$	$2.2 \pm 0.6 \times 10^5$	$2.1 \pm 0.9 \times 10^5$	$1.0 \pm 0.4 \times 10^4$	1.2×10^{10}	200
	Area	41.580	13.020	5.400			
MA		$1.1 \pm 0.3 \times 10^3$	$1.1 \pm 0.3 \times 10^3$	$7.5 \pm 6.4 \times 10^4$	*	1.9×10^8	5.1
	Area	6.908	11.275	2.316			
PA		$2.1 \pm 0.9 \times 10^4$	$2.9 \pm 3.1 \times 10^5$	$3.4 \pm 3.0 \times 10^5$	$6.5 \pm 4.1 \times 10^4$	1.2×10^{10}	44.4
	Area	11.850	19.800	18.400			
ME		$3.0 \pm 4.2 \times 10^3$	$1.9 \pm 2.1 \times 10^3$	$8.0 \pm 1.0 \times 10^4$	0		
	Area	75.500	15.450	12.050		1.2×10^9	4.8
HÅ		$8.9 \pm 4.2 \times 10^4$	$3.1 \pm 3.1 \times 10^6$	$2.9 \pm 2.7 \times 10^6$	*		
	Area	35.000	22.000	16.000		1.2×10^{11}	240

* No sampling.

Table 5. Egg abundance of different species at Stn. 3.

	Species							
	E. affinis	*T. longicornis*	*A. teclae*	*A. clausi*	*C. hamatus*	*Podon* spp.	*Evadre nordmanni*	Rotifera
SV	$7.8 \pm 3.7 \times 10^5$		$4.6 \pm 2.2 \times 10^6$		+		+	+
ST		$1.2 \pm 0.7 \times 10^5$						
HY	$1.4 \pm 0.4 \times 10^5$	+		$3.5 \pm 0.8 \times 10^4$		+	+	+
MA	$7.5 \pm 6.4 \times 10^4$			+			+	
PA				$3.4 \pm 3.0 \times 10^5$		+		+
ME	$0.8 \pm 1.0 \times 10^5$			+		+	+	+
HÅ		$2.9 \pm 2.7 \times 10^6$	+				+	

+ = egg densities $< 10^3$ pr. m^2.

found in ST. Typically, this pond has a very large and synchronous spring population of *Temora longicornis*, undoubtly originating from resting eggs. In spring 1992 during a warm period, pumping of colder water from outside, used as a temperature regulator, resulted in a dramatic dilution of the *T. longicornis* population. This probably occurred prior to the seasonal production of resting eggs.

The higher egg densities in most of the enclosed systems compared to the open sea, was probably due to the species composition, with a dominance of species with resting egg production, a higher zooplankton production (more females) due to the operational procedures (Naas *et al.*, 1991) and probably also lower predation in the sediments due to anoxia and predator removal by rotenone (cf. Næss, 1991a).

Only a few species produced resting eggs in the localities and high egg densities were found only in species belonging to the studied, Temoridae or Acartidae, each represented by two species. These were among the dominant species in all of seven localities despite the different geographical and climatic conditions. In addition to the species listed in Table 2, seasonal populations of many other neritic calanoids were present, e.g. *Pseudocalanus minutus*, *Centropages typicus*, *Acartia longiremis*, *Paracalanus parvus*. These probably did not produce resting eggs.

Table 6. Hatchablity (%) of eggs of dominant species on selected stations and in upper parts of the sediment. – denotes missing values, open space = eggs not present.

		Species							
		E. affinis		*T. longicornis*		*A. teclae*		*A. clausi*	
		0–1 cm	1–3 cm	0–1 cm	1–3 cm	0–1 cm	1–3 cm	0–1 cm	1–3 cm
SV	Stn. 1	–	52.2			–	79.2		
	2	–	45.8			79.2	75.0		
	3	79.2	58.3			50.0	70.8		
HY	Stn. 1	50.0	30.4					12.5	65.0
	3	45.8	54.2					0.0	0.0
PA	Stn. 2							66.6	75.0
	3							–	70.8
ME	Stn. 3	72.7	–						
HÅ	Stn. 3			70.8	91.7				

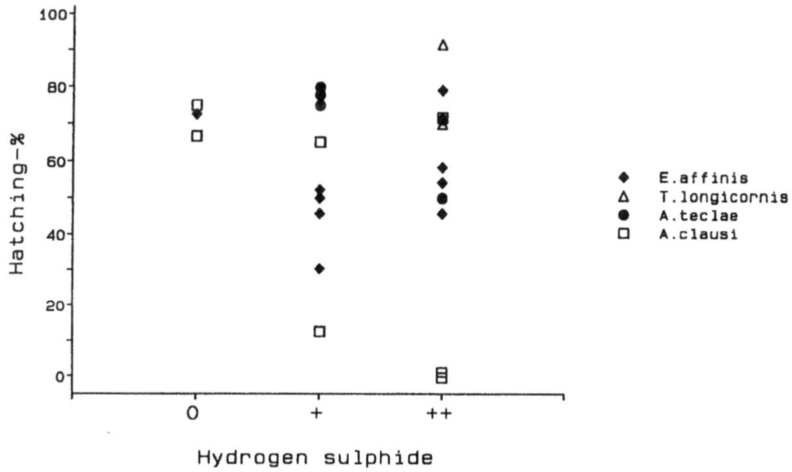

Fig. 2. Hatching success of benthic eggs of most common species relative to hydrogen sulfide concentration. 0 denotes oxic sediment, + = low concentration of H_2S, ++ = high concentration of H_2S.

Because of the large individual differences and unsystematic hatching data from some sediment cores, the observed hatching percentages (Table 6) were not used as correlative in the calculation of nauplii potential. Consequently, the presented data for this (Table 4) were higher than real values. However, even an estimated hatching success as low as 50% will give a considerable nauplii potential in at least three of the localities.

In the present study anoxia and presence of hydrogen sulphide was a general characteristic of deeper parts of most of the ponds during parts of the season. Generally, pore water of muddy sediments is known to be anoxic below 5 mm from the sediment water interface (e.g. Revbech *et al.*, 1980). Uye *et al.* (1984) found a negative correlation between egg viability and anoxic conditions in the sediments. A plot of hatching success versus the presence of anoxia and hydrogen sulphide in the sediments (Fig. 2), verified no clear connection. However, some inferences could be drawn. The few hatching observations of eggs from oxygenated sediments all showed relatively high hatching abilities. In anoxic sediments containing hydrogen sulphide, eggs showed both high and low hatching success. Some differences between species were observed. Considering the high viability of eggs from *E. affinis*, *T. longicornis* and *A. teclae* exposed to high levels of hydrogen sulphide, eggs of these species must be able to withstand this toxin for several months. The data for *A. clausi* indicate a lesser tolerance which is in agreement with the findings of Uye *et al.* (1984). Interspecific differences in the tolerance of resting eggs to low oxygen levels and anoxia were also shown by Uye *et al.* (1984) and Lutz *et al.* (1992), but hydrogen sulphide has been assumed to be toxic (Uye *et al.*, 1984). All eggs used for hatching analysis in the present study were several months old. The oldest were probably those of the two Temoridae species, which typically disappear from the plankton during early summer in these locations.

Acknowledgements

The study was initiated by a proposal from the operators of the pond systems and they are all acknowledged for assistance and information. I thank Aengus Parsons for improving the English. Torbjørn Lønøy is thanked for field and laboratory assistance and Pia Kupka Hansen for equipment and instructions regarding redox measurements.

References

Fenchel, T. & B. B. Hemmingsen, 1974. Manual of Microbial Ecology. Akademisk Forlag.

Grice, G. D. & N. H. Marcus, 1981. Dormant eggs of marine copepods. Oceanogr. mar. Biol. A. Rev. 19: 125–140.

Kajak, Z., 1971. Benthos of standing water, Chap. 3. In W. T. Edmondson & G. G. Winberg (eds), A manual on methods for the assessment of secondary productivity in fresh waters. IBP Handbook no. 17. Blackwell Scientific Publications, Oxford: 25–65.

Kasahara, S., S. Uye & T. Onbé, 1975. Calanoid copepod eggs in sea-bottom muds. II. Seasonal cycles of abundance in the populations of several species of copepods and their eggs in the inland Sea of Japan. Mar. Biol. 31: 25–29.

Lutz, R. V., N. H. Marcus & J. P. Chanton, 1994. Hatching and viability of copepod eggs at two stages of embryological development: anoxic/hypoxic effect. Mar. Biol. 119: 199–204.

Lutz, R. V., N. H. Marcus & J. P. Chanton, 1992. Effects of low oxygen concentrations on the hatching and viability of eggs of marine calanoid copepods. Mar. Biol. 114: 241–247.

Marcus, N. H., 1984. Recruitment of copepod nauplii into the plankton: importance of diapause eggs and benthic processes. Mar. Ecol. Prog. Ser. 15: 47–54.

Naas, K. E., T. van der Meeren & D. L. Aksnes, 1991. Plankton succession and larval responses to manipulations in a marine basin for larval fish rearing. Mar. Ecol. Prog. Ser. 74: 161–173.

Næss, T., 1991a. Tolerance of marine calanoid resting eggs: effects of freezing, desiccation and rotenone exposure – a field and laboratory study. Mar. Biol. 111: 455–459.

Næss, T., 1991b. Marine calanoid resting eggs in Norway: abundance and distribution of two copepod species in the sediment of an enclosed marine basin. Mar. Biol. 110: 261–266.

Næss, T. & Ø. Bergh, 1994. Calanoid copepod resting eggs can be surface-disinfected. Aquacult. Engin. 13: 1–9.

Onbé, T., 1978. Sugar flotation method for sorting the resting eggs of marine cladocerans and copepods from sea-bottom sediment. Bull. Jap. soc. Scient. Fish. 44: 1411.

Revsbech, N. P., J. Sorensen & T. H. Blackburn, 1980. Distribution of oxygen in marine sediments measured with microelectrodes. Limnol. Oceanogr. 25: 403–411.

Uye, S., 1985. Resting egg production as a life history strategy of marine planktonic copepods. Bull. mar. Sci. 37: 440–449.

Uye, S., M. Yoshiya, K. Ueda & S. Kasahara, 1984. The effect of organic sea-bottom pollution on survivability of resting eggs of neritic copepods. Crustaceana (Suppl.) 7: 390–403.

Hydrobiologia **320**: 169–178, 1996.
V. R. Alekseev & G. Fryer (eds), Diapause in the Crustacea.

Role of parthenogenetic natality and emergence from diapausing eggs in the dynamics of some rotifer populations

Elena A. Mnatsakanova & Leonard V. Polishchuk[1,] *
Department of Hydrology, Geographical Faculty and [1] Department of Vertebrate Zoology and General Ecology, Biological Faculty, M. V. Lomonosov Moscow University, Moscow 119899, Russia

Key words: zooplankton, birth rate, diapausing eggs, population dynamics, rotifers

Abstract

There are few quantitative data on the role of emergence from diapausing eggs in population dynamics of natural populations of zooplankton species; to our knowledge, all these concern copepods and 'cladocerans'. We present here direct estimates of emergence from bottom resting eggs for another important category of freshwater zooplankton, namely rotifers. Three populations of rotifers of the genus *Brachionus* were studied in a lake. During the study period 10 population increases, each corresponding to an individual sampling interval, were detected. For each interval, emergence from immediately hatching, parthenogenetic eggs calculated on the basis of the Edmondson-Paloheimo model and emergence from diapausing bottom eggs determined in short-term experiments were estimated and compared to each other. We found that three of the population increases observed are entirely explained by parthenogenetic natality. In contrast, emergence from diapausing eggs can, on its own, account for none of population increases. For two population increases, however, it accounts for that part of population growth which remains unexplained by the parthenogenetic natality. For rotifer populations studied, emergence from diapausing eggs is generally less important than parthenogenetic births, when both are regarded as an immediate cause of population growth. This is in sharp contrast to the data available for some crustaceans (De Stasio, 1990) where the role of emergence from diapausing eggs in population dynamics has been clearly shown.

Introduction

There is increasing, although mostly indirect, evidence that emergence from bottom diapausing eggs or cysts may be of great importance for population dynamics and seasonal succession in plankton systems (Bogoslovsky, 1969; Marcus, 1984; Kawabata & Ohta, 1989; De Stasio & Hairston, 1992; Yampolsky & Kalabushkin, 1992; Ishikawa & Taniguchi, 1994). However, direct estimates of the emergence in the field are still rare and for planktonic animals are mostly confined to crustaceans (De Stasio, 1989, 1990). As to another important category of freshwater zooplankton such as rotifers, little evidence exists of the role of their bottom resting eggs in population dynamics of active planktonic stages. Information on this process may be important not only for the study of rotifers itself but also for comparison with crustaceans. This may provide better understanding of the whole zooplankton community because copepods, 'cladocerans', and rotifers often coexist.

Several major processes are responsible for local population growth in zooplankton. These are recruitment from immediately hatching eggs which are often attached to the mother's body (in 'cladocerans' and rotifers such eggs are produced by parthenogenesis and referred to below as parthenogenetic eggs), from diapausing bottom eggs, and due to advection. Since all these factors may operate simultaneously they should be studied on the same time scale. We know of no instance, however, where this requirement is entirely fulfilled. Here we present data both on births from planktonic parthenogenetic eggs and on emergence from bottom diapausing eggs obtained concurrently.

* Author for correspondence.

We have no data on advection, so its role will be judged only indirectly.

The normal way to calculate parthenogenetic natality in zooplankton is based on the Edmondson (1968) and Paloheimo (1974) model (for review see Polishchuk, 1986). It provides instantaneous, per capita estimates of the birth rate. In contrast, emergence from diapausing eggs as an empirical characteristic is expressed in terms of finite, absolute values. To make contributions of these two sources of population recruitment comparable, we slightly modify the standard approach to birth rate estimation.

In this paper we present direct estimates of the rate of emergence from the bottom diapausing eggs for three rotifer species. We estimate contributions of population recruitment due to parthenogenetic natality and due to emergence from diapausing eggs to the population increases observed. Finally, we compare our findings with those available in the literature for copepods and 'cladocerans'. Part of the material concerning the emergence from diapausing eggs has already been presented (Mnatsakanova, 1990). For the present work, the data were recalculated and corrected.

Materials and methods

Study area and sampling procedure

The study was carried out at Lake Chirtovo not far from Nizhny Novgorod, in the flood-lands of the Oka River. The lake is subdivided into three highly extended arms (west, north-eastern, and south) with a total area of 62 ha (Fig. 1). The lake is part of a system of slowly moving stretches of water and has an inflow to its west arm and outflow from the north-east. Through the inflow it is fed with treated sewage from a piggery with high concentration of N (N-NH_4, 59 mg l^{-1}) and P (P-P_2O_5, 15.5 mg l^{-1}, see Voroshilov *et al.*, 1985). Samples were taken in the south bay which is clear of the main stream and can be considered semi-isolated (Voroshilov *et al.*, 1985). It is 100 m wide and 2 m maximum depth. The water is thermally uniform, with temperature ranging from 14 to 24 °C during the study period. Zooplankton species composition varies from year to year partly due to whether fish are introduced with the Oka River during spring flood. In the year studied, zooplankton was dominated by brachionid rotifers, although *Daphnia*, *Moina* and several copepod species were also present.

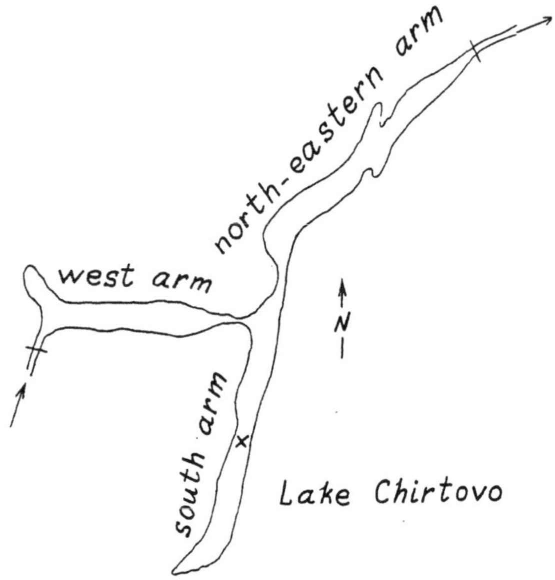

Fig. 1. A map of Lake Chirtovo. Arrows denote the inflow to and the outflow from the lake. The cross indicates the study area.

Zooplankton samples were collected from 23 May to 21 July 1987 every 3 d, with a gap in the observations from 19 June to 4 July. On each occasion samples were taken at about 11 a.m. Four hauls from different stations situated not far from one another (Fig. 1) were made with a plankton net (open area 0.05 m^2, mesh size 100–200 μm) towed from the bottom to the surface. The four samples were pooled, transported to the laboratory and treated immediately. Subsamples of sufficient volume to provide at least 100 animals of each species studied were counted in a Bogorov plankton tray under a dissecting microscope. Formalin was added by drops directly in the tray to prevent egg loss. Counts were made of the number of animals per litre (N) and the number of female parthenogenetic eggs per litre (E) for three rotifer species: *Brachionus calyciflorus*, *B. urceus*, and *B. angularis*. The ratio of eggs to animals, E/N, called the egg ratio was then calculated. Unsmoothed data were used throughout. Resting eggs which were also counted when observed, were not included in E. Males, because of their small size, must have passed through the net and were never seen in the samples. Male-producing parthenogenetic eggs when observed were therefore excluded from calculations of E.

Water temperature was measured every day or every second day. On that basis the egg development time (D, in days) was calculated according to the equation presented by Bottrell *et al.* (1976).

An important source of statistical error involved in population abundance estimates is spatial heterogeneity of zooplankton. Unfortunately, in 1987 we have no data allowing it to be assessed. Such data are available on 18 August 1985 when 9 net samples were taken at different stations simultaneously (within an hour) and each sample was counted separately. All these stations were within the area where samples were taken in 1987. Zooplankton at that time consisted almost entirely of *B. calyciflorus*. Using a program written in BASIC we computed the standard error of population size estimates for all $\binom{9}{4} = 126$ distinct combinations selecting 4 samples out of 9. Then the relative error, that is, the ratio of SE to the mean size, was calculated. It averaged 18% and was applied to the population size estimates in 1987.

Technique for hatching experiments

For taking bottom sediments, we used a corer designed by Sorokin (Chirkova, 1975) which is similar to the Kajak corer (Downing, 1984). The corer has a transparent plastic tube with a 33 mm inner diameter which penetrates the sediments 7–10 cm. On withdrawal, the tube was half or a little less full of sediments, with the above-sediment water filling its upper part. When hoisted to the boat, the tube was immediately closed at the bottom and detached from the frame. To transfer the sediments undisturbed, a glass tube of a smaller diameter of 30 mm was forced through the corer's tube. The corer's tube was then fastened to the frame, and the process was repeated to yield 1 to 4 sample tubes on each sampling occasion. Samples were taken from 5 to 19 June and from 3 to 22 July 1987 every day or every second day from the same area as zooplankton samples.

The glass tubes with the sediments were immediately transported to the laboratory. After 20 min standing, the water in the tubes was carefully removed with a siphon to prevent disturbing the sediment. In order to remove every zooplankter from the remainder of the above-sediment water some 100 ml of carbon dioxide saturated soda-water was poured into the tube. Highly concentrated soda-water kills zooplankters (Mudretsova-Wiss, 1933; pers. obs.). After 10 to 30 min standing, the soda-water was siphoned off and fresh, filtered lake water was poured into the tube.

The tubes were allowed to stand undisturbed in the lab or, on several occasions, in the lake. Temperature in the lab was no different from that in the lake. After 1 to 2 d, the above-sediment water was drawn off and any rotifer hatchlings present were counted by species. We never used the tubes once examined for further examination of hatching; instead, fresh samples of sediments were regularly taken and emergence of newly hatched rotifers was recorded.

All hatching rates symbolized below as V were expressed as animal numbers per litre per day. We first calculated the number of animals found in a tube to the number per square metre per day and then converted this to a litre basis, taking a lake depth in the study area to be 2 m.

Contribution of input from the sediments and parthenogenetic births to population increase

In zooplankton, recruitment from parthenogenetic eggs is commonly estimated from the Edmondson (1968) and Paloheimo (1974) model which yields an estimate for the instantaneous per capita birth rate b,

$$b = (1/D)\ln(1 + q)\ , \qquad (1)$$

where D is the egg development time (d), $q = E/N$ is the egg ratio. Two other characteristics routinely calculated in combination with birth rate are the instantaneous per capita rate of population change r and death rate d,

$$r = (\ln N_t - \ln N_0)/t\ , \qquad (2)$$

$$d = b - r\ , \qquad (3)$$

where N_0 and N_t are the population abundances at the beginning and at the end of sampling interval t, respectively. Equation 1 is supposed to be based on the exponential model of population growth (for general discussion see Polishchuk, 1986), although when $t = D$ this assumption is shown to be unnecessary (Voronov, 1991). In our study, as it is typical when studying rotifers, $t = 3$ d is several times larger than D which is about 1 d at the temperatures observed in the lake, so the assumption concerning the exponential growth remains essential.

An important assumption underlying Equations 1 and 3 is that the population is closed, that is, there occurs no immigration or emigration. This assumption is, of course, violated when a population experiences an inflow from diapausing eggs deposited in the sediments which is actually the case for the rotifer populations studied. We will show, however, that the contribution of emergence from diapausing eggs to the population growth is relatively small, so Equation 1 can, we believe, be used to estimate birth rate.

A problem with comparison between the contribution of emergence from resting eggs and that of parthenogenetic births to population growth is that the hatching experiments give a value of emergence which corresponds to an absolute population increase such as $N_t - N_0$ or $(N_t - N_0)/t$, whereas birth rate from Equation 1 corresponds to the instantaneous per capita population growth rate. Contribution of emergence from resting eggs relative to an absolute population increase can be expressed as

$$V\,t(N_t - N_0)^{-1}\,, \tag{4}$$

where $V\,t$ represents an absolute number of individuals entering the planktonic population from diapausing eggs over sampling interval $(0, t)$. Given the exponential population growth model, the number of individuals coming from parthenogenetic eggs over interval $(0, t)$ is given by

$$B = \int_0^t b\,N\,\mathrm{d}t = b\,r^{-1}(N_t - N_0)\,. \tag{5}$$

Contribution of parthenogenetic natality found by analogy with Eq. 4, takes the form

$$b\,r^{-1}\,. \tag{6}$$

The normal practice is to employ the egg ratio in Equation 1 specific to the beginning of sampling interval. When sampling interval equals the egg development time, this practice is certainly valid since it is justified by the arguments underlying the Edmondson-Paloheimo model (Polishchuk, 1986; Voronov, 1991). However, when $t > D$, it is probably better to use the average q and D. We use here the arithmetic means of q and of D over each sampling interval.

Using Equations 1, 2, 4, and 6, we calculated the contribution of emergence from diapausing eggs and that of parthenogenetic natality for each sampling interval where the populations were found to increase. Only population increases are analysed here because emergence manifests itself, as a factor of population dynamics, most clearly when the population is increasing. Only summer increases are considered because data on emergence from diapausing eggs are available beginning on 5 June.

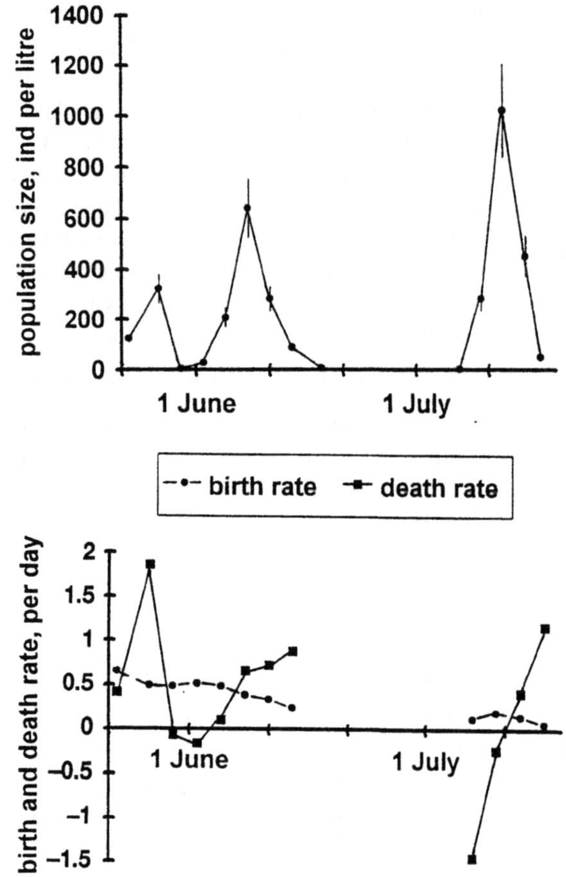

Fig. 2. Seasonal dynamics of *Brachionus calyciflorus* in summer 1987. *Upper panel* – Population size. Error bars indicate ± SE, and, if not shown, are smaller than the symbol size. *Lower panel* – Birth and death rates. See text for detail.

Results

Seasonal changes in population sizes and contribution of parthenogenetic births to population growth

At the beginning of summer we observed two peaks in *Brachionus calyciflorus*, in late May and early June. Maximum population size reached 640 ind l^{-1} (Fig. 2). Following the gap in observations, a sharp peak of population abundance of about 1000 ind l^{-1} occurred in mid July. *B. angularis* and *B. urceus* each exhibited only one population peak of about 1500 ind l^{-1} in early June (Figs 3 and 4). Also, seasonal changes in birth and death rates calculated by Equations 1 and 3 are shown in Figs 2 to 4.

For the three rotifer species studied, ten summer population increases, each relating to an individual

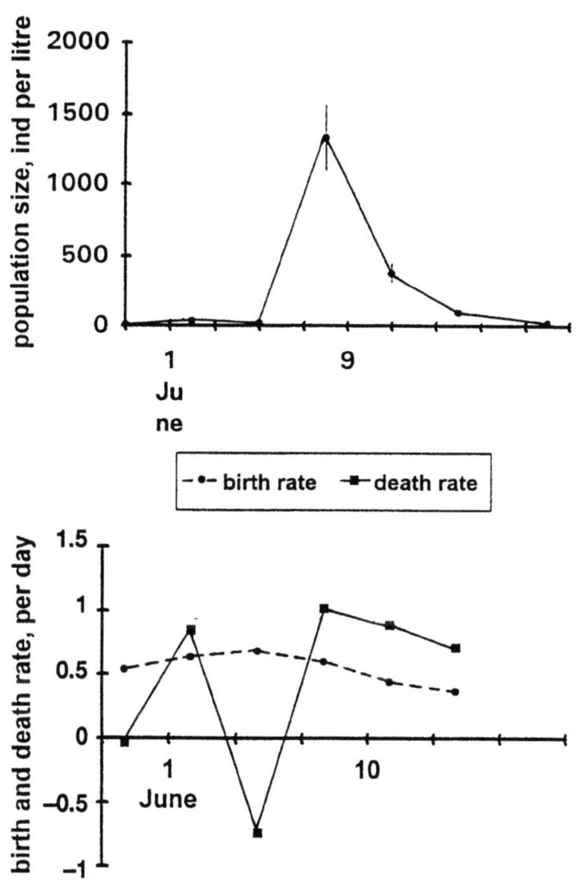

Fig. 3. As Fig. 2, but for *Brachionus angularis*.

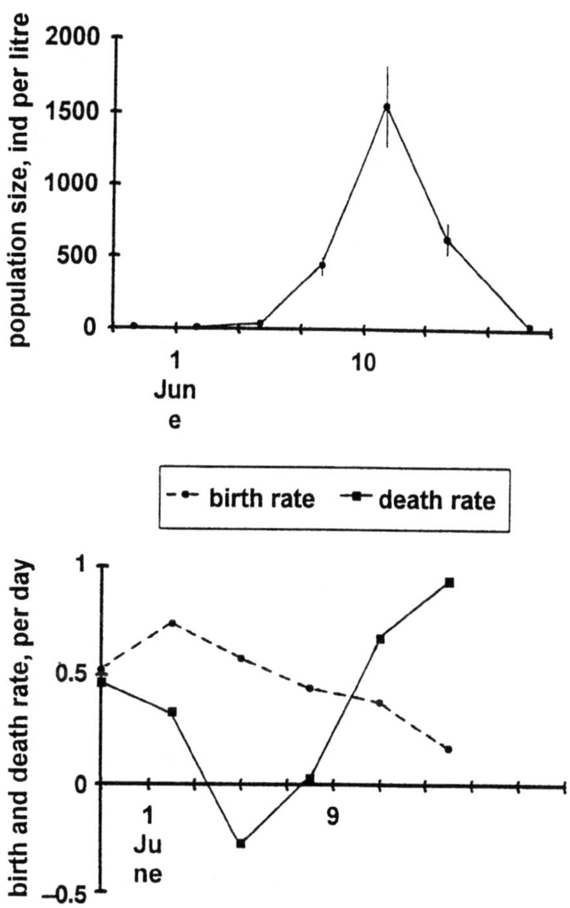

Fig. 4. As Fig. 2, but for *Brachionus urceus*.

sampling interval, have been analysed (Table 1). We consider only those increases which appear to be real, given the range of population size variation over individual sampling intervals and possible errors of population size estimates. Contributions of parthenogenetic natality to population increases vary from 8% to more than 100% (three intervals out of 10, Table 1). The latter means that $b > r$ so that the increases where contributions exceed 100% can be entirely attributed to parthenogenetic natality. Excluding figures higher than 100%, parthenogenetic natality is responsible on average for 62% of the magnitude of population increases.

Contribution of emergence from diapausing eggs to population growth

Emergence from diapausing eggs was rare (there were zero to 5 hatchlings per tube per day for *Brachionus calyciflorus*, for example), and so it is hard to discuss temporal changes in the emergence rate. Emergence rates in the tubes left in the laboratory did not significantly differ from the rates obtained in the tubes concurrently suspended in the lake. In three cases with synchronous determination of emergence rates in the laboratory and in the lake, differences between average rates are 44%, 9%, and 12% (all the data for *B. calyciflorus*). This appears to be rather small taking into account that the maximum and minimum emergence rates observed in individual tubes which were run concurrently and held in the laboratory sometimes differ by a factor of 10. Here we use only the average emergence rate for periods of continuous observation.

B. calyciflorus hatched throughout the whole study period, while *B. angularis* and *B. urceus* hatched only in June. This correlates with the fact that the two latter species were not found in the plankton in July. Mean emergence rate ranges from 0.6 to 2.1 ind l^{-1} d^{-1},

Table 1. Population size and contribution of parthenogenetic births to population increase for three rotifer species. Notation: N_0 = actual population size at the beginning of sampling interval, N_t = actual population size at the end of sampling interval, SE = standard error, D = mean egg development time, q = mean egg ratio (E/N), b = birth rate, $b = (1/D)\ln(1+q)$, r = rate of population increase, $r = (\ln N_t - \ln N_0)/t$, where t = 3 d is sampling interval, Cont = contribution of parthenogenetic births to population increase, Cont = br^{-1} 100%. If contribution is less than 100% the observed population increase cannot be explained by the recruitment from parthenogenetic eggs.

Sampling interval	$N_0 \pm$ SE (ind l^{-1})	$N_t \pm$ SE (ind l^{-1})	D (day)	q	b (d^{-1})	r (d^{-1})	Cont (%)
		Brachionus calyciflorus					
30 May – 2 June	5.4 ± 1.0	27.5 ± 5.0	1.6	1.2	0.48	0.54	89
2 – 5 June	27.5 ± 5.0	206.1 ± 37.1	1.2	0.9	0.51	0.67	76
5 – 8 June	206.1 ± 37.1	640.0 ± 115.2	1.0	0.6	0.48	0.38	128
7 – 10 Jule	2.6 ± 0.5	282.7 ± 50.9	0.9	0.1	0.12	1.57	8
10 – 13 July	282.7 ± 50.9	1019.4 ± 183.5	0.8	0.2	0.20	0.43	46
		Brachionus angularis					
30 May – 2 June	6.5 ± 1.2	35.8 ± 6.4	1.6	1.3	0.54	0.57	94
5 – 8 June	18.9 ± 3.4	1337.4 ± 240.7	1.0	1.0	0.72	1.42	51
		Brachionus urceus					
2 – 5 June	10.3 ± 1.8	34.9 ± 6.3	1.2	1.5	0.74	0.41	180
5 – 8 June	34.9 ± 6.3	445.2 ± 80.1	1.0	0.8	0.61	0.85	72
8 – 11 June	445.2 ± 80.1	1548.6 ± 278.7	0.9	0.5	0.44	0.42	106

Table 2. Rate of emergence from bottom diapausing eggs expressed on a litre basis. Notation: n = number of observations, V = emergence rate, SD = standard deviation, CV = coefficient of variation. Emergence rate presents the mean value for the periods indicated.

Period	n	V (ind l^{-1} d^{-1})	SD (ind l^{-1} d^{-1})	CV (%)
		Brachionus calyciflorus		
5 – 19 June	21	1.64	1.21	73
3 – 22 July	12	0.59	0.37	63
		Brachionus angularis		
5 – 19 June	14	0.92	0.59	64
		Brachionus urceus		
5 – 19 June	17	2.13	2.15	101

with a coefficient of variation between 60 and 100% (Table 2).

Such a low hatching rate as compared with the population increases observed (Table 3) implies that recruitment due to emergence from diapause eggs can only play a part in population growth if population size is low. Indeed, all three sampling intervals where emergence makes a sizeable contribution of 10 to 25% to population increase comprise the very early stage of the population growth (Table 3). In the other seven intervals where the populations increase, the contribution of emergence is less then 3%. A mean contribution of emergence from diapausing eggs, for all ten sampling intervals, comprises 6.5%. Emergence from diapausing eggs on its own accounts for none of the population increases. It might be important, however, as an additional factor when parthenogenetic natality alone fails to explain population increase.

Consider the role of emergence from diapausing eggs for those sampling intervals where the contribution of parthenogenetic natality is less than 100%, that is, the birth rate is lower than the actual rate of population growth. We estimated the extent to which emergence from diapausing eggs can close the gap between actual population increase and the number of individuals hatched from parthenogenetic eggs. That is, we calculated the quantity $V\,t\,[(N_t - N_0) - B]^{-1}$ where B is taken from Equation 5. In two cases out of seven, emergence from diapausing eggs contributes more than 100% to that gap (Table 3), thus being the sec-

Table 3. Contribution of emergence from bottom diapausing eggs to population increase or to the gap between population increase and the number of parthenogenetic births over sampling interval. Notation: N_0 and N_t = population sizes at the beginning and end of sampling interval. Cont 1 = contribution of emergence from diapausing eggs to population increase; Cont 1 = $V\,t(N_t - N_0)^{-1}$ 100%, where V is emergence rate from Table 2 and t = 3 d is sampling interval. B = number of individuals coming from parthenogenetic eggs over sampling interval; $B = b r^{-1}(N_t - N_0)$, where b is birth rate, r is the rate of population increase, both are taken from Table 1. G = the gap between population increase and the number of individuals coming from parthenogenetic eggs, $G = (N_t - N_0) - B$. Cont 2 = contribution of emergence from diapausing eggs to gap G, Cont 2 = $V\,t\,G^{-1}$ 100%. PR means that population increase can be entirely attributed to parthenogenetic births, that is, $B > N_t - N_0$ and $b > r$.

Sampling interval	$N_t - N_0$ (ind l^{-1})	Cont 1 (%)	B (ind l^{-1})	G (ind l^{-1})	Cont 2 (%)
		Brachionus calyciflorus			
30 May – 2 June	22.1	22	19.7	2.5	200.8
2 – 5 June	178.6	2.8	136.5	42.1	11.7
5 – 8 June	433.9	1.1	554.4	PR	–
7 – 10 Jule	280.1	0.6	21.3	258.8	0.7
10 – 13 July	736.8	0.2	339.1	397.7	0.4
		Brachionus angularis			
30 May – 2 June	29.3	9.5	27.5	1.8	153.3
5 – 8 June	1318.6	0.2	670.9	647.7	0.4
		Brachionus urceus			
2 – 5 June	24.6	26	44.4	PR	–
5 – 8 June	410.3	1.6	295.8	114.5	5.6
8 – 11 June	1103.4	0.6	1169.7	PR	–

ond most important factor of population growth. We consider it the second on those two sampling intervals, as compared with parthenogenetic natality, because parthenogenetic natality taken alone explains 89 and 94% of the amount of the population increases, whereas emergence from diapausing eggs taken alone, only 22 and 10% (Tables 1 and 3). For five remaining intervals parthenogenetic births and emergence from resting eggs even taken together are insufficient to explain the population increases observed.

Discussion

We have analysed population increases for three species of brachionid rotifers over ten sampling intervals. Parthenogenetic natality on its own explains population increases over three intervals, whereas emergence from diapausing eggs on its own explains none at all. On average, parthenogenetic natality is responsible for 62% of the amount of population increases while emergence from diapausing eggs for 6.5% only. Notice that for parthenogenetic natality it is a conservative estimate because values larger than 100% were excluded when the average was calculated. Hence for the populations studied, parthenogenetic natality is a more important factor of population growth than emergence from diapausing eggs.

A role of emergence from diapausing eggs in population growth is revealed in two intervals where parthenogenetic natality taken alone cannot explain the population increases while both these factors operating simultaneously can. Nevertheless on five sampling intervals out of ten, population increases remain uninterpreted. We speculate that one more factor such as horizontal transport of animals within the lake affected the population growth.

It may be that our observations yield biased estimates of emergence rate. Temperature which is known to be among the factors inducing rotifer resting eggs to hatch (May, 1987), was approximately the same in the lake and laboratory. Yet other conditions in the experimental tubes may differ from those in the lake: diapausing eggs in the tubes are affected by increased aeration due to sediment disturbance which cannot be completely avoided, by additional light when tubes are being taken from the bottom and left standing in the laboratory (though we observed no marked difference when holding the tubes in the lake), and by a 30-min exposure to carbon dioxide. For crustaceans, all these influences are known to stimulate emergence from diapausing eggs (Alekseev, 1990), although for *Diaptomus sanguineus*, a freshwater calanoid copepod, low oxygen shock is reported to stimulate emergence (Hairston & Munns, 1984; De Stasio, 1989 but see Lutz *et al.*, 1992 for stimulating effect of high oxygen concentrations on hatching from diapausing eggs in marine calanoid copepods). If our experimental perturbation stimulates hatching from diapausing eggs, the observed emergence rates tend to be overestimated, and an actual recruitment from diapausing eggs becomes still less important for population growth.

Although it is widely appreciated that emergence from diapausing eggs may play a significant role in ensuring the existence of zooplankton populations on both ecological and evolutionary time scales (Hairston & De Stasio, 1988; De Stasio & Hairston, 1992; Marcus *et al.*, 1994), we know of no instance of estimating the rate of hatching of diapausing eggs from lake sediments in natural populations of rotifers (for review of the state of the art see Hollowday, 1993). Generally speaking, much more is known about the density of resting eggs in sediments and cues inducing hatching in the laboratory (for rotifers see e.g. Smirnova, 1985; May, 1987 and the literature therein) than about the natural input from resting eggs to the pelagial.

Pioneer research by De Stasio (1989, 1990) has shed new light on the role of emergence from diapausing eggs in seasonal dynamics of some planktonic freshwater crustaceans. Nauplii of calanoid copepods *Diaptomus sanguineus* and *Epischura nordenskioldi* from Bullhead Pond (Rhode Island, USA) appear in the plankton in late autumn every year. The start of emergence from diapause eggs, after a period of lack of emergence, was detected directly. Moreover, all nauplii that emerge in late autumn are from the bottom diapausing eggs as there have been no adult females in the plankton during a 2- (for *Epischura*) or 6-month (for *Diaptomus*) period before the first nauplii appear. Emergence from diapausing eggs thus determines the start of the annual population cycle. De Stasio (1990) also found that the timing of peak abundance of *Ceriodaphnia reticulata*, an anomopod 'cladoceran', in the plankton coincides well with peak emergence from the diapausing eggs. For another anomopod, *Eubosmina longispina*, emergence from diapausing eggs seems rare, with no correspondence between population dynamics in the plankton and emergence from the sediments. Note that De Stasio (1990) correlates the timing of planktonic and bottom events but does not compare numerically the magnitude of recruitment from those sources.

For the *Brachionus* species studied here, the situation in June and July appears similar to that for *Eubosmina*. Although emergence from diapausing eggs occurred during the whole study period, comparison of the contributions of parthenogenetic natality and diapausing eggs leads us to the conclusion that parthenogenetic, immediately hatching eggs play a bigger part in seasonal population growth than diapausing eggs.

When studying zooplankton population dynamics, one often comes up against the so-called "negative death rate" (Polishchuk, 1982). It arises when the difference between birth rate and the actual rate of population growth which normally serves as an estimate of the death rate (see Equation 3) turns out to be negative. For our data set we observed negative death rates on seven sampling intervals (Figs 2 to 4). These intervals are exactly those on which the contribution of parthenogenetic natality to population increase is less than 100% (Table 1). This is because $b/r < 1$ entails $d < 0$, and vice versa.

Negative death rate is commonly attributed to an input of individuals into the population (Polishchuk, 1982). Here we examine one possible source of such input, emergence from diapausing eggs deposited in sediments. We find that while the input from sediments does actually occur, in the majority of cases it is insufficient to explain that part of population growth which remains unexplained by the parthenogenetic natality. Hence some additional source of population recruitment must have been involved. Moreover, based on the percentage of sampling intervals where parthenogenetic natality and emergence from diapausing eggs taken together do now allow us to explain population increases (five out of ten intervals), we suggest that an unknown factor may play a rather large role in the population dynamics.

An important, while not well studied, agent in population recruitment of zooplankton is water movements and horizontal transport of animals (Threlkeld, 1987).

It is well documented for a natural re-establishment event for *Bosmina longirostris* in Lake Tahoe which resulted from wind-induced advection from a semi-isolated embayment (Threlkeld, 1981). We did not measure the water movement at the locality studied. Thus we can only suggest that in our water body horizontal transport may also have an effect on the population dynamics.

Since in nature diverse factors operate concurrently a better understanding can be attained when different aspects of population dynamics are studied simultaneously. Our study is the first to quantify the relative importance of input from bottom diapausing eggs and from planktonic, immediately hatching, eggs to population growth in zooplankton. However, half of the population increases observed cannot be accounted for by either of the processes considered. More comprehensive studies are needed to get more insight into the population dynamics of planktonic animals, whether they be crustaceans or rotifers. When both immediately-hatching and diapausing eggs are involved, the same principles apply to both groups.

Acknowledgements

E. A. M. is grateful to Drs Tatyana S. Maltsman and Yuri I. Voroshilov for much help in logistics and field work. We thank Profs Nelson G. Hairston, Jr. and Geoffrey Fryer for valuable advice and correcting earlier versions of the manuscript and an anonymous reviewer for comments. The final part of the work was supported by grant M7L-000 from the Soros Foundation (International Science Foundation) and by grant 94-04-12760 from the Russian Foundation for Fundamental Research.

References

Alekseev, V. R., 1990. Diapause in Crustaceans: Ecological and Physiological Aspects. Nauka, Moscow, 144 pp. (in Russian).

Bogoslovsky, A. S., 1969. Studying resting eggs of rotifers. Part 3. Bull. Mosk. Obshch. Ispyt. Prirody. Otd. Biol. 74(3): 60–79 (in Russian).

Bottrell, H. H., A. Duncan, Z. M. Gliwicz, E. Grygierek, A. Herzig, A. Hillbricht-Ilkowska, H. Kurasawa, P. Larsson & T. Weglenska, 1976. A review of some problems in zooplankton production studies. Norw. J. Zool. 24: 419–456.

Chirkova, Z. N., 1975. Microzoobenthos. In F. D. Mordukhay-Boltovskoy (ed.), Methods of Studying Ecosystems of Inland Waters. Nauka, Moscow: 178–184 (in Russian).

De Stasio, B. T., Jr., 1990. The role of dormancy and emergence patterns in the dynamics of a freshwater zooplankton community. Limnol. Oceanogr. 35: 1079–1090.

De Stasio, B. T., Jr., 1989. The seed bank of a freshwater crustacean: copepodology for the plant ecologist. Ecology 70: 1377–1389.

De Stasio, B. T., Jr. & N. G. Hairson, Jr., 1992. Environmental variability and the persistence of multiple emergence strategies. Bull. math. Biol. 54: 313–334.

Downing, J. A., 1984. Chapter 4. Sampling the benthos of standing water. In J. A. Downing & F. H. Rigler (eds), A Manual on Methods for the Assessment of Secondary Productivity in Fresh Waters. Blackwell, Oxford: 87–130.

Edmondson, W. T., 1968. A graphical model for evaluating the use of the egg ratio for measuring birth and death rates. Oecologia 1: 1–37.

Hairston, N. G., Jr. & B. T. De Stasio, Jr., 1988. Rate of evolution slowed by a dormant propagule pool. Nature (Lond.) 336: 239–242.

Hairson, N. G., Jr. & W. R. Munns, Jr., 1984. The timing of copepod diapause as an evolutionarily stable strategy. Am. Nat. 123: 733–751.

Hollowday, E. D., 1993. Recent world-wide studies in Rotiferology. Quekett J. Microscopy 37: 73–78.

Ishikawa, A. & A. Taniguchi, 1994. The role of cysts on population dynamics of *Scrippsiella spp.* (*Dinophyceae*) in Onagawa Bay, northeast Japan. Mar. Biol. 119: 39–44.

Kawabata, Z. & M. Ohta, 1989. Cyst distribution and excystment conditions for the dinoflagellate *Peridinium penardii* (Lemm.) Lemm. in a reservoir. Freshwat. Biol. 21: 437–444.

Lutz, R. V., N. H. Marcus & J. P. Chanton, 1992. Effects of low oxygen concentrations on the hatching and viability of eggs of marine calanoid copepods. Mar. Biol. 114: 241–247.

Marcus, N. H., 1984. Recruitment of copepod nauplii into the plankton: Importance of diapause eggs and benthic processes. Mar. Ecol. Prog. Ser. 15: 47–54.

Marcus, N. H., R. Lutz, W. Burnett & P. Cable, 1994. Age, viability, and vertical distribution of zooplankton resting eggs from an anoxic basin. Evidence of an egg bank. Limnol. Oceanogr. 39: 154–158.

May, L., 1987. Effect of incubation temperature on the hatching of rotifer resting eggs collected from sediments. Hydrobiologia 147: 335–338.

Mnatsakanova, E. A., 1990. Estimation of rotifer hatching from resting eggs in natural conditions. In L. A. Kutikova (ed.), Rotifera. Proceedings of the Third All-Union Rotifer Symposium. Zoological Institute, Leningrad: 33–41 (in Russian).

Mudretsova-Wiss, K. A., 1933. Effect of Carbonic Acid, Hydrogen Sulphide, Methane, and Lack of Oxygen on Aquatic Organisms. Gosstroyizdat, Moscow, 48 pp. (in Russian).

Paloheimo, J. E., 1974. Calculation of instantaneous birth rate. Limnol. Oceanogr. 19: 692–694.

Polishchuk, L. V., 1982. The "negative mortality" of planktonic animals and the sampling schedule. Zh. obshch. Biol. 43: 411–418 (in Russian).

Polishchuk, L. V., 1986. Dynamic Characteristics of Populations of Planktonic Animals. Nauka, Moscow, 128 pp. (in Russian).

Smirnova, I. A., 1985. Recruitment of rotifers due to hatching from resting eggs. In Biological Resources in Waters under Anthropogenic Stress. Naukova Dumka, Kiev: 47–49 (in Russian).

Threlkeld, S. T., 1981. The recolonization of Lake Tahoe by *Bosmina longirostris*: Evaluating the importance of reduced *Mysis relicta* populations. Limnol. Oceanogr. 26: 433–444.

Threlkeld, S. T., 1987. *Daphnia* population fluctuations: Patterns and mechanisms. In R. H. Peters & R. de Bernardi (eds), Daphnia. Mem. Ist. ital. Idrobiol. 45: 367–388.

Voronov, D. A., 1991. On the interpretation of some planktonology equations. Oecologia 88: 303–304.

Voroshilov, Yu. I., V. S. Zhitkov, T. S. Maltsman & G. L. Margolina, 1985. The transformation of the biocenosis of a running-water lake under the effect of treated sewage water from a pig breeding farm. Vodnye Resursy No. 4: 148–154 (in Russian).

Yampolsky, L. Yu. & B. A. Kalabushkin, 1992. Genetical variability in an intermittent population of *Daphnia magna*. Genetika 28: 41–50 (in Russian).

Hydrobiologia **320**: 179–189, 1996.
V. R. Alekseev & G. Fryer (eds), Diapause in the Crustacea.

The biology and morphology of the marine harpacticoid copepod *Heteropsyllus nunni* Coull, during encystment diapause *

Judith Williams-Howze
Belle W. Baruch Institute for Marine Biology and Coastal Ecology, University of South Carolina, Columbia, SC, USA

Key words: diapause, harpacticoid, copepod, morphology

Abstract

Heteropsyllus nunni Coull, a meiobenthic harpacticoid copepod is the only marine crustacean to undergo a state of diapause within a cyst. A 12 month field study indicated *H. nunni* adults reached peak population densities in winter, with nauplii maturing in the spring, becoming adults by April or May.

At the last stage of development, a mature but unmated adult, they begin to prepare for encystment diapause. The copepods remain within their cyst in a state of diapause for 3–4 months during the summer only. Studies on the effects of temperature and photoperiod suggested that these two environmental cues are not crucial for induction or termination of diapause. Low temperature delayed development and time to encystment, while high temperatures accelerated development, making the time to encystment shorter. There were more males than females in the cysts in laboratory experiments. Upon excystment, the copepods mate, and females begin egg production within one week. Adults that have excysted and mated die after a few weeks of active reproductive effort. Nauplii go on to mature and begin the univoltine diapause/reproductive cycle.

The copepods prepare for dormancy in two ways: they begin to produce and store two types of secretory products to be used in cyst construction; then they produce large quantities of lipid to be used as a nutrient supply throughout diapause. Histochemistry of the cyst-building material indicated the lower urosome is full of two chemically different products. Dorsally, there is a storage sac of proteinaceous material. The ventral sac of secretory product is a mucopolysaccharide. The copepod builds the spherical cysts in a matrix of small and large sand grains. The cysts fit tightly around the ventral portion of the animal in the its flexed position: however, there is a large space between the cyst and the sides of the copepod.

Biochemical analysis of the cyst showed it is composed of an amino acid complex similar to collagenous material. Scanning electron microscopy revealed a complex of large cuticular pores located in the lower urosome and caudal rami. There are specific pores for secretion of the two cyst-building products.

Introduction

Many invertebrates (aquatic and terrestrial) have developed adaptations to avoid periodic (cyclic or acyclic) harsh environmental conditions, such as dry seasons, extreme high or low temperatures, and food shortages. Diapause and quiescence are two adaptations used to avoid these environmental catastrophes/perturbations. While often used interchangeably the two are distinctly different. Quiescence is characterized as short-termed, irregular and controlled by the adverse factors. For example, a cold shock might send an invertebrate into a state of quiescence, i.e., the animal enters and remains in a state of torpor until the environmental catastrophe (e.g. cold shock, heat wave) ends and then resumes normal physiological functions. Quiescence is reversible, not fixed to a specific stage of development and may be induced repeatedly in the same individual (Andrewartha 1952; Tauber *et al.*, 1986).

Diapause disrupts the normal metabolic program away from its developmental pathway at a specific

* Contribution No. 1072 from the Belle W. Baruch Institute for Marine Biology and Coastal Research.

ontogenetic stage. It is not controlled by the direct action of sporadic environmental factors (as is quiescence); rather it is cued in advance by some predictable cyclic change in the environment (Andrewartha, 1952; Danks, 1987). Diapause is by definition neurohormonally driven (Danks, 1987) and subsequently involves a more complicated developmental process committing the organism to a greater metabolic investment (Tauber, *et al.*, 1986). In the silkworm *Bombyx* there is a specific diapause hormone which is part of the larger neuropeptide family PBAN (pheromone biosynthesis activating neuropeptides) (Nagasawa, 1992).

As a reproductive strategy diapause is an alternative to normal development, and is favoured when the expectation of fitness accruing from active growth and reproduction is less than that from survival in diapause (Cohen, 1970).

Insects are the most extensively studied of the invertebrates exhibiting diapause (Danks, 1987; Tauber & Tauber, 1970; Tauber *et al.*, 1986). Copepods also exhibit diverse forms of diapause (Elgmork, 1980; Marcus, 1980; Coull & Grant, 1981; Hairston, 1987). Different groups of copepods undergo dormancy in different ways. Of the major free-living orders, the largely planktonic Calanoida (Centropagoidea) produce primarily resting eggs (chitin covered/desiccant resistant), although some species of *Calanus* and *Neocalanus* (Megacalanoidea) diapause as a stage V copepodid in deep waters (Miller *et al.*, 1991). Diapause eggs are heavier than the subitaneous (immediately hatching) eggs and sink to the sediment. They are crucial in the reproductive success of these copepods as they aid in predator avoidance (Hairston, 1987) and as a mechanism for surviving desiccation (Taylor *et al.*, 1990).

In the Cyclopoida, the copepod enters into a state of dormancy at the fourth or fifth copepodid stage (C-IV/V); rarely are resting eggs produced (reported only for *Mesocyclops edax* (S. A. Forbes) (Wyngaard, 1988). Diapausing cyclopoids sink to the bottom and remain obscured by mud/detrital coverings; some such as *Diacyclops thomasii (= Cyclops bicuspidatus thomasii)* Forbes are covered in a cyst-like mucus structure (Cole, 1953; Fryer & Smyly, 1954; Nilssen, 1980; Elgmork, 1980).

Diapausing harpacticoids construct and reside in cysts, typically at the adult stage, but diapause is not common in the Harpacticoida. Most encysting species are freshwater and encyst during summer months (Sarvala, 1979; Coull & Grant, 1981; Nalepa, 1985). Until the discovery of encysted *Heteropsyllus nunni* Coull (Family Cletotidae) by Coull & Grant (1981) in the marine environment, it was believed that only freshwater species of the family Canthocamptidae underwent diapause within a cyst.

Timing for entering diapause is critical to reproductive success in copepods (Cohen, 1967, 1970; Taylor, 1980; De Stasio, 1990). The initiation of diapause has most often been associated with cues of photoperiod, temperature or a combination of both. The production of calanoid diapausal eggs appears to be controlled by photoperiod (Marcus, 1980, 1982a, 1982b) and temperature combined (Marcus, 1987; Hairston & Olds, 1984; Cooley, 1978). Diapause egg production also may differ geographically (Hairston & Olds, 1984; Marcus, 1987). In cyclopoids, photoperiod has been shown to be the cue for induction of diapause (Watson & Smallman, 1971a, 1971b; Elgmork & Nilssen, 1978). The dormant cyclopoid C-IV or C-V stage overwinters or oversummers depending on the geographical effect on the induction response (Elgmork, 1955; George, 1973; Cooley, 1978; Elgmork & Langeland, 1980).

Little is known about the effect of environmental cues initiating diapause in harpacticoids. Sarvala (1979) determined that a particular combination of photoperiod and temperature were necessary to induce encystment and excystment in the freshwater *Canthocampus staphylinus* (Jurine). Changes in the light/temperature regime arrested development of the copepodids, inhibited egg production in mature females, and induced the copepods to produce prediapause oil droplets. Exposure to low temperature inhibited encystment in *C. staphylinus* which normally are encysted in sediments during summer months.

Comprehensive morphological and physiological studies of copepods before or during diapause are few. Most focus on one physiological aspect such as nutrient storage. Cyclopoid copepodids are reported to accumulate oil before diapause, have low metabolic rates, and crystalline material within the intestines during diapause (Elgmork, 1962, 1973; Wierzbicka, 1970; Watson & Smallman, 1971a). Marine calanoid copepods (Megacalanoida) undergo diapause during copepodid stage IV or V in deep cold waters of the Atlantic and Arctic circle. During dormancy, species of *Calanus* have been described as having reduced body weight, lower levels of nitrogen and carbon (Tande, 1982), higher lipid and lower body weight (Bamstedt & Ervik, 1984; Bamstedt *et al.*, 1990), as well as reduced gut epithelium and lowered digestive enzymes during diapause (Hallberg & Hirche, 1980). No studies describing morphological and physiological changes during diapause exist for freshwater or marine harpacticoids.

Heteropsyllus nunni, is a recurring member of the intertidal meiobenthos on South Carolina sandflats (Coull & Grant, 1981). Data exist for numbers of animals in the free-living state versus encysted state over a 10 year period (Coull, unpublished), but little is known about the biology of this animal related to encystment. Presented here is a summary of the results of a long-term field study and a laboratory experiment testing environmental cues for encystment/excystment (data from Williams-Howze & Coull, 1992). The internal, external morphology and physiological processes associated with encystment diapause in *H. nunni* are also presented.

Materials and methods

Population biology of H. nunni

During 1988 through 1989 an intertidal sand flat (Oyster Landing) at North Inlet, South Carolina, USA (33 ° 19.0′ N, 79 ° 11.6′ W. was studed by Williams-Howze & Coull (1992). They took random cores by hand at low tide with a 2.54 cm diameter core tube in the upper 10 cm of sediment. Eight samples were taken monthly for one year (September through October the following year). All samples were immediately preserved with 10% buffered Formalin with Rose Bengal added. In the laboratory, copepods were extracted via elutriation, the number, sex, and life-history stage of *H. nunni* were recorded for each core sample. The following year experiments were conducted to test specific environmental cues on encystment/excystment (see Williams-Howze & Coull (1992) for detailed description). Live *H. nunni* were extracted from the sand, separated from other meiofauna, and placed in large petri dishes containing filtered artificial seawater (ASW). All copepods were held in an incubator at 18 °C, with 16/8 h day-night cycle until sufficient nauplii hatched. Sand substrate for culture was obtained from the sand flat at Oyster Landing) in a 500 ml flask, covering the sand with F/2 medium solution (Guillard, 1972), autoclaving again, then inoculating the sand and medium with 10 ml of *Phaeodactylus tricornutum*. The sand/algae substratum was removed from the flask with a sterile pipette, washed with filtered seawater and placed in sterile plastic petri dishes with sterile-filtered ASW, salinity 29–30‰. Nauplii were held in 35×10 mm petri dishes and all other stages were held in 60×20 mm dishes; there were at least 50 nauplii per dish. All dishes with copepods were established during the same day. Experiments were conducted in two large incubators, one at ambient winter temperature (10 °C), and one at early summer (20 °C) temperatures for North Inlet. Within each incubator, 'black boxes' were used to isolate copepods in the dark from the rest of the incubator (which had light). The four treatment conditions were cold-short day (15 h dark/6 h light), hot-short day, and cold-long day (15 h light/9 h dark), hot-long day. At the end of the experiment (which lasted 23 weeks), all dishes were removed from incubators and 10% formalin with Rose Bengal was added to each to preserve and stain the copepods. Every individual was counted and life- history stage and sex determined.

Full or empty cysts were counted and copepods removed from full cysts and sexed. Comparisons of monthly abundance in the field, sex ratios, and free-living to encysted animals were statistically analyzed using the GLM (General Linear Model) procedure and Turkey's multiple comparison procedure (significance levels were set at alpha <0.05). Free-living and encysted abundance within the four experimental treatments was analyzed separately (Williams-Howze & Coull, 1992).

Morphological studies

Live *H. nunni* were collected during winter months, extracted from the sand and cultured as for the laboratory experiment. Four ontogenetic stages were chosen for morphological studies; pre-encystment, encystment, post- encystment and reproductive. By analyzing these, the morphological changes before, during and after diapause could be determined. Some copepods were extracted, preserved in formalin, and placed in a glycerol/water solution for examination and photography. Specimens for scanning (SEM) and transmission electron microscopy (TEM) were fixed in cacocylate buffered glutaraldehyde (2%) and post-fixed in osmium tetroxide. For TEM, specimens were embedded in either Epon or Spurr's plastic resin; for SEM, dehydrated, critically point dried, and coated with 20 nm of gold palladium. For light microscopy, serial plastic sections (1–2 μm) were stained with either periodic acid-Schiff's reagent (PAS), PAS with methylene blue, for carbohydrates, (Hayat, 1975) or a tribasic (trichrome) stain (Grimley, 1967). Some copepods were embedded in Lowicryl, thick sectioned and

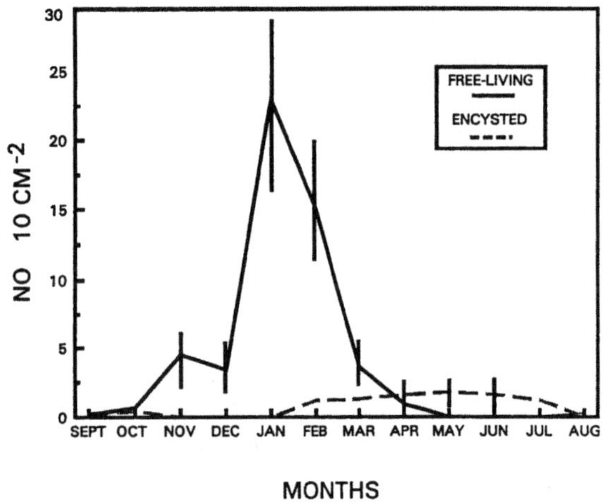

Fig. 1. Mean number of encysted and free-living copepods from the field study. Errors bars are one standard deviation of mean (from Williams-Howze & Coull, 1992).

stained with Coomassie Blue stain, for protein determination (Cawood, 1978). For amino-acid analysis of cyst material, cysts were collected from the culture dishes soon after formation, hydrolyzed in 6N HCL, dissolved in buffer and analyzed on a Beckman 6300 amino-acid analyzer by Dr Yoshinore Ishikawa, Department of Chemistry, University of South Carolina.

Results

Free-living *H. nunni* reached maximum abundance in winter and were low in number, then absent as summer progressed (Fig. 1). In December and January, females (particularly gravid) greatly outnumbered males. Free-living males were most numerous just after emergence from cysts (October–November); sex ratio was about 3 males to 1 female.

The mean number of full cysts was not significantly different between months because the number of cysts in the cores was extremely low (Williams-Howze & Coull, 1992). Cysts were most abundant in summer when free-living *H. nunni* were absent. In the laboratory the frequency of encystment (at least one cyst in a dish) was surprisingly high (69%) in all four treatments (Williams-Howze & Coull, 1992): Hot- SD = 11/16; Hot-LD = 11/16; Cold-SD = 11/16; Cold- LD = 12/16 (sixteen dishes for each treatment), indicating that copepods encysted in most of the five life-history stages originally placed within the dish.

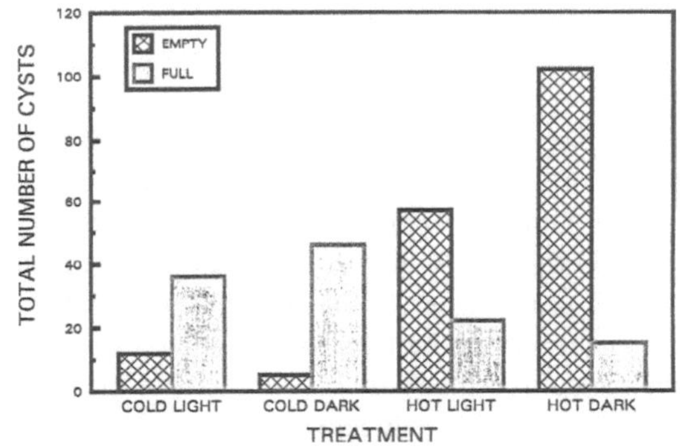

Fig. 2. Number of full and empty cysts, by treatment, at the end of the temperature/photoperiod experiment (total time — 23 weeks). (from Williams-Howze & Coull, 1992).

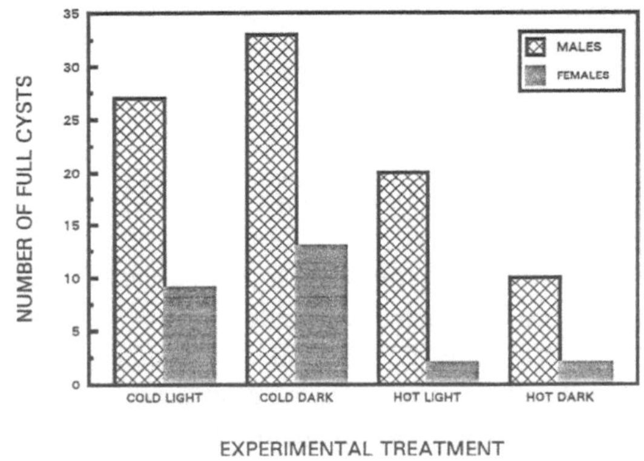

Fig. 3. Number of males compared to females, removed from full cysts, by treatment, at the end of the temperature/photoperiod experiment (total time — 23 weeks) (from Williams-Howze & Coull, 1992).

All the copepods encyst as unmated adults or stage 6. When placed in culture, all life-history stages matured to adult before encysting. The time to first encystment for the Hot treatments was 37 days and 67–77 days for the Cold treatments; thus, while encystment was delayed by the cold, it was not inhibited. This delay resulted in more occupied cysts in cold than in hot treatments, and more full than empty cysts at the end of the experiment (Fig. 2). More males than females were found within the cysts at the end of the experiment (Fig. 3).

Morphological studies

H. nunni has three morphologically distinct stages related to encystment diapause pre-encystment, encystment and post-encystment. The last consists of two substages, a premoult, and a reproductive stage.

Both sexes pass through a free-living stage in the sandy sediments prior to encystment (Fig. 4A). During this stage, which lasts four months, the copepods begin to produce and store large amounts of material to be used in cyst construction. Cyst-building material is stored in two compartments that fill the entire lower urosome (Fig. 4B). The dorsal sac contains a homogeneous looking mass, while the ventral sac is globular in appearance (Fig. 4C). Each sac of material has a separate secretory pore to the exterior (Fig. 4D). The composition of the two secretory products is very different as shown by histochemical staining. The ventral product stained positively (dark purple) with the periodic acid-Schiff's reagent, indicating the presence of mucopolysaccharides, and the dorsal secretory product was positive (dark blue) for protein after staining with Coomassie Blue (Fig. 5A, B). If during culture of *H. nunni*, the water is agitated, the copepods stick to the bottom of the dish with a visible secretion from the ventral sac. Only by using their mouthparts, and flexing backward (apparently either biting the sticky adhesive, or using some enzymatic labral secretion) can the copepod free itself and swim away. Fig. 4D, shows this secretion being extruded from the lateral pore as a result of being placed in fixative. The adhesive is so sticky that once the copepod is stuck to the dish, trying to help free them mechanically tears them in two! Even after fixation in formalin, the urosome with the adhesive extruded sticks to forceps making the copepod almost impossible to remove.

The dorsal sac is connected to a pore in the apical caudal ramus, and the ventral sac to a lateral pore in the caudal ramus (Fig. 5C,D). The sacs and pores are bilateral — a secretory canal and pore for each sac on each caudal ramus.

H. nunni generally builds its cyst on large grains of sand (Fig. 6A), but sometimes in the concave portion of small shell fragments. The following pattern of cyst-building was observed in culture dishes.

The copepods selected 3–4 large sand grains and appeared to 'tack' secretion onto them, while circling continuously backward. Small sand grains were then cemented to the large ones until the copepod was surrounded by sand and barely visible. The copepods appeared to use some labral secretion in cyst construction in that the caudal rami pores were in direct contact with the mouthparts when the animal was flexed dorsally backwards (Fig. 6A). *H. nunni* moved slowly, circling approximately one-eight of a turn, then stopping and working the mouthparts before moving another one-eight of a turn. This cycle continued for several hours. Any mechanical disturbance (touching with forceps or moving the dish) caused cessation of activity and sometimes movement to a new site. Hours into the building process, the copepods had buried themselves under sand grains and could no longer be seen. Unfortunately, it is during this time that the extrusion of the cyst material and final construction occurs. If left undisturbed, the copepods would complete cyst-building overnight (approx. 12–14 hours). Once the cyst was formed, the copepods continued in a backward circular motion, apparently working the interior surface of the still pliable cyst. Initially, the cyst with the copepod inside is somewhat pink in colour. Some of this coloration is due to the pink-orange colour of the pre-diapause lipid stores. After another 24–48 hours, the cyst is darker, less pliable, and the copepod ceases activity.

Using SEM, the cyst appears like fibrous paper (Fig. 6B). Copepods within the cysts are always flexed in a dorsally backward position (Fig. 6A,B). Encysted copepods remain in this position throughout their diapause (approximately 5 months). The apical pore on each caudal ramus has a diameter of about 4 μm. (Fig. 6C,D). The lateral pore is less visible, often being covered by the lateral setae just above the apical setae (Fig. 6C). The copepod stored vast quantities of lipid prior to encystment to be utilized as an energy supply during diapause.

The lower intestinal tract shows lipid spheres being assimilated for storage (Fig. 7A). TEM of urosomal cells shows the rough endoplasmic reticulum, a dividing cell, and stores of globular mucopolysaccharide (mucin) substance (Fig. 7B).

Following diapause, the copepods chew their way out of the cyst. A short-lived (72 h) pre-moult stage, is followed by a reproductive stage lasting 3–4 months. The copepods that emerge moult, mate, produce 3–4 eggs batches per female, then die. The material used in cyst-building is greatly reduced within the urosome following excystment, but small amounts of both were still present to some degree.

Analysis of the cyst material shows it to be high in acidic amino acids and to consist of proline, glycine, hydroxyproline, and hydroxylysine similar in proportion to that of collagen. Cysts have elevated levels of glycine, common in adhesive proteins of mussels, but a lower level of lysine.

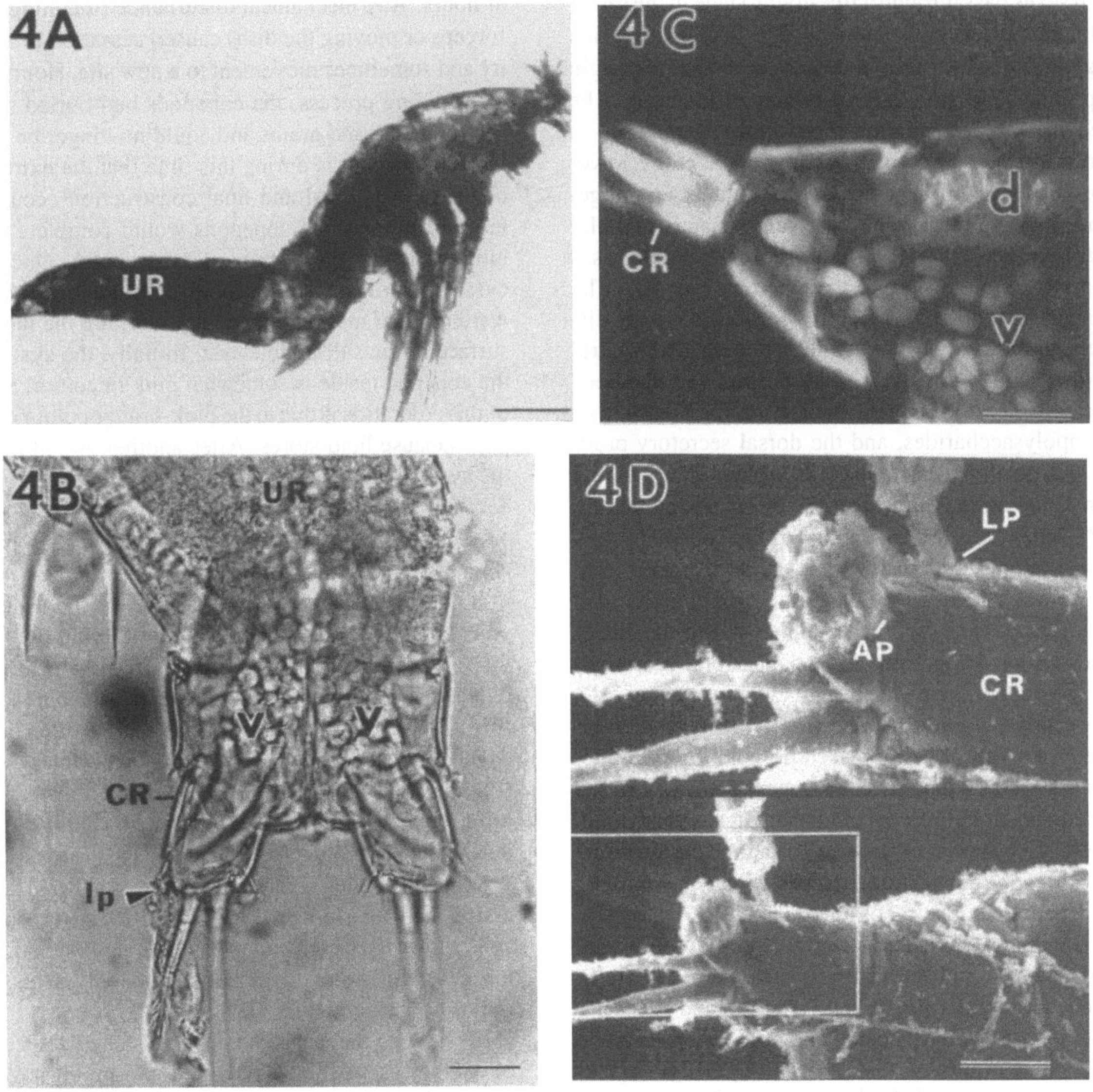

Fig. 4. A. *Heteropsyllus nunni* free-living female (not encysted). Scale bar = 90 μm. B. Last urosomal segments (UR) showing the two large sacs of ventral secretory product (v) within the caudal rami (CR). Arrow indicates location of large apical pore with secretory product being extruded out of the pore (lp). Scale bar = 50 μm. C. Confocal laser scanning micrograph of lower urosome and caudal rami (CR) showing the two sacs of secretory product, dorsal (d) and ventral (v). Scale bar = 20 μm. D. Scanning electron micrograph showing the caudal ramus (CR) and secretory product being extruded out of both the lateral pore (LP) and apical pore (AP). Scale bar = 25 μm.

Discussion

The biology of *H. nunni* is unique in copepod diapause. No other marine copepod undergoes diapause within a cyst, and diapause appears to be under genetic control as experiments to inhibit diapause failed. Copepods that encyst do so as mature, but unmated adults. In *Canthocamptus staphylinus*, another harpacticoid, that encysts in freshwater, encysted females with spermatophores attached have been recorded (Sarvala, 1979). Some females of the cyclopoid copepod, *Cyclops strenuus* Fischer, enter diapause mated with spermatophore attached (Naess & Nilssen, 1991).

The laboratory environmental cues experiments showed that while temperature and photoperiod affect development, the chosen experimental settings did not alter the set diapause timing for encystment or excystment. Encystment occurred in all treatment con-

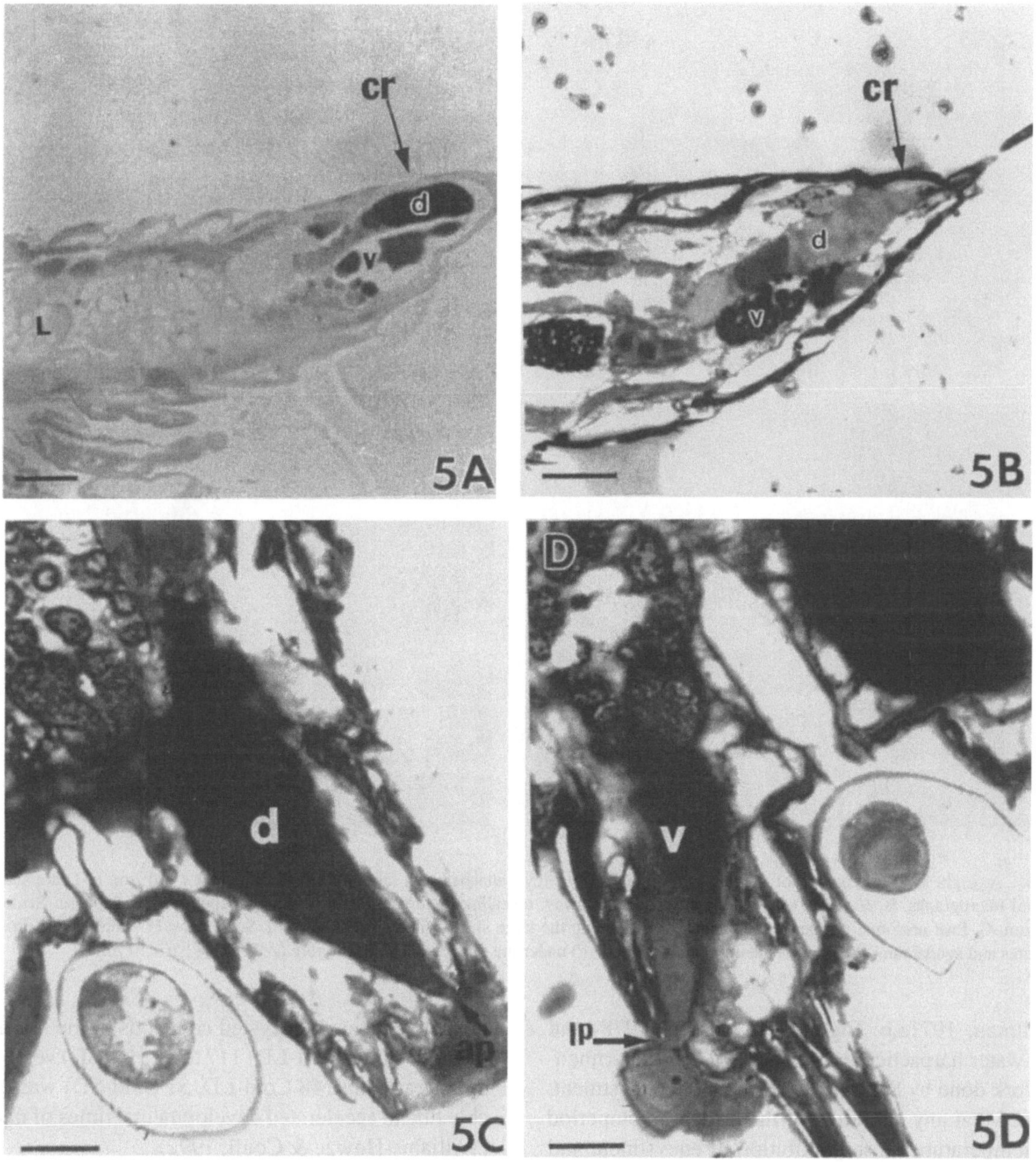

Fig. 5. Thick plastic sections of *H. nunni*. A. Sagittal section through lower urosome, showing stained secretory sacs in the caudal rami (cr). Dorsal (d) secretory product stained positively for protein using Coosamie Blue. L = lipid. B. Sagittal section through lower urosome showing both dorsal (d) and ventral (v) secretory products. Ventral secretory product stained positively for mucopolysaccharides using the PAS reaction. C. Longitudinal section through a caudal ram showing dorsal secretory product (d) being extruded out of the apical pore (ap). D. Longitidinal section through a caudal ramus showing ventral secretory product (v) being extruded out of the lateral pore (lp). Scale bars = 50 μm.

ditions, and was not inhibited by a cold, dark environment (10 °C, 9 h light, 15 h dark) (Williams-Howze & Coull, 1992). This is in direct contrast to previous findings which suggested that specific photoperiod and temperature requirements must be met to induce or terminate diapause or diapause egg production, e.g., for calanoids (Marcus, 1980, 1982a, 1982b, 1987; Hairston *et al.*, 1990), cyclopoids (Watson &

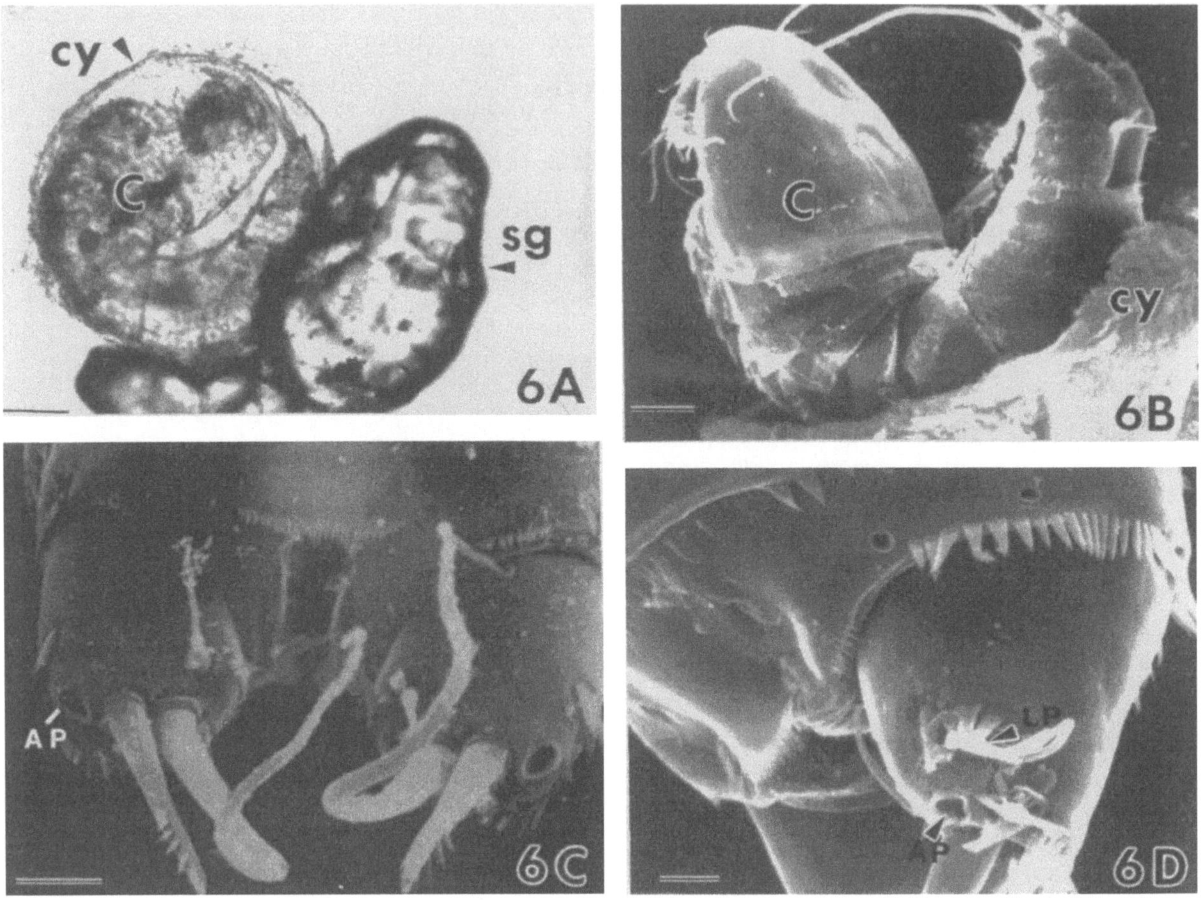

Fig. 6. A. Light level micrograph of *H. nunni* (c) within its cyst (cy), attached to a sand grain (sg). Scale bar = 110 μm. B–D are scanning electron micrographs. B. *H. nunni* (c) with the cyst (cy) torn away, revealing the dorsal flexure of the copepod during diapause. Scale bar - – 56 μm. C. Last urosomal segments and caudal rami showing the large size of the apical pore (AP). Scale bar = 15 μm. D. Last urosomal segments and caudal rami showing the lateral pore location (LP) under the setae, and the apical pore (AP). Scale bar = 10 μm.

Smallman, 1971a,b; Elgmork & Nilssen, 1978) and freshwater harpacticoids (Sarvala, 1979). Experimental work done by Sarvala on harpacticoid encystment, showed that any alteration of the critical photoperiod and temperature caused inhibition of encystment and significant morphological/reproductive changes. This was emphatically not the case for *H. nunni*. Temperature had a greater effect on the copepod's physiology than did photoperiod. The most significant effect was on the nauplii, since naupliar development to adult, then to encystment took twice as long in cold treatments (67–77 days) as it did in hot (20 °C), only 37 days. In the field reproduction and naupliar growth is in winter/spring (February–March); therefore, cold treatments were probably closer to 'normal' for early naupliar development than the hotter conditions. The greater number of encysted copepods in the hot treatments (81 total, Hot-LD) 117 total, Hot-SD) versus the cold treatments (48 Cold-LD, 51 Cold-SD) was probably due to accelerated developmental times of nauplii (Williams-Howze & Coull, 1992).

Initially, it was thought that the entire population of *H. nunni* encysted during the summer, however, in culture experiments, not all the adults encysted. In most treatment dishes there were mating and reproducing copepods throughout the entire 23 week experimental period (Williams-Howze, 1992). This was unexpected because no free-living forms have even been found during summer at North Inlet, (Coull & Grant, 1981). Coull & Grant (1981) hypothesized that the free-living population either moved to another area or all copepods were in diapause within their cysts.

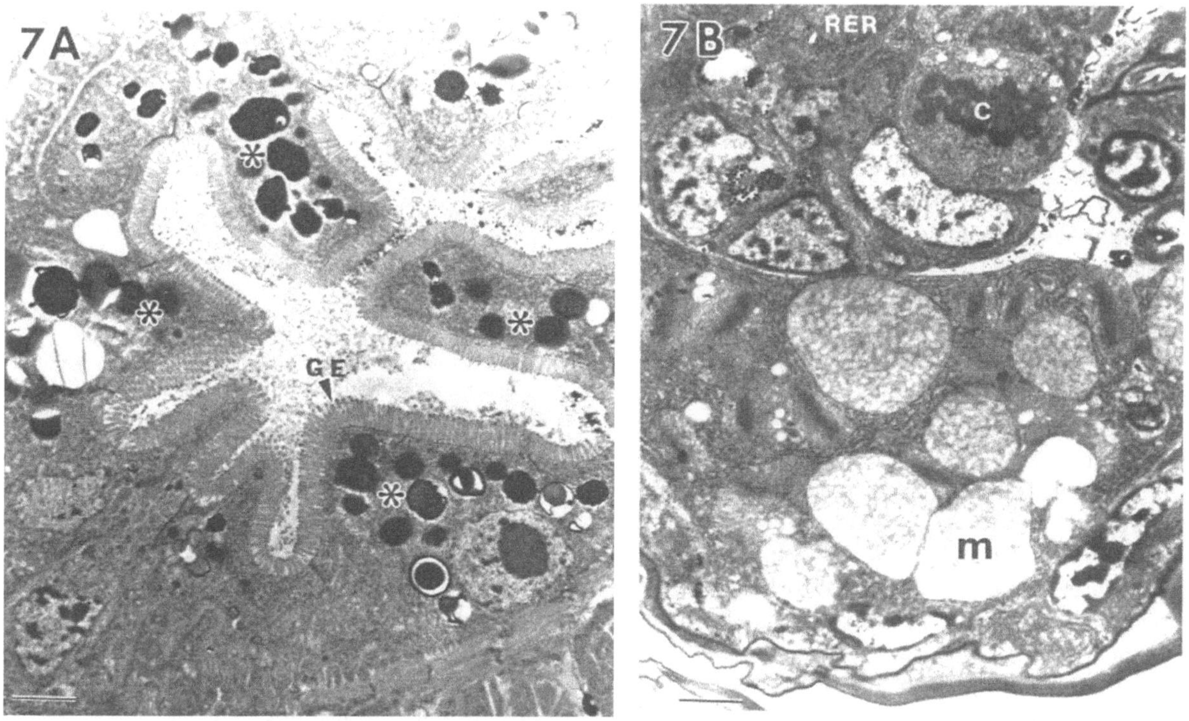

Fig. 7. Transmission electron micrographs of *H. nunni*. A. Cross section through the intestine showing the microvilli capped gut epithelium (GE). Lipid spheres are indicated by asterisks. Scale = 3 μm. B. Section is through the lower urosome area showing active cell proliferation (c), rough endoplasmic reticulum (RER) and stores of mucopolysaccharide substance (m). Scale bar = 2 μm.

Other possible causes for initiation of diapause, or encystment in *H. nunni*, could be predation (Sarvala, 1979; Maier, 1990) or overcrowding (Gehrs & Martin, 1974). These factors have not been tested, but should be at least considered as potential inducers of diapause. Predation by bottom feeding juvenile fishes during the summer in the shallow intertidal zone could possibly be a cause for summer diapause in *H. nunni*, but the soft, pliable cyst offers no protection from either desiccation or consumption. The large peak in population in a relatively small area of intertidal sand may also be a cue to undergo diapause (i.e., overcrowding), possibly indicating that food may be limited if the population does not diapause. These are speculations as experimental work on such factors has not been done.

More males than females were consistently found in the cysts, and in the field study males predominated during the time of emergence from the cysts (Sept.–Oct.). This is apparently not unusual in copepod diapause where the male to female ratio is often in favour of the male (George, 1973; Maier, 1989, 1990).

The morphology of *H. nunni* is highly specialized. Of particular interest are the two large storage sacs of secretory material which fill the entire lower urosome. The ventral sac of globular material occurrs in all ontogenetic stages and is present (though reduced) even after cyst construction. This adhesive allows all ontogenetic stages to stick to the substratum until they release themselves.

Why these copepods anchor themself when not building a cyst is unknown. They may do so to prevent tidal flushing away from the sand flat. While the pores are large, the possession of secretory pores is not uncommon in harpacticoids that build some type of structure. Williams-Howze *et al.*, (1987) studied the functional morphology of *Pseudostenhelia wellsi* Coull and Fleeger, that builds, and lives within, a mucoid tube. It has large pores on the last urosome segment and the caudal rami, and secretes mucoid material from them when tube-building.

Nehring (1993) found the same mucus tube construction via urosomal pores in *Stenhelia palustris* Brady, a closely related species. Other harpacticoids, particularly of the genus *Diarthrodes* (*D. nobilis* Baird, *D. cystoecus* Fahrenbach and *D. aegidus* Brian) secrete large quantities of mucus to form mucus bags in which they live (Fahrenbach, 1962; Hicks & Grahame, 1979; and personal observation). *D. nobilis* secretes its mucus through a large cuticular vent on its cephalothorax, as does *D. cystoecus*. The external morphology of *D. aegidus* has not been described.

The cyst material of *H. nunni* is different from the mucus structures of *Diarthrodes*. It is fibrous and similar to collagen in amino-acid structure. The cyst proteins are similar to byssal or adhesive proteins of mussels and to those of, for example, annelid tube cement, egg shells of turbellarians and dogfish and skate egg capsules (Harris, 1979; Waite, 1990). The cyst material is being analyzed for the presence of DOPA (3,4 dihydroxyphenylalanine), a common component of quinone-tanned structures such as byssal threads and egg shell cases (Waite & Little, 1989). These data will be presented in a paper comparing the amino-acid composition of the cysts with other extracoporeal secreted products such as the permanent adhesive proteins of a byssal thread. The cysts of *H. nunni* are very durable (like byssal threads) and do not decompose readily. They have been found still intact and attached to sand grains in old core samples.

References

Andrewartha, H. G., 1952. Diapause in relation to the ecology of insects. Biol. Rev. 27: 50–107.

Bamstedt, U. & A. Ervik, 1984. Local variation in size and activity among *Calanus finmarchicus* and *Metridia longa* (Copepoda: Calanoida) overwintering on the west coast of Norway. Plankton Res. 6: 843–857.

Bamstedt, U. & J. Hakanson, J. Frenner-Larsen, P. Bjornsen, O. Geertz-Hansen & P. Tiselius, 1990. Copepod nutritional condition and pelagic production during autumn in Kosterfjorden, western Sweden. Mar. Biol. 104: 197–208.

Cawood, A. H., U. Potter & H. G. Dickinson, 1978. An evaluation of Coomassie Brilliant Blue as a stain for quantitative microdensitometry of protein in section. Histochem. Cytochem. 26: 645–650.

Cohen, D., 1967. Optimizing reproduction in a randomly varying environment when a correlation may exist between the condition at the time a choice has to be made and the subsequent outcome. J. theor. Biol. 16: 1–14.

Cohen, D., 1970. A theoretical model for the optimal timing of diapause. Am. Nat. 104: 389–400.

Cole, G. A., 1953. Notes on copepod encystment. Ecology 34: 208–211.

Cooley, J. M., 1978. The effect of temperature on the development of diapausing and subitaneous eggs in several freshwater copepods. Crustaceana 35: 27–34.

Coull, B. C. & J. Grant, 1981. Encystment discovered in a marine copepod. Science 212: 342–344.

Danks, H. V., 1987. Insect Dormancy: An Ecological Perspective Biol. Sur. Can. Mono. 1: 439 pp.

DeStasio, B. T., 1990. The role of dormancy and emergence patterns in the dynamics of a freshwater zooplankton community. Limnol. Oceanogr. 35: 1079–1090.

Elgmork, K., 1955. A resting stage without encystment in the annual life cycle of the freshwater copepod *Cyclops strenuus strenuus*. Ecology 3: 739–743.

Elgmork, K., 1962. A bottom resting stage in the planktonic freshwater copepod *Cyclops scutifer* Sars. Oikos 13: 306–310.

Elgmork, K., 1973. Bottom resting stages of planktonic copepods in meromictic lakes. Ver. int. Ver. Limnol. 18: 1474- -1478.

Elgmork, K., 1980. Evolutionary aspects of diapause in freshwater copepods. In W. C. Kerfoot (ed.), Evolution and Ecology of Zooplankton Communities. The University Press of New England, Hanover (N.H.): 411–417.

Elgmork, K. & A. Langeland, 1980. *Cyclops scutifer* Sars — one and two-year cycles with diapause in the meromictic lake Blankvatn. Arch. Hydrobiol. 88: 178–201.

Elgmork, K. & J. P. Nilssen, 1978. Equivalence of copepod and insect diapause. Verh. int. Ver. Limnol. 20: 2511- -2517.

Fahrenbach, W., 1962. The biology of a harpacticoid copepod. La Cellule 62: 303–376.

Fryer, G. & W. J. P. Smyly, 1954. Some remarks on the resting stages of some freshwater cyclopoid and harpacticoid copepods. Ann. Mag. nat. Hist. 7: 65–72.

Gehrs, C. W. & B. D. Martin, 1974. Production of resting eggs by *Diapatomus clavipes* Schact (Copepoda: Calanoida). Am. Midl. Nat. 91: 487–489.

George, D. G., 1973. Diapause in *Cyclops vicinus*. Oikos 24: 136–142.

Grimley, P. M., 1967. A tribasic stain for thin sections of plastic embedded O_SO_4 fixed tissue. Stain Tech. 39: 229–232.

Guillard, R. L., 1972. Culture of phytoplankton for feeding marine invertebrates. In W. L. Smith & M. H. Chaney (eds), Culture of Marine Animals. Plenum Press, New York: 20– 60.

Hairston, N. G., Jr., 1987. Diapause as a predator- avoidance adaptation. In W. C. Kerfoot & A. Sih (eds), Predation: Direct and Indirect Impacts on Aquatic Communities. The University Press of New England, (N.H.): 282–290.

Hairston, N. G., Jr. & E. J. Olds, 1984. Population differences in the timing of diapause: Adaptation in a spatially heterogeneous environment. Oecologia 61: 42–48.

Hallberg, E. & H. J. Hirche, 1980. Differentiation of mid-gut epithelium in adults and overwintering copepodids of *Calanus finmarchicus* (Gunnerus) and *C. helgolandicus* (Claus. J. exp. mar. Biol. Ecol. 48: 283– 295.

Harris, B., 1979. The mechanical behavior of composite materials. Symp. soc. exp. Biol. 34: 37–74.

Hayat, M. A., 1975. Positive Staining for Electron Microscopy. Van Nostrand/Reinhold, New York, 343 pp.

Hicks, G. R. F. & J. Grahame, 1979. Mucus production and its role in the feeding behavior of *Diarthrodes nobilis* (Copepoda: Harpacticoida). J. mar. biol. Ass. U. K. 59: 321–33-.

Hirche, H., 1985. Overwintering of *Calanus finmarchicus* and *Calanus helgolandicus*. Mar. ecol. Prog. Ser. 11: 281–290.

Maier, G., 1989. The seasonal cycle of *Thermocyclops crassus* (Fisher 1853) (Copepoda: Cyclopoida) in a shallow eutrophic lake. Hydrobiologia 178: 43–58.

Maier, G., 1990. Spatial distribution of resting stages, rate of emergence from diapause and times to adulthood and to the appearance of the first clutch in 3 species of cyclopoid copepods. Hydrobiologia 206: 11–18.

Marcus, N. H., 1980. Photoperiod control of diapause in the marine calanoid copepod *Labidocera aestiva*. Biol. Bull. 159: 311–318.

Marcus, N. H., 1982a. The reversability of subitaneous and diapause egg production by individual females of *Labidocera aestiva* (Copepoda: Calanoida). Biol. Bull. 162: 39–44.

Marcus, N. H., 1982b. Photoperiod and temperature regulation of diapause in *Labidocera aestiva* (Copepoda: Calanoida). Biol. Bull. 162: 45–52.

Marcus, N. H., 1987. Differences in the duration of egg diapause of *Labidocera aestiva* (Copepoda: Calanoida) from the Woods Hole, Massachesetts, region. Biol. Bull. 173: 169–177.

Miller, C. B., T. J. Cowles, P. H. Wiebe, N. J. Copely & H. Grigg, 1991. Phenology in *Calanus finmarchicus*; hypothesis about control mechanisms. Mar. biol. prog. Ser. 72: 79–91.

Naess, T. & J. P. Nilssen, 1991. Diapausing fertilized adults. Oecologia 86: 369–371.

Nagasawa, A. 1992. Newro peptides of the silkworm, *Bombyx mori*. Experientia 48: 425–440.

Nalepa, T. F., 1985. Occurrence of a resting stage in cyclopoid and harpacticoid copepods in nearshore Lake Michigan. J. Great Lakes Res. 11: 59–66.

Nehring, S., 1993. Tube-dwelling meiofauna in marine sediments. Int. Revue ges. Hydrobiol. 521–534.

Nilssen, J. P., 1980. When and How to Reproduce: A Dilema for Limnetic Cyclopoid Copepods. In W. C. Kerfoot (ed.), Evolution and Ecology of Zooplankton Communities. The University Press of New England, Hanover (N.H.): 418–424.

Sarvala, J., 1979. A parthenogenetic life cycle in a population of *Canthocamptus staphlinus* (Copepod: Harpacticoid). Hydrobiologia 62: 113–129.

Tande, K. S., 1982. Ecological investigations on the zooplankton community of Balsfjorden, Northern Norway: generation cycles, and variations in body weight and body content of carbon and nitrogen related to overwintering and reproduction in the copepod *Calanus finmarchicus* (Gunnerus). J. exp. mar. biol. Ecol. 62: 129–142.

Tauber, T. M. & C. A. Tauber, 1970. Photoperiod induction and termination of diapause in an insect: response to changing day lengths. Science 167: 170.

Tauber, M. J., C. A. Tauber & S. Mazaki, 1986. Seasonal Adaptations of Insects. Oxford University Press. pp. 411.

Taylor, F., 1980. Optimal switching to diapause in relation to the onset of winter. Theor. pop. Biol. 18: 125– 133.

Taylor, B. E., G. Wyngaard & D. L. Mahoney, 1990. Hatching of *Diapatomus stagnalis* from a temporary pond after a prolonged dry period. Arch. Hydrobiol. 117: 271–278.

Waite, J. H. & K. Little, 1989. The glue protein of ribbed mussels (*Geukensia demissa*): a natural adhesive with some features of collagen. J. comp. Physiol. B. 159: 517- -525.

Waite, J. H., 1990. The phylogeny and chemical diversity of quinone-tanned glues and varnishes. Comp. Biochem. Physiol. 97: 19–29.

Watson, N. & B. N. Smallman, 1971a. The role of photoperiod and temperature in the induction and termination of an arrested development in two species of freshwater cyclopoid copepods. Can. J. Zool. 49: 855–862.

Wierzbicka, M., 1970. Contents of lipids in resting stages of copepoda, Cyclopoida. Polk. Arch. Hydrobiol. 17: 289–293.

Watson, N. & B. N. Smallman, 1971b. The physiology of diapause in *Diacyclops navus* Herrick (Crustacea, Copepoda). Can. J. Zool. 49: 1449–1454.

Williams-Howze, J., 1992. Ecology and Functional Morphology of *Heteropsyllus nunni* Coull (Copepoda: Harpacticoida) with a review of diapause in the Cyclopoida, Calanoida and Harpacticoida. Ph.D. Thesis, University of South Carolina, pp. 189.

Williams-Howze, J. & B. C. Coull, 1992. Are temperature and photoperiod necessary cues for encystment in the marine benthic harpacticoid copepod *Heteropsyllus nunni* Coull? Biol. Bull. 182: 109–116.

Williams-Howze, J., J. H. Silverman & J. W. Fleeger, 1987. Internal morphology related to tube-building in the meiobenthic copepod *Pseudostenhelia wellsi*. J. crust. Biol. 7: 171–181.

Wyngaard, G., 1988. Geographical variation in dormancy in a copepod: evidence from population crosses. Hydrobiologia 167/168 (Dev. Hydrobiol. 47): 367–374.

Hydrobiologia **320**: 191–196, 1996.
V. R. Alekseev & G. Fryer (eds), Diapause in the Crustacea.

Distribution and summer diapause of *Canthocamptus staphylinus* (Jurine) (Copepoda: Harpacticoida) in Lake Ladoga

Eugene A. Kurashov
Institute of Limnology, Sevastyanova 9, 196199 St. Petersburg, Russia

Key words: harpacticoids, diapause, temperature, Lake Ladoga

Abstract

The spatial distribution of *Canthocamptus staphylinus* (Jurine) in Lake Ladoga was examined. *C. staphylinus* is one of the most numerous species of harpacticoids in the lake. It can dwell in almost all littoral biotopes. In profundal biotopes strongly exposed to anthropogenic influence (pollution and eutrophication) only this species of harpacticoid was found. A peculiarity of *C. staphylinus* biology is that in summer individuals are in the resting stage as cysts. In summer active individuals were found in littoral habitats at a water temperature of 20 °C, but their number was small in comparison with cold periods of the year. In the littoral zone the relative number of individuals in cysts was clearly related to water temperature. Twelve degrees temperature appears to be critical for development of littoral populations of *C. staphylinus*.

Introduction

Harpacticoids of the family Canthocamptidae are typical representatives of lake meiobenthos. Data on harpacticoid faunistic composition, abundance and biology were extremely rare until 1983, since when different aspects of meiobenthic ecology have been studied by Kurashov (Kurashov, 1987, 1989, 1992; Kurashov & Slepukhina, 1988; Kurashov & Gorichenskiy, 1992). *Canthocamptus staphylinus* (Jurine) was reported in Lake Ladoga for the first time by Scorikov (1911). Some data exist on its occurrence in the different regions of the lake (Dengina & Sokolova 1968), but the biology, distribution and quantitative characteristics of its populations have not been specifically studied. A peculiarity of *C. staphylinus* biology is that in summer individuals are in diapause as cysts (Donner, 1928; Smyly, 1957).

The present study considers the quantitative distribution of *C. staphylinus* over the different biotopes in Lake Ladoga and describes some features of its biology in the littoral zone of the lake.

Materials and methods

The paper considers: (1) the results from samples collected at 46 littoral stations in July 1989 and 1990, in various regions of Lake Ladoga (Fig. 1). A general description of Lake Ladoga and its modern ecological state is given in (Viljanen & Drabkova, 1992). Six types of macrophyte associations were investigated: (i) reed beds on silt bottom and bottom with root mat; (ii) reed beds on sand; (iii) polydominant vegetation with *Polygonum amphibium* (L.), *Cicuta virosa* (L.), and Eleocharis acicularis (L.); (iv) associations with the prevalence of *Equisetum fluviatile* (L.); (v) with the prevalence of *Potamogeton spp.*; (vi) with the prevalence of *Carex spp.*; (2) the data obtained in 1983–93 during studies of profundal and littoral biotopes both in undisturbed zones of gulfs in the northern part of Ladoga and in the zones affected by industrial (pulp and/or paper mills) and domestic sewage (Schuchij Bay, sites near Pitkyaranta, Sortavala, Laskela, Impilahti; Jakkimvarskij Bay); (3) the data on seasonal dynamic of *C. staphylinus* obtained during the sampling period from March to October 1983 at two littoral stations in Jakkimvarskij Bay. Station 1 was situated in polydominant vegetation with the prevalence of *Pota-*

Fig. 1. Lake Ladoga, the main areas of investigation of *C. staphylinus* distribution and location of stations 1 and 2 in Jakkimvarskij Bay. 1 – Volkhov Bay; 2 – area near Pitkyaranta; 3 – Impilahti Bay and Haukalahti Bay, 4 – Sortavala's skerries 5 – Jakkimvarskij Bay, 6 – area including Lehmalahti Bay, 7 – Priozersk and Schuchij Bay, 8 – Tajpolovskij Bay.

mogeton spp. and *E. fluviatile*. Station 2 was situated in reed beds on a sand bottom.

Meiobenthos was sampled with an MB-TE sample corer with a cross-section area of 12.6 cm^2 (Travyanko & Evdokimova, 1968), preserved with 4% formaldehyde and sifted with a 0.08 mm nylon net. From 3 to 5 samples were taken at each station, later they were merged together and treated as a single integrated sample.

Comparison of numerical data was carried out using the statistical Student's *t*-test (Elliott, 1977).

Results

C. staphylinus was the dominant harpacticoid in the many littoral biotopes of Lake Ladoga. During the period of summer encystment the highest values of number and biomass (33 970 and 44 520 ind. m^{-2}; 955 and 709 mg wet weight m^{-2}) were found at the unpolluted sites, namely in the association of *Phragmites australis* (Cav.) and in association with the prevalence of *Carex spp.* and *E. fluviatile* (L.) (Haukalahti Bay, region of northern skerries). In July the main part of the population of *C. staphylinus* in these biotopes was encysted (95 and 71% respectively). In the reproductive period the largest number and biomass (50 880 ind. m^{-2} and 2020 mg wwt m^{-2}) of *C. staphylinus* were observed in Jakkimvarskij Bay at station 1. Here the polydominant vegetation had a prevalence of *Potamogeton spp.* and *E. fluviatile*. The share of encysted individuals at that time was 31%.

On average, the number and biomass of *C. staphylinus* (free-swimming and diapausing individuals) were significantly higher ($P<0.05$) than the number and biomass of other littoral harpacticoid copepods (*Attheyella crassa* (G. O. Sars), *Paracamptus schmeili* (Mrazek), *Moraria brevipes* (G. O. Sars) and *Bryocamptus minutus* (Claus)) taken together in the reed beds and macrophyte associations with prevalence of *E. fluviatile* (Table 1). In the macrophyte associations without *P. australis* as a whole, quantitative indices for *C. staphylinus* were lower (statistically not reliable, $P>0.05$) than for other harpacticoids.

Polluted littoral biotopes and littoral biotopes located at some distance from the sources of pollution did not differ significantly $P>0.05$) in number and biomass of *C. staphylinus* (Table 1). At the same time, number and biomass of *C. staphylinus* in polluted biotopes were significantly higher than these indices for all the rest of the harpacticoids ($P<0.05$; $P<0.01$). Densities of *C. staphylinus* and other Harpacticoida in the undisturbed biotopes were not significantly different ($P>0.05$).

Open littoral habitats without macrophytes were the least prefered environment for *C. staphylinus*. In these conditions its abundance was much less than among macrophytes (Table 1).

The share of encysted individuals of *C. staphylinus* in the different littoral biotopes differed insignificantly ($P>0.05$) (Table 1).

Outside the littoral zone *C. staphylinus* occurs almost exclusively in areas affected by polluted sewage and in biotopes recovering after radical reductions in pulp and paper mill discharges of waste waters (Schuchij Bay, zone near Laskela) (Table 2).

Encystment of *C. staphylinus* within the different littoral biotopes is subject to variation. In littoral habitats in Jakkimvarskij Bay encysted individuals were found even in March (Fig. 2). At station 1 (poly-

Table 1. Mean values of number (N = ind. m^{-2}), and biomass (B = mg wwt m^{-2}) of *C. staphylinus* and other harpacticoid copepods and the share of encysted individuals of *C. staphylinus* (S = %) in different littoral biotopes in Lake Ladoga in July. Biotopes are: I - reed beds on silt bottoms and bottoms with root mat; II - reed beds on sandy bottom; III - polydominant vegetation with *Polygonum amphibium, Cicuta virosa*, and *Eleocharis acicularis*; IV - associations with the prevalence of *Equisetum fluviatile*; V - associations with the prevalence of *Potamogeton* spp.; VI - associations with the prevalence of *Carex* spp.; VII - reed beds on the whole; VIII - all macrophyte associations excluding reed beds; IX - polluted biotopes; X - undisturbed biotopes; XI - littoral without macrophytes; ** - no exact information.

	C. staphylinus			Other Harpacticoida	
	N	B	S	N	B
Biotopes	$X \pm$ SE	$X \pm$ SE	$X \pm$ SE	$X \pm$ SE	$X \pm$ SE
I	13350 ± 2450	331.8 ± 64.9	60 ± 8	4070 ± 1200	37.3 ± 9.1
II	8280 ± 2710	228.9 ± 76.6	58 ± 13	2800 ± 700	27.3 ± 6.5
III	5490 ± 1970	126.3 ± 54.1	55 ± 13	13270 ± 6640	126.6 ± 63.3
IV	11190 ± 3180	253.5 ± 67.9	71 ± 10	4500 ± 3210	33.8 ± 22.6
V	2870 ± 2730	88.5 ± 84.1	32 ± 24	38360 ± 34030	341.0 ± 319.0
VI	13200 ± 5920	284.4 ± 103.9	70 ± 13	12650 ± 6070	130.6 ± 58.8
VII	12340 ± 1940	314.1 ± 51.3	60 ± 7	4450 ± 910	41.5 ± 6.6
VIII	10480 ± 2380	237.5 ± 45.6	63 ± 7	26140 ± 10900	242.1 ± 101.9
IX	8540 ± 2330	233.3 ± 65.9	57 ± 8	2780 ± 1280	25.6 ± 10.2
X	11140 ± 1840	260.6 ± 40.3	67 ± 6	11500 ± 4630	106.7 ± 43.1
XI	1200 ± 400	42.7 ± 15.6	**	13250 ± 5070	148.9 ± 66.6

Table 2. Number (N = ind. m^{-2}) and biomass (B = mg wwt m^{-2}) of *C. staphylinus* beyond the littoral zone in Lake Ladoga.

Location	Depth in meters	Date	N	B	Source of pollution
Schuchij Bay	2.8	July 1992	24450	632.1	Pulp mill
Schuchij Bay	2.5	July 1993	41860	1286.6	Pulp mill
Volkhov Bay	8.0	July 1992	400	14.1	River Volhov
Pitkyaranta	12.0	August 1992	780	33.3	Pulp mill
Pitkyaranta	32.0	August 1992	2400	76.7	Pulp mill
Pitkyaranta	24.0	July 1990	2860	76.2	Pulp mill
Lehmalathi Bay	13.0	August 1992	260	8.6	Technogenic pollution
Laskela	11.0	October 1991	8000	250.7	Wood enterprise, Pulp mill
Laskela	10.0	July 1993	35200	939.7	Wood enterprise
Sortavala	10.0	July 1984	1590	68.4	Industrial and domestic sewage

dominant vegetation with prevalence of *Potamogeton spp.* and *E. fluviatile*) free-swimming individuals were captured throughout the summer even at a water temperature of 20 °C. At station 2 (reed beds on sand bottom) in July no free-swimming individuals were found (Fig. 2), i.e. the total population was encysted. Females with egg-sacs were still found at the beginning of June, when the water temperature already reached 11–12 °C. Subsequently, with increasing water temperature, reproduction ceased. Reproduction resumed in September after the water cooled to 11–12 °C.

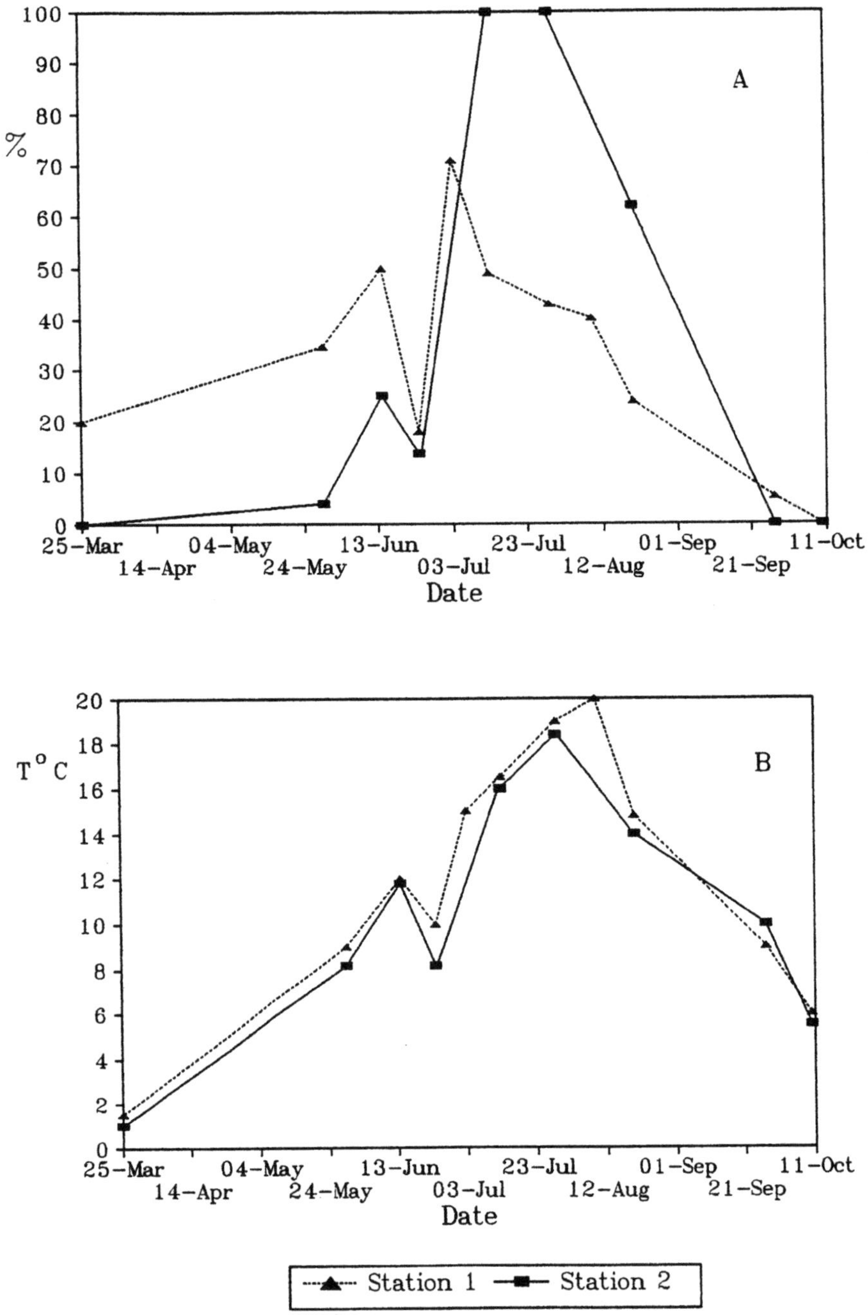

Fig. 2. Percentage of encysted individuals of *C. staphylinus* (A) and seasonal variation of water temperature (T °C) (B) at station 1 and station 2 in Jakkimvarskij Bay.

Discussion

Distribution of *C. staphylinus* in the profundal biotopes of Lake Ladoga appears to be directly related to eutrophication, pollution and organic loading. These data indicate that *C. staphylinus* prefers naturally high-productive biotopes (macrophyte associations) and profundal biotopes with a high input of allochthonous organic matter. Similar results have been reported by Sarvala (1979) and Sarkka (1992a, 1992b).

Human impact (pollution and eutrophication) increasing total productivity appears to be favourable for colonization of profundal biotopes of bays in the northern skerries of Lake Ladoga by *C. staphylinus*. In the unpolluted profundal biotopes (Haukkalahti Bay as an example) in Lake Ladoga *C. staphylinus* is absent. Food shortage seems to prevent it from colonizing the oligotrophic areas of the profundal zone (Sarvala, 1979). Notably, it colonizes only biotopes without oxygen deficit (Sarvala, 1979). Taking into consideration these facts and data on its distribution in Finish lakes (Sarvala, 1979; Sarkka, 1987a, 1987b, 1992a, 1992b) it may be deduced that *C. staphylinus* is an indicator of moderate pollution and eutrophication when it presents in the profundal and littori-profundal zones. In contrast, pollution has little or no effect on its distribution in the littoral zone of Lake Ladoga. This is not so in the other Harpacticoida, most of which typically indicate oligotrophic conditions (Kurashov & Gorichenskiy, 1992; Sarkka, 1992a).

The seasonal population dynamics observed in the littoral zone of lake Paajarvi (southern Finland) (Sarvala, 1979) and Lake Constance (Germany) (Frenzel, 1980) is similar to that in Lake Ladoga's littoral habitats. However, it should be pointed out that encystment of the total population is not an obligatory event in the seasonal development of populations of *C. staphylinus* within the different littoral biotopes in Lake Ladoga. Data supporting this view (Table 1, Fig. 1) were obtained in July, when total encystment is most likely to occur. The degree of encystment in the littoral habitats is dependent probably on a number of combined factors. Among other things, in littoral habitats *C. staphylinus* is prey for *Acanthocyclops viridis* Jurine (Kurashov, 1989) and perhaps other predators. As reported by Sarvala (1979), in permanent water bodies summer encystment is an efficient way of avoiding predation. Consequently, the intensity of predation is liable to influence the frequency of encystment.

In the littoral zone of Lake Ladoga the relative number of encysted individuals is clearly related to water temperature (Fig. 2). In littoral populations encystment is thought to be induced by long days (daylight) and temperatures of over 12 °C (Sarvala, 1979). It is remarkable that during a short period in late June when the water temperature has fallen to 9 °C at station 2 and to 10,4 °C at station 1, the relative number of encysted individuals decreased at both of these stations too. Similar temperature shocks acting as an awakening factor in the field have been observed in *Cyclops strenuus* Fischer by Elgmork (1959). Influx of cold river water into a small warm pond led to emergence and thus to reduced percentage of diapausing copepodids. These observations support the view that temperature is a principal factor governing development of *C. staphylinus*. It seems likely that twelve degrees temperature is actually critical in many cases for development of littoral populations of *C. staphylinus*. However, temperature and daylight are not the sole factors initiating encystment. Muckle (1967) has shown that in the profundal zone (40–140 m) of Lake Constance *C. staphylinus* also forms cysts; however, neither light (more precisely darkness) nor temperature (4.1–6,2 °C) are likely to induce the encystment at such depths. Research on encystment of *C. staphylinus* within the profundal habitats of Lake Ladoga needs further investigations.

Acknowledgments

This research was supported by Research Support Scheme of the Open Society Institute (grant N ° 990/93).

References

Dengina, R. S. & M. F. Sokolova, 1968. On the species composition of zooplankton of Lake Ladoga. In S. V. Kalesnik (ed.), Biological resources of Lake Ladoga. Zoology. Nauka, Leningrad: 117–129.(in Russian)

Donner, F., 1928. Die Harpacticiden der Leipziger Umgebung und der Schneeberger Erzbergwerke. Int. Rev. ges. Hydrobiol. Hydrogr. 20: 221–353.

Elgmork, K., 1959. Seasonal occurrence of *Cyclops strenuus strenuus* in relation to environment in small water bodies in southern Norway. Folia limnol. scand. 11: 1–196.

Elliott, J. M., 1977. Some methods for the statistical analysis of samples of benthic invertebrates. Freshwat. Biol. Association sci. publ., N 25: 1–160.

Frenzel, P., 1980. Die Populationsdynamik von *Canthocamptus staphylinus* (Jurine) (Copepoda, Harpacticoida) im Litoral des Bodensees. Crustaceana 39: 282–286.

Kurashov, E. A., 1987. Quantitative development, interrelation of components of meiobenthic community and influence on it some environment factors in Lake Ladoga' coastal zone. Proceedings of GOSNIORH, Leningrad 264: 138–146 (in Russian).

Kurashov, E. A., 1989. Experimental research of feeding of littoral predators *Acanthocyclops viridis* Jurine and *Atractides (Megapus) gracilis* Sokolow on meiobenthos in Lake Ladoga. Ecologiya 1: 42–49 (in Russian).

Kurashov, E. A., 1992. Role of meiobenthos in the lake ecosystem. In N. A. Petrova & A. Yu. Terzhevik (eds), Ladoga Lake: criteria of ecosystem state. Nauka, Leningrad: 187–204. (in Russian).

Kurashov, E. A. & T. D. Slepukhina, 1988. The use of the system and the experimental approaches for study of the relationships in the benthic communities of the Ladoga Lake. In L. A. Kutikova (ed.), Hydrobiological investigations of the sea and fresh waters. USSR Academy of Sciences, Zoological Institute: 60–67. (in Russian).

Kurashov, E. A. & A. Y. Gorichenskiy, 1992. The use of meiobenthic community for indication of ecological state of environment in lakes Ladoga and Onega. A comparative analysis of some methods and approaches. Russian J. Aquatic Ecol. 1: 125–136.

Muckle, R., 1967. Die Temperaturverhaltnisse des Bodensee-Obersees in den Jahren 1961 bis 1963. Internationale Gewasserschutzkommission fur den Bodensee, Bericht 5: 9–46.

Scorikov, A. S., 1911. Zoological study of water of Lake Ladoga as drinking. In Lake Ladoga as source of water-supply of Sankt-Petersburg. Sanitary part. S. Petersburg: 589–709. (in Russian)

Sarkka, J., 1987a. Meiobenthos of the profundal depths of lake Paijanne. Biol. Res. Rep. Univ. Jyvaskyla 10: 95–104.

Sarkka, J., 1987b. Meiobenthos of a lake chain affected by pulp mill effluent. Aqua Fennica 17: 35–41.

Sarkka, J., 1992a. Lacustrine profundal meiobenthos as an environmental indicator. Hydrobiologia 243/244 (Dev. Hydrobiol. 79): 333–340.

Sarkka, J., 1992b. Effects of eutrophication and organic loading on the occurence of profundal harpacticoids in a lake in southern Finland. Environm. Monit. Assessment 21: 211–223.

Sarvala, J., 1979. A parthenogenetic life cycle in a population of *Canthocamptus staphylinus* (Copepoda, Harpacticoida). Hydrobiologia 62: 113–129.

Smyly, W. J. P., 1957. Observations on the life-history of the harpacticoid copepod, *Canthocamptus staphylinus* (Jurine). Ann. Mag. nat. Hist. 12: 509–512.

Travyanko, V. S. & L. V. Evdokimova, 1968. Benthometer MB-TE. Hydrobiol.J. 4: 94–96.

Viljanen, M. & V. Drabkova (eds), 1992. Intercalibration of research methods in Lake Ladoga. University of Joensuu. Karelian Institute. Working Papers 8: 35–40.

Hydrobiologia **320**: 197–208, 1996.
V. R. Alekseev & G. Fryer (eds), Diapause in the Crustacea.

Diapause in the life cycle of *Calanoides carinatus* (Kroyer), (Copepoda, Calanoida)

E. G. Arashkevich, A. V. Drits & A. G. Timonin
P. P. Shirshov Institute of Oceanology, Russian Academy of Science, 23 Krasikova Street, 117218 Moscow, Russia

Key words: Calanoides carinatus, diapause, biochemical composition, respiration, gonad development

Abstract

The distribution and stage composition of *Calanoides carinatus* (Kroyer) are described together with changes in protein and lipid content, respiration rate and gonad development in copepodite stages IV and V (CV) and adults in the Northern Benguela current (17 °S, 23 °S and 25 °S).

During active upwelling the population consisted of two parts: the surface part over the shelf was represented by all developmental stages, while the deep part offshore was dominated (90–95%) by diapausal CVs. In the surface CVs the surplus assimilated energy was allocated to structural growth and maturation or to synthesis of reserve lipids. CVs with large oil sacs and high lipid content descended into deeper layers and formed diapausal stock; they were characterized by a dramatic decrease of respiration rate. Increase of gonad size in association with decrease of oil sac volume in diapausal CV suggests that reserve lipids were expended not only for respiration but also for gonad development. The moulting of diapausal CVs into adults took place in deep water.

These results are discussed in relation to the life cycle of the *C. carinatus* population and the factors causing the formation and termination of the diapausal phase.

Introduction

It is well known that most temporal and polar zooplankton species of the family Calanidae have a diapause phase in their life cycle (see review Conover, 1988) as an adaptation to the seasonal periodicity of environmental factors. Tropical and subtropical species inhabiting upwelling regions off the western shores of America and Africa are also exposed to variability in temperature and food availability, in association with changes in upwelling intensity. Like mid- and high latitude copepods their life cycles also include a diapausal phase (Longhurst, 1967; Thiriot, 1978; Heinrich, 1993).

Calanoides carinatus (Kroyer) is the most characteristic species in upwelling areas along the entire African west coast (Thiriot, 1978) and off Somalia on the African east coast (Smith, 1982, 1984). The life history of *C. carinatus* in the Eastern Tropical Atlantic was described by Binet & Suisse de Sainte Claire (1975), Petit & Courties (1976) and Petit (1982). According to their scheme, during the lack of coastal upwelling, usually associated with the austral autumn, the resting fifth stage copepodids (CVs) inhabit deep water until the onset of the following upwelling season. The upwelled water brings them to the surface in the shelf zone, where during the productive period *C. carinatus* produces 4–6 generations. The final generation at the end of the upwelling season descends to deep layers as CV. This scheme has been accepted by many authors and is assumed to be typical for this species (Smith, 1982; Alldredge *et al.*, 1984; Miller *et al.*, 1991; Verheye *et al.*, 1991; Heinrich, 1993). However in the Benguela ecosystem (off the west coast of Southern Africa) the *C. carinatus* population inhabits the upper layers over the shelf throughout the year (Unteruberbacher, 1964; Andronov, 1976; Verheye *et al.*, 1991, 1992; Timonin *et al.*, 1992). Simultaneously, diapausal CVs exist in deep waters offshore during different phases of upwelling activity (Borchers & Hutchings; 1986; Timonin, 1990; Timonin *et al.*, 1992). These data suggest that the diapausal stock forms not only at the end of upwelling period (as a reaction to unfavourable environmental conditions), but during other periods of the year as well.

It is well known that the formation of the diapausal stock in Calanidae is accompanied by changes in biochemical body composition, respiration and excretion rate, feeding behaviour and mobility (Hirche, 1983; Alldredge *et al.*, 1984; Head & Harris, 1985; Conover, 1988; Drits *et al.*, 1993). These physiological characteristics could be considered as indicators of a diapausal state in calanoid copepods. The present study was aimed at the analysis of the distribution and stage composition of *Calanoides carinatus* and with assessing data on protein/lipid content, respiration rate and gonad development in relation to dormancy in its life cycle.

Material and methods

Sampling and treatment

Sampling and experiments were carried out during a cruise on the R/V *Vitjaz* to the south-west African coast in September–October 1990. At each station prior to plankton sampling, hydrological investigations were carried out. Vertical profiles of temperature, salinity, oxygen and nutrient concentration were drawn with CTDO Mark III (NBIS), and 10-day mean sea surface temperature maps during the month preceding the cruise were analysed. Water samples for chlorophyll *a* determination were also collected. The chlorophyll *a* concentration was estimated with a spectrophotometer 'Specol-10'.

Zooplankton was collected along transects at 17 °S, 23 °S and 25 °S (Fig. 1), the positions of which were selected according to locations of upwelling centres (Shannon, 1985). Along each 110–150 mile transect, 7 or 8 stations 5–20 miles apart were occupied. Plankton was sampled with Judey-OM net (mouth area 0.5 m^2, mesh size 180 μm) hauled vertically from 0–25, 25–50, 50–100, 100–200, 200–300, 300–400, 400–500 and 500–1000 m layers. Recovery speed was 1.2–1.5 m s^{-1}. The volume of water filtered was also calculated. During sampling, the net was lowered to the required depth and closed electronically during ascent. The net-closing system had a depth sensor attached (Flint *et al.*, 1978). Sometimes additional samples were taken to collect sufficient numbers of copepods for physiological experiments. The catches were diluted with cool, 1-μm filtered sea-water and taken to the ship's laboratory. Live, uninjured *Calanoides* were immediately sorted with a binocular microscope, counted, placed in 200 ml jars (40–60 ind. per jar) filled with 1-μm filtered sea-water at near-ambient temperature and kept in a refrigerator for 0.5–1.5 hours. These were used for biochemical analysis, ETS activity measurements and respiration experiments. The rest of the sample was fixed in 4% formaldehyde, buffered with hexamine, for determination of abundance, stage composition and gonad development.

Stage composition and abundance

The abundance of *C. carinatus* was estimated from subsamples or total samples. All specimens were subdivided by sex and copepodite stage, counted and measured. All data were standardized on the basis of the amount of water filtered by the net and expressed as the number of individuals m^{-2} or m^{-3}.

Body biochemical composition

Biochemical assays were performed on freshly collected copepods, i.e. without any fixation or narcotization procedure. Totals of 20 CIV, 10 CV and 10 adults were selected to prepare a homogenate. First, the total body length and oil sac length (l) and diameter (d) were measured with a binocular microscope scale at × 32 magnification. The oil sac volume was estimated as $1/6\ l\ 3.14\ d^2$. During the sorting procedure the specimens of the same copepodite stage were classified into several categories according to the total body length and oil sac size. Whenever possible from 2 to 5 homogenates of each copepodid stage were prepared from the same catch. Lipid and protein determinations were performed on each homogenate. Protein content was measured by the Lowry technique (Lowry *et al.* 1951), with human serum albumin as the standard. Total lipids were extracted in 2:1:1 (by volume) mixture of chloroform, methanol and ethanol. They were then analysed using the sulphophosphovanillin reaction described by Barnes & Blackstock (1973) with olive oil as the standard.

Respiration

Respiration rate was estimated by measuring the electron transport system (ETS) activity of freshly collected copepods as described by Owens & King (1975). On some occasions, the rate of respiration (Q) was determined also by incubating 10 CV copepods in 100 ml

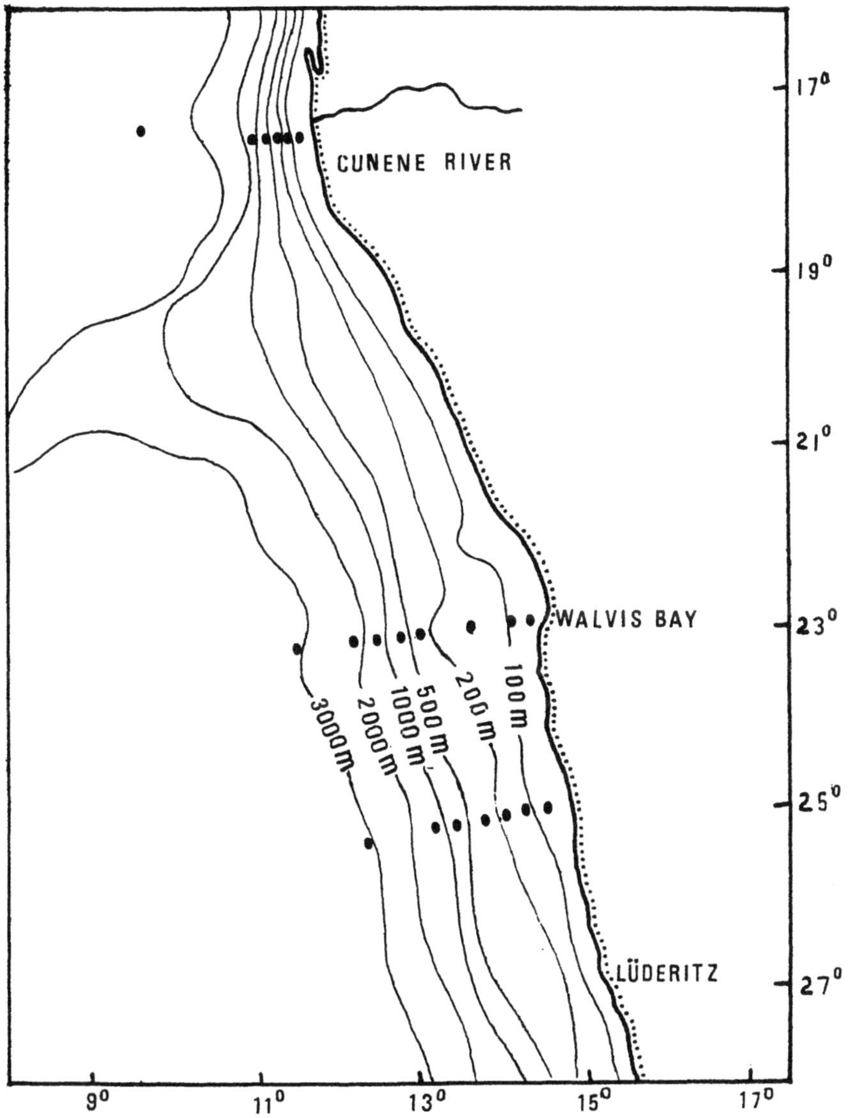

Fig. 1. The area studied and the location of the plankton stations along transects sampled in September–October 1990.

of 1-μm filtered seawater in glass bottles for 24 h at a temperature of 17 °C. Dissolved oxygen concentration was determined by titration according to the Winkler method. The ETS/Q ratio calculated was 2.2 (SD = 0.55, $n = 6$), similar to the values of 1.86–2.65 reported for different species of zooplankton at a range of temperatures from −0.5 to 25 °C (Owens & King, *op. cit*; Bamstedt, 1980; Bidigare *et al.* 1982; Flint *et al.* 1991; Drits *et al.* 1993).

To investigate the influence of temperature on ETS activity two sub-samples from the same homogenate were exposed to different temperatures (Table 1). The mean value, 2.09, of Q_{10} coefficient is similar to the value of 2.27, which was found by Ivleva (1981). We used this coefficient to compare respiration rates measured at different temperatures.

Gonad development

The degree of gonad development was characterized by gonad size. For gonad size measurement we used the staining technique proposed by Tande & Hopkins (1981). The copepods were stained in a 2% Borax carmine solution in filtered sea-water, washed in distilled water, dehydrated in ethanol and cleared in cedar-wood oil. Before staining, the oil sac size in each

Table 1. Calanoides carinatus. ETS-activity ($\mu l\ O_2\ ind^{-1}\ h^{-1}$) at different temperatures (t°C).

Depth, m	Stage	t°C	ETS-activity	Q_{10}
0-50	CV	10	0.61	
		16	0.95	2.01
10-40	CV	8	1.32	
		18	2.32	1.76
0–70	CV	10	0.99	
		18	1.74	2.01
10-50	female	5	0.52	
		15	1.21	2.29
0-50	female	10	0.88	
		20	2.2	2.49
200-500	CV	8	0.5	
		18	0.95	1.90
300-400	CV	10	0.14	
		16	0.23	2.29
300-500	CV	7	0.10	
		15	0.17	1.94
400-600	CV	10	0.21	
		18	0.38	1.76
Mean Q_{10} ± standard deviation				2.09±0.23

specimen was measured with a binocular microscope scale at × 32 magnification. 75 individuals of CV, 30 females and 15 males were examined from surface samples and 130 CV, 30 females and 15 males – from deep samples.

Results

Hydrology and distribution

Typical for the austral spring period, intensification of upwelling (Shannon, 1985) was observed throughout the study area. However the features of upwelling development along each transects were different.

In the northern (17 °S) part of the study area upwelling was strong. The water column was isothermal within 25-miles of the coast. Taking into account the very low concentration of chlorophyll (0.3–0.6 mg m^{-3}) nearshore we assume that 'physical' upwelling had not yet led to enhanced biological production.

A different hydrological regime was observed in the southern part of the region. At 23 °S the peak of 'physical' upwelling was completed. Associated with the temperature stratification was abundant phytoplankton evidenced by a high concentration of chlorophyll (1–15 mg m^{-3}) in the 15-mile nearshore zone. At 25 °S the hydrological regime was similar to that at 23 °, but characterized by more advanced phases of upwelling. Here transformed upwelled waters extended farther offshore and high concentrations of chlorophyll extended offshore to 25-miles.

The vertical distribution and stage composition of *C. carinatus* along the different transects is shown in Figs 2–4. The different hydrological situations between transects were reflected in the abundance and stage composition of the surface part of *C. carinatus* population over the shelf. On the 17 °S transect the population in the shelf waters consisted mostly of older copepodid stages. Total abundance was relatively low and did not exceed 500 ind. m^{-2}. A slight maximum occurred near the shelf break. At 23 °S adult females and CIs dominated over the shelf. Numerically the population of *C. carinatus* was more abundant (up to 5000 ind. m^{-2}) than at 17 °S. All development stages were present in the surface of the shelf area at the 25 °S. The abundance of *C. carinatus* reached 93 000 ind. m^{-2} (mainly younger copepodids) over the mid-shelf.

Unlike the surface population, the abundance of deep-living *C. carinatus* inhabiting the 300–1000 m layer was similar for the three transects varying from 4000 to 8500 ind. m^{-2}. The deep-water stock was dominated by CV (90–95% total abundance) with a small number of adults. CIVs were found at only one station at 23 °S near the shelf break.

Biochemical body composition

Protein and lipid contents of the older stages and adults of *C. carinatus* are listed in Table 2. All data obtained from the three transects were combined into three zones: shelf, slope and offshore. Two size groups (2.1 ± 0.1 and 2.4 ± 0.1 mm total length) were distinguished among the surface copepodid stage Vs. The protein and lipid contents of smaller specimens tended to be lower than those of larger animals. The larger size group of CV specimens was further subdivided according to their oil sac volume:

(1) Copepods with a large oil sac volume (>0.030 mm^3) were characterized by high lipid (72–96 μg ind.$^{-1}$) and protein (35–52 μg ind.$^{-1}$) content. In the upper layer, these copepods were found over the shelf. In the slope zone they inhabited the 100–500 m layer; offshore they were found in 300–500 (1000) m layer.

Table 2. Calanoides carinatus. Protein and lipid body content in older stages and adults at different depths. Values Indicate protein and lipid content as μg ind.$^{-1}$, oil sac volume as mm^3, body total length as mm.

Depth (m)	Stage	Body length	Oil sac volume	Protein content	Lipid content
Shelf					
	CIV	1.7±0.1	No oil sac	12.3±3.6	12.7±2.1
	CV	2.1±0.1	No oil sac	20.1±2.4	29.7±9.3
0–100	CV	2.4±0.1	0.007±0.003	30.0±0.2	33.8±8.9
	CV	2.4±0.1	0.043	37.5	95.9
	female	2.7±0.1	0.001±0.001	59.7±10.9	24.2±4.8
Slope					
	CV	2.1±0.1	No oil sac	17.8±2.8	16.4±6.9
0–100	CV	2.4±0.1	0.010±0.005	33.2±3.8	41.0±0.7
	female	2.7±0.1	0.002±0.002	57.6±4.4	28.2±1.9
	CV	2.3	0.006±0.002	28.2±1.1	32.6±0.9
100–300	CV	2.4±0.1	0.035±0.005	34.4±0.9	68.7±8.2
	female	2.6	0.002±0.002	33.4±5.8	28.2±7.4
	CV	2.3	0.008±0.002	22.5	29.5
	CV	2.4±0.1	0.045±0.005	42.6±9.9	91.7±22.3
300–500	CV	2.4±0.1	0.018±0.003	35.8±2.3	42.8±3.3
	CV	2.4±0.1	No oil sac	18.8±3.0	20.7±0.8
	female	2.5	0.006	14.2	37.6
	male	2.5	0.015	44.9	59.6
Offshore					
100–300	CV	2.4±0.1	0.020±0.002	24.4±1.9	48.0±5.3
	female	2.5	0.005	16.5	33.9
	CV	2.3	0.007±0.003	20.4±3.3	31.0±7.1
	CV	2.5	0.042	52.5	92.4
300–500	CV	2.4±0.1	0.010±0.001	33.1±0.6	41.7±2.3
(1000)	CV	2.4±0.1	No oil sac	22.1±1.4	20.4±1.1
	female	2.5	0.007±0.002	17.8±2.7	30.6±10.4

(2) Copepods with an oil sac volume of 0.010–0.020 mm^3 were characterized by an intermediate lipid content of 30–50 μg ind.$^{-1}$ and protein of 20–35 μg ind.$^{-1}$. These were found throughout the water column from the shore to the end of the transects.

(3) Copepods without a visible oil sac had a lipid content lower than 30 μg ind.$^{-1}$. These were found over the shelf and deeper than 300 m over the slope and in the oceanic waters. In the latter case, not only their lipid, but also protein content was extremely low (19–22 μg ind.$^{-1}$). They seemed close to death judging by their opaque bodies and broken appendages.

Adult females at the surface had a high protein content and low lipid content. The biochemical body composition of females from 300–500 m layer was similar to that of deep CVs with similar oil sac volume. In oceanic waters these females were found not only in the 300–500 m layer but also between 100 and 300 m.

As we chose copepods for biochemical and physiological analyses according to their oil sac sizes we considered the relationship between oil sac volume and measured lipid body content. Copepods having no visible oil sac contained a noticeable amount of lipids ('structural' lipids) (Table 2). The ratio of the

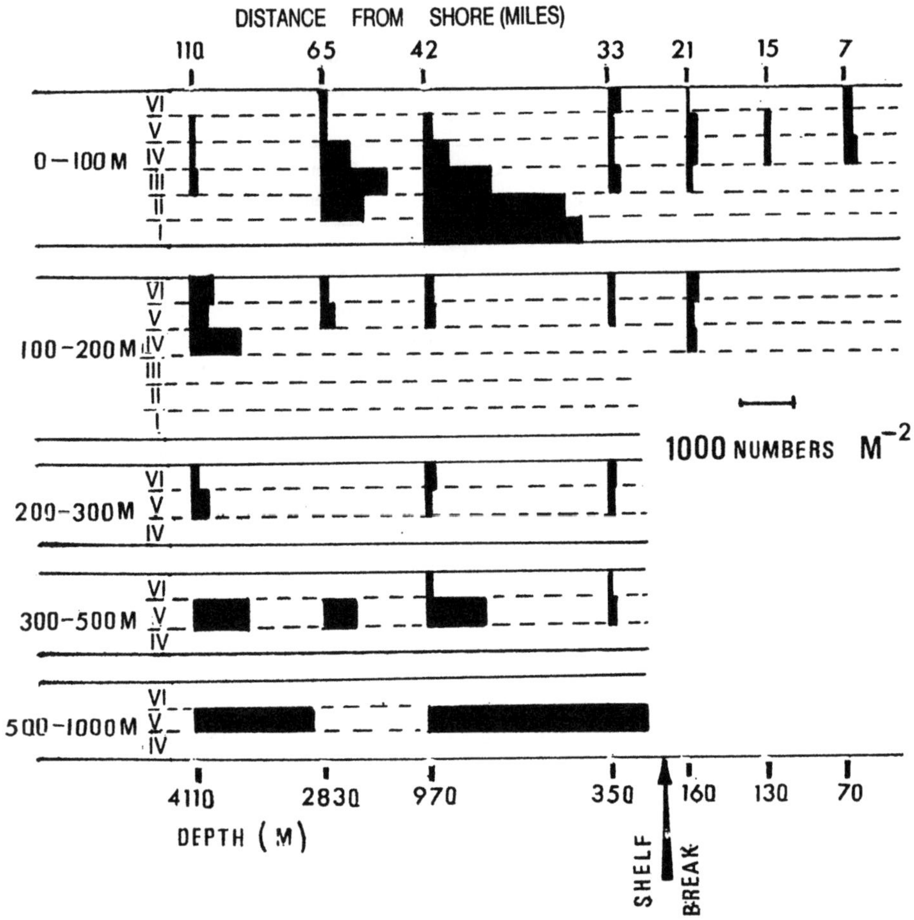

Fig. 2. Vertical distribution and stage composition of *Calanoides carinatus* along the transect at 17 °S (Oct. 15–17). Roman numerals – different developmental stages; stage VI includes both females and males.

weight of these 'structural' lipids (SLW) to protein body weight (PW) was calculated to be 0.90 ± 0.16 ($n = 14$) To obtain an oil sac lipid (reserve lipids) weight (RLW), SLW was subtracted from the total lipid weight (TLW). RLW was plotted against mean oil sac volume (OSV) (Fig. 5). RLW was proportional to OSV and the relationship is described by following equation:

$$(\text{TLW} - \text{SLW}) = \text{RLW} = 4.465 =$$
$$1,183\ \text{OSV}(r^2 = 0.891, n = 47). \qquad (1)$$

Gonad development

The genital system of copepodid stage V of *C. carinatus* resembles that of *Calanus finmarchicus* described by Tande & Hopkins (1981). In the early phase of development, before sex differentiation, it consists of a small gonad, which lies dorsally in the prosome in the first free thoracic segment, and two genital ducts. The length of the undifferentiated gonads is usually 0.10–0.15 mm. Sometimes two small diverticula can be seen at the anterior end of the gonad. In the late phase of CV gonad development males and females could be distinguished. The female system consists of an ovary, two oviducts and ovarian diverticula. The male system comprises a testis and a thin genital duct on the left side of the body. The length of the testis does not exceed 0.25 mm. The length increase in the gonads of female CVs up to 0.75–0.80 mm is due to the development of the ovary and the elongation of diverticula into the head region.

Variation in gonad size in relation to oil sac volume in CVs and adults from different depths is shown in Fig. 6. Gonad size increased as oil sac volume decreased. All CVs with a large oil sac had either small (<0.2 mm) gonads or gonads which were not apparent.

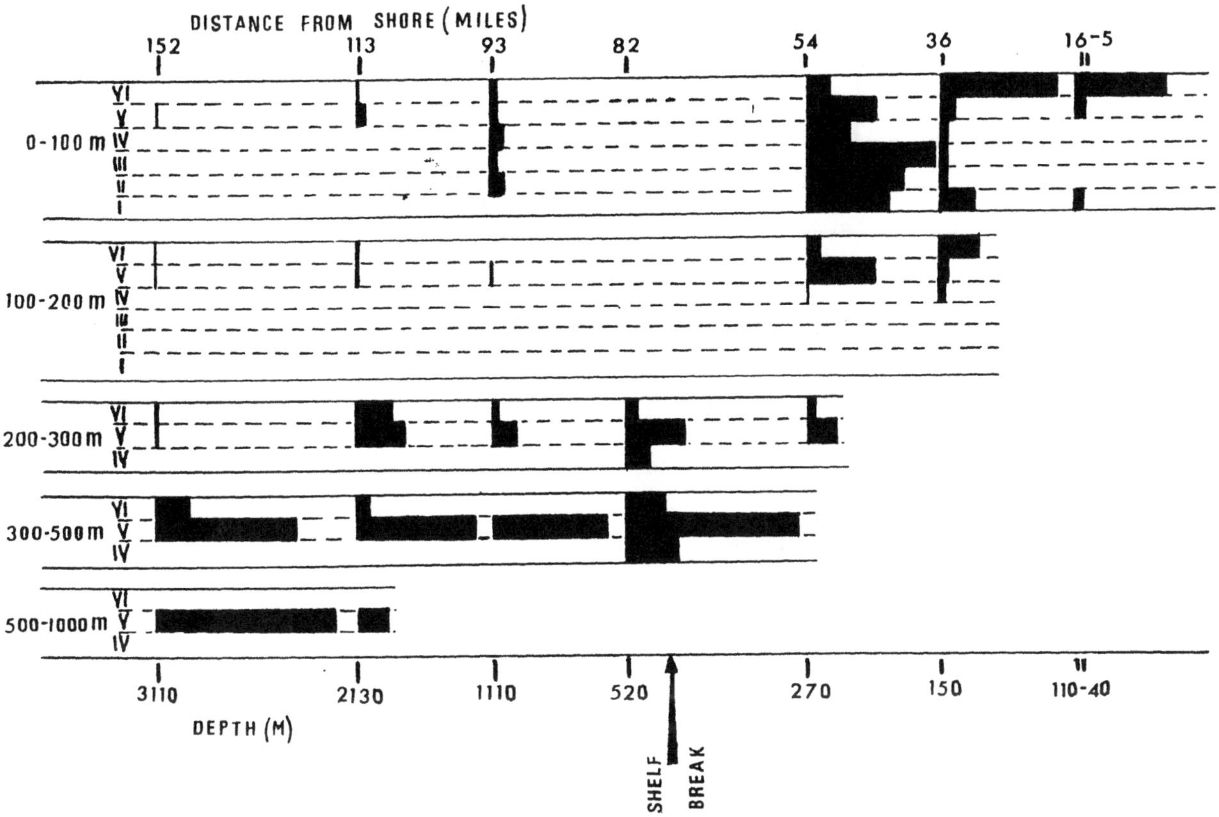

Fig. 3. Vertical distribution and stage composition of *Calanoides carinatus* along the transect at 23 °S (Sept. 28–Oct. 1). Designations as in Fig. 2.

Gonads in surface CVs with a small oil sac showed a wide size range with a predominance of the larger size group. In the deep layer gonad size of CVs with small oil sac volume reached those of adult females from the same depth. Deep-living CVs without any visible oil sac had no gonads; the gonads appear to be resorbed. All females examined from the upper layer had more advanced gonads than those from the deeper layer. Both surface and deep males had larger oil sacs than females.

Respiration

The data on respiration rate of CV and females from different depths are shown in Table 3. The respiration rate of deep females, estimated from ETS-activity, was only about 20% of that of surface females, while specific respiration rate of deep and surface females converted to 10 °C was practically the same. The respiration rate of deep-living CVs decreased dramatically, levelling off at a value 10 to 20% of that of surface animals. The lowest respiration rate was measured in CVs with large oil sacs from 300–500 m depth. In deep CVs with a small oil sacs the respiration rate was two times high-

Table 3. *Calanoides carinatus.* Respiration rate (Q μl O_2 ind^{-1} h^{-1}) estimated from ETS activity and specific respiration (Q_{sp} μl O_2 mg^{-1} protein day^{-1} at 10°C) of females and Stage CV at different depths.

Stage	Total length (mm)	Oil sac volume (mm^3)	Q	Q_{sp}
0–100 m (temperature = 13°C)				
CV	2.0±0.1	0.002±0.002	0.41±0.03 (3)	0.43±0.03
CV	2.2±0.1	0.019±0.007	0.55±0.05 (4)	0.33±0.04
female	2.7±0.1	0.001±0.001	0.59±0.14 (9)	0.18±0.04
300–500 m (temperature = 8°C)				
CV	2.4±0.1	0.044±0.008	0.05±0.02 (6)	0.03±0.01
CV	2.3±0.1	0.011±0.002	0.10±0.02 (4)	0.12±0.02
female	2.5	0.009	0.12 (1)	0.15

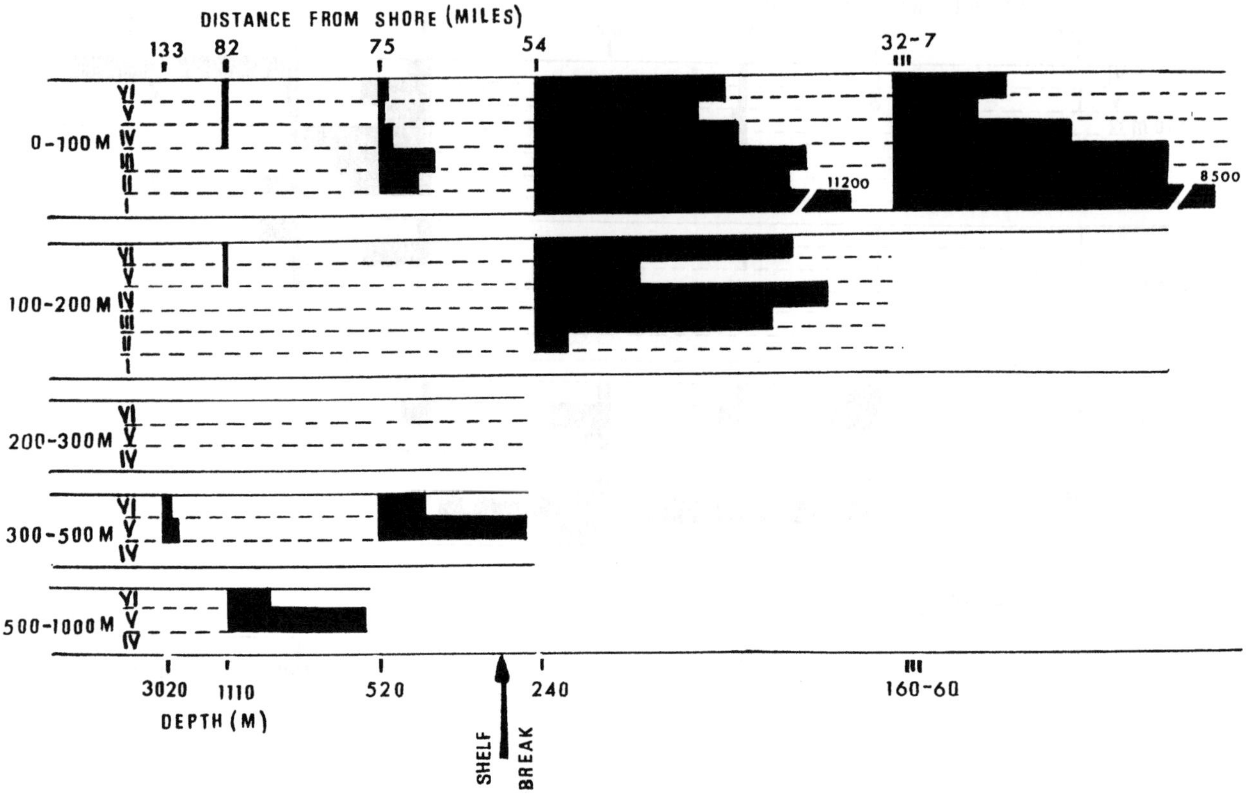

Fig. 4. Vertical distribution and stage composition of *Calanoides carinatus* along the transect at 25 °S (Oct. 4–7). Designations as in Fig. 2.

er; their specific respiration rate increased by a factor of 4 and almost reached that of deep females.

Discussion

Our data on *C. carinatus* distribution show that during the phase of active upwelling the population consists of two parts; the surface part in the shelf and slope zones is represented by all developmental stages, the deep part offshore is dominated by CVs.

A similar pattern has been observed during other seasons in the North Benguela region (Timonin, 1990; Timonin *et al.*, 1992). At the end of upwelling activity (January, 1986), intense development, reproduction, and growth of *C. carinatus* took place in shallow shelf waters. Offshore, copepodid stage Vs dominated depths greater than 300 m. During the quiescent phase (May 1985) 'active' adults and sometimes young stages occupied surface waters inshore while CVs were simultaneously present in deep water near the edge of the shelf and offshore. In the Southern Benguela Verheye *et al.* (1991) reported the presence of diapausing *C. carinatus* in deeper layers in June when upwelling was minimal.

A comparison of *C. carinatus* abundance during different upwelling phases showed considerable variation over the shelf. The number of copepods varied from 4000 ind. m^{-2} in the non-upwelling phase (Timonin, 1990) to 93 000 ind. m^{-2} during the active upwelling (present study) and up to 370 000 ind. m^{-2} in the upwelling relaxation phase (Timonin, 1990). At the same time the abundance of deep-living copepods near the slope in different seasons was relatively constant varying from 4000 to 9000 ind. m^{-2} (Timonin, 1990).

The absence of considerable variation of deep CV abundance at different phases of upwelling is contrary to the traditional viewpoint (Binet & Suisse de Sainte Claire, 1975; Petit & Courties, 1976; Petit, 1982; Smith, 1982) that formation of diapausal stock occurs once a year at the end of upwelling period. The distribution data themselves are insufficient for understanding the physiological processes underlying the formation and termination of diapause. Analysis of physiological parameters enabled us to follow these processes in

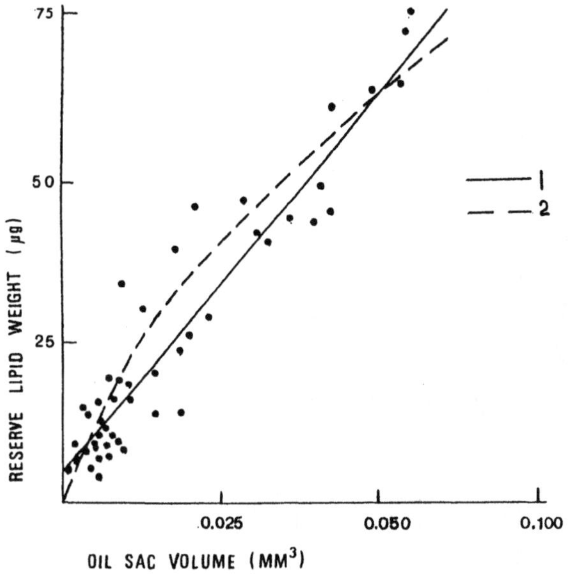

Fig. 5. *Calanoides carinatus*. Reserve lipid weight versus oil sac volume, our data (1) compared with the relationship obtained by Escribano & McLaren (1992) for *Calanus glacialis* (2).

the *C. carinatus* population during the active phase of upwelling.

Timonin *et al.* (1992) found that at high concentrations of phytoplankton the daily ration of *C. carinatus* over the shelf (40–58% body carbon) was much higher than carbon lost through respiration (6.2–7.5% body carbon). Thus the surplus amount of assimilated energy could be allocated to 'structural' growth and fast gonad development or to accumulation of reserve lipids.

Our data indicate that both processes could be distinguished within the surface *C. carinatus* population. In CVs with low lipid content and advanced gonad development close to that of adults the energy seems to be directed into rapid growth and maturation instead of reserves. Apparently, these CVs will moult into adults followed by production of the next generation. CV with high lipid content and undeveloped small gonads transformed surplus energy into reserve lipids and could be considered as forming a diapausal part of the *C. carinatus* population. These 'fat' copepods with undeveloped gonads descend into deeper layers; over the shelf-break they were found in 100–300 m layers; over the slope and further offshore they were caught deeper than 300 m.

The fact that the population of *C. carinatus* existing under the same conditions (temperature, light regime and food abundance) separated into two parts, 'active' with persistent growth and reproduction, and diapausal with suppressed development, suggests an endogenous mechanism of diapause induction. Data on neurosecretion and hormones in active and diapausal copepods obtained by Carlisle & Pitman (1961) for *Calanus finmarchicus* and *C. helgolandicus* point to this possibility.

Hormone secretion is likely to have a genetic basis. In this case each generation of *C. carinatus* would contribute to the formation of diapausal stock, as was shown by McLaren *et al.* (1989) and McLaren & Corkett (1986) for a *C. finmarchicus*.

Diapausal CVs which descended in to deep waters were characterized by a dramatic decrease in respiration rate. Also our observations showed that these copepods had empty guts and low mobility. Similar changes in the physiology of the diapausal copepods in deep layers are known for calanids inhabiting high and temporal latitudes (Hirche, 1983; Head & Harris, 1985; Flint *et al.*, 1991; Drits *et al.*, 1993). High lipid content and low respiration rate enable these copepods to survive for a long time without feeding.

The presence of CVs with low lipids and advanced gonads in deep layers suggest that reserve lipids were expended not only for respiration but also for gonad development. The respiration rate of these CVs being twice as high as that of recently descended copepods was close to the respiration rate of females from the same layer (300–500 m). The biochemical composition, oil sac volume and gonad size of these CVs were also similar to those of the deep adults. These data demonstrate that in *C. carinatus* as in *Calanus finmarchicus* (Miller *et al.*, 1991), *Neocalanus tonsus* (Ohman, 1987) and *N. plumchrus* and *N. cristatus* (Conover, 1988) the moulting of deep-living CVs into adults takes place at depth. Increase of respiration in deep CVs, gonad development, moulting into adults, each of these is an evidence of the termination of diapause.

At present little is known about the factors terminating the diapausal phase in *C. carinatus*. Miller *et al.* (1991) postulated that *C. carinatus* has a purely light-stimulated arousal mechanism. An *a priori* guess of these authors is that light is a direct stimulus; when CVs are moved to the surface layer by upwelled water, they mature, initiate feeding and spawn. However, the postulated mechanism is hardly possible. First, according to Brenning (1985) it is impossible to bring out copepods from 400–1000 m layer by upwelled flows because upwelling does not come from such depths. Second, we show that maturation and moulting of diapausal CV occurs in deep layers. It is more likely that

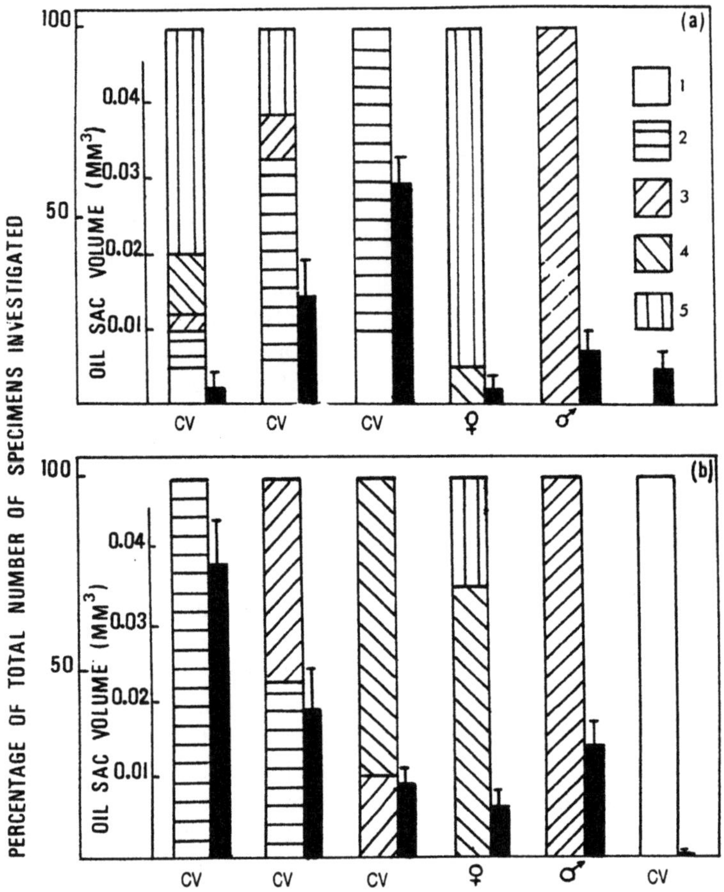

Fig. 6. Calanoides carinatus. Variation in gonad size in relation to oil sac volume in CV and adults. (Gonad size: 1 – gonads not apparent, 2 – <0.2 mm, 3 – 0.2–0.5 mm, 4 – 0.5–0.8 mm, 5 – >0.8 mm). Oil sac volume, mm^3; a)– 0–100 m layer, b)– 300–500 (1000) m layer.

endogenous causes could trigger arousal and maturation of diapausal *C. carinatus* at depth.

Based on respiration rates of diapausal CV and differences in biochemical composition between deep adults and recently descended CV with high protein and lipid content the duration of diapausal period could be estimated. The differences in mean protein and lipid content between CV and deep females were 18.4 and 68.7 μg respectively, i.e. 67.6 μg C (carbon equivalent of lipids is assumed to be 85%, of protein – 50% (Mayzaud, 1986). Given the measured respiration rate of 0.6 μg C ind.$^{-1}$ d^{-1} (respiratory quotient=0.75, after Mayzaud, *op. cit*), this quantity of organic carbon would be sufficient to sustain potential females in copepodid stage V for 112 days. Similar calculations for potential males in copepodid stage V yield a sustainable period of 55 days. The shorter duration of diapause of potential males of *C. carinatus* is in accordance with results of Tande & Hopkins (1981) for *Calanus finmarchicus* in Balsfjorden. They showed that moulting of overwintering CV into males occurred earlier than into females.

The approach used in this study to select animals for biochemical and physiological analyses enabled us to recognize apparently dying specimens among deep-living CVs. The reasons causing these copepods to entirely expend their lipid reserves, preventing gonad development are not clear. Probably they descended with lipid reserves insufficient for gonad development and moulting into adults.

The final stage of development is an ascent of copepods from deep layers to the surface. We can exclude their passive transport by upwelled water as it has already been mentioned that the water never upwells from 400–1000 m depths (Brenning, 1985). We believe that deep adults could ascend actively to the upper layers. During active upwelling by ascending to the 300–200 m layer they could find themselves moved towards shore by upwelled water and thus join the 'active' part of the population. The presence of recently moulted females in our samples from 300–100 m depth confirms this suggestion.

Acknowledgments

The authors are especially grateful to Dr M. Ohman and Dr N. Marcus for their thorough review of the manuscript and valuable comments.

This research was supported by the Program 'Ecological Security of Russia' Grant No 91.4.5.0, Grant of Russian Fond for Fundamental Research (RFFR) Project No 93-05-14044 and Grant of George Soros International Science Foundation (ISF) Project No MP 6000.

References

Alldredge, A. L., B. H. Robinson, A. Fleminger, J. J. Torres, J. M. King & W. M. Hamner, 1984. Direct sampling and *in situ* observation of a persistent copepod aggregation in the mesopelagic zone of Santa Barbara Basin. Mar. Biol. 80: 75–81.

Andronov, V. N., 1976. Some data about the development of *Calanus carinatus (Copepoda, Calanoida)* in South-East Atlantic. Trudy AtlantNIRO 60: 117–130 (In Russian).

Bamstedt, U., 1980. ETS activity as an estimator of respiratory rate of zooplankton populations. The significance of variations in environmental factors. J. exp. mar. Biol. Ecol. 42: 267–283.

Barnes, H. & J. Blakstock, 1973. Estimation of lipids in marine animals and tissues: detailed investigation of the sulphophosphovanilin method for 'total' lipids. J. exp. mar. Biol. Ecol. 12: 103–118.

Bidigare, R. R., F. D. King & D. G. Biggs, 1982. Glutamate dehydrogenase (GDH) and respiratory electron-transport-system (ETS) activities in Gulf of Mexico zooplankton. J. Plankton Res. 4: 895–911.

Binet, D. & E. Suisse de Sante Claire, 1975. Le copepode planctonique *Calanoides carinatus*. Repartition et cycle biologique au large de la Cote d Ivoire. Cah. ORSTOM, Ser. oceanogr. 13: 15–30.

Borchers, P. L. & L. Hutchings, 1986. Starvation tolerance, development time and egg production of *Calanoides carinatus* in the southern Benguela Current. J. Plankton Res. 8: 855–874.

Brenning, U., 1985. Structure and development of calanoid populations (*Crustacea, Copepoda*) in the upwelling regions off North West and South West Africa. Beitr. Meersk. 52, S: 3–33.

Carlisle, D. B. & W. J. Pitman, 1961. Diapause, neurosecretion and hormones in copepoda. Nature 190 (4778): 827–828.

Conover, R. J., 1988. Comparative life histories in the genera *Calanus* and *Neocalanus* in high latitudes of the northern hemisphere. Hydrobiologia 167/168 (Dev. Hydrobiol. 47): 127–142.

Drits, A. V., A. F. Pasternak & K. N. Kosobokova, 1993. Feeding, metabolism and body composition of the Antarctic copepod *Calanus propinquus Brady* with special reference to its life cycle. Polar Biol. 13: 13–21.

Escribano, R. & I. A. McLaren, 1992. Influence of food and temperature on lengths and weights of two marine copepods. J. exp. mar. Biol. Ecol. 159: 77–88.

Flint, M. V., A. G. Timonin, M. V. Geptner & M. S. Polyakov, 1978. Use of the electronic closing device for planktonic nets with a depth recorder ('Murena'). Oceanologija 18: 365–372 (in Russian).

Flint, M. V., A. V. Drits & A. F. Pasternak, 1991. Characteristic features of body composition and metabolism in some interzonal copepods. Mar. Biol. 111: 199–205.

Head, E. J. H. & R. Harris, 1985. Physiological and biochemical changes in *Calanus hyperboreus* from Jones Sound NWT during the transition from summer feeding to overwintering condition. Polar Biol. 4: 99–106.

Heinrich, A. K., 1993. Comparative ecology of the plankton oceanic communities. Nauka, Moscow, 159 pp. (In Russian).

Hirche, H.-J., 1983. Overwintering of *Calanus finmarchicus* and *Calanus helgolandicus*. Mar. Ecol. Prog. Ser. 11: 281–290.

Ivleva, I. V., 1981. Temperature of environment and exchange rate of aquatic animals. Naukova Dumka, Kiev. 231 pp.(In Russian)

Longhurst, A. R., 1967. Vertical distribution of zooplankton in relation to the eastern Pacific oxygen minimum. Deep Sea Res. 14: 51–63.

Lowry, O. N., N. J. Rosenbrough, A. L. Farr & R. J. Randall, 1951. Protein measurements with the folin phenol reagent. J. Biol. Chem. 193: 265–275.

Mayzaud, P., 1976. Respiration and nitrogen excretion of zooplankton. 4. The influence of starvation on the metabolism and the biochemical composition of some species. Mar. Biol. 37: 47–58.

McLaren, I. A. & C. J. Corkett, 1986. Life cycles and production of two species of copepods on the Scotian shelf, eastern Canada. Syllog. (Nat. Mus. Can.). 58: 362–367.

McLaren, I. A., M. J. Tremblay, C. J. Corkett & J. C. Roff, 1989. Copepod production on the Scotian Shelf based on life-history analyses and laboratory rearings. Can. J. Fish. aquat. Sci. 14: 560–583.

Miller, C. B., T. J. Cowles, P. H. Wiebe, N. J. Copley & H. Grigg, 1991. Phenology in *Calanus finmarchicus*; hypotheses about control mechanisms. Mar. Ecol. Progr. Ser. 72: 79–91.

Ohman, M. N., 1987. Energy sources for recruitment of the subantarctic copepod *Neocalanus tonsus*. Limnol. Oceanogr. 32: 1317–1330.

Owens, T. G. & F. D. King, 1975. The measurement of respiratory electron-transport-system activity in marine zooplankton. Mar. Biol. 30: 27–36.

Petit, D., 1982. *Calanoides carinatus* (copepode pelagique) sur le plateau continental congolais. 3. Abondance, tailles et temps de generation. Relations avec la chlorophylle. Oceanogr.trop. ORSTOM 17: 155–175.

Petit, D. & C. Courties, 1976. *Calanoides carinatus* (Copepode pelagique) sur le plateau continental congolais. 1. Apercu sur la repartition batymetrique des stades; generations durant la saison froide 1974. Cah. ORSTOM, Ser.oceanogr. 14: 177–199.

Shannon, L. V., 1985. The Benguela ecosystem. 1. Evolution of the Benguela, physical features and processes. In M. Barnes (ed.), Oceanography and Marine Biology. An Annual Review 23. The University Press, Aberdeen: 105–182.

Smith, S. L., 1982. The northwestern Indian Ocean during the monsoons of 1979: distribution, abundance, and feeding of zooplankton. Deep Sea Res. 29: 1331–1353.

Smith, S. L., 1984. Biological indications of active upwelling in the northwestern Indian Ocean in 1964 and 1979, and a comparison with Peru and Northwest Africa. Deep Sea Res. 31: 951–967.

Tande, K. S. & C. C. E. Hopkins, 1981. Ecological investigations of the zooplankton community of Balsfjorden, northern Norway: the genital system in *Calanus finmarchicus* and the role of gonad development in overwintering strategy. Mar. Biol. 63: 159–164.

Thiriot, A., 1978. Zooplankton communities in the West African upwelling area. In R. Boje & M. Tomczak (eds), Analysis of Upwelling Systems. Springer, Berlin: 32–60.

Timonin, A. G., 1990. Composition and zooplankton distribution in the Benguela upwelling area off Namibia. Oceanology 30: 651–655 (translated from Russian).

Timonin, A. G., E. G. Arashkevich, A. V. Drits & T. N. Semenova, 1992. Zooplankton dynamics in the northern Benguela ecosystem, with special reference to the copepod *Calanoides carinatus*. In A. I. L. Payne, K. H. Brink, K. H. Mann & R. Hilborn (eds), Benguela Trophic Functioning. S. Afr. J. mar. Sci. 12: 545–560.

Unteruberbacher, N. K., 1964. Zooplankton studies in the waters off Walvis Bay with special reference to the Copepoda. Invest. Rep. mar. Res. Lab. S. W. Afr. 11: 42 pp. + Plates 2–36.

Verheye, H. M., L. Hutchings & W. T. Peterson, 1991. Life history and population maintenance strategies of *Calanoides carinatus* (Copepoda: Calanoida) in the southern Benguela ecosystem. S. Afr. J. mar. Sci. 11: 179–191.

Verheye, H. M., L. Hutchings, J. A. Huggett & S. J. Painting, 1992. Mesozooplankton dynamics in the Benguela ecosystem, with emphasis on the herbivorous copepods. In A. I. L. Payne, K. H. Brink, K. H. Mann & R. Hilborn (eds), Benguela Trophic Functioning. S. Afr. J. mar. Sci. 12: 561–584.

Hydrobiologia **320**: 209–222, 1996.
V. R. Alekseev & G. Fryer (eds), Diapause in the Crustacea.

Growth and fecundity of *Daphnia* after diapause and their impact on the development of a population

Kęstutis Arbačiauskas & Zita Rasuolė Gasiūnaitė
Institute of Ecology, Akademijos St. 2, Vilnius 2600, Lithuania
(Present address of ZRG: University of Klaipėda, Sportininkų St. 13, Klaipėda 5813, Lithuania)

Key words: Daphnia, diapause, growth, maturity, clutch size, intrinsic rate of increase

Abstract

Laboratory and field investigations revealed that the life history traits of exephippial and parthenogenetic generations of *Daphnia* differ substantially. Daphniids hatching from resting eggs grow faster and their definitive body sizes are bigger than of hatchlings from subitaneous eggs. Size at maturity for exephippial animals is significantly larger. In spite of this, they mature a few days earlier than parthenogenetic females. In this study, the difference was 3–4 days for the laboratory experiments and 1–3 days for the field. Fecundity of the exephippial generation is markedly higher. Here, the clutch size for this generation was up to 3.5–4.0 times as large as for the parthenogenetic generation. Moreover, obtained results suggest that the relationship between clutch size and body length for both generations differ significantly.

Estimates of the intrinsic rate of increase for field *Daphnia* populations demonstrated that life history traits of exephippial animals lead to a two or threefold higher rate of increase in the conditions of invertebrate predation pressure. Under moderate fish pressure, obtained *r* values for the daphniids hatching from resting eggs were larger than those from subitaneous. High growth rate of exephippial females is disadvantageous only under the conditions of severe pressure by fish. Obtained results suggest that hatchlings from diapausing eggs enable an acceleration of population increase by several times during the beginning of the development of a population with periodical re-establishment from resting eggs.

Introduction

Diapause is an adaptation that enables many animals to survive harsh environmental conditions. It is prevalent in crustaceans living in variable habitats which undergo diapause in various life cycle stages (see Alekseev, 1990). 'Cladoceran' species have embryonic diapause. These cyclical parthenogens, under special circumstances, predominately during unfavorable conditions, initiate sexual reproduction resulting in the production of resting eggs (ephippia). Resistant diapausing eggs are able to survive conditions, which frequently eliminate adult animals, and remain viable for extended time periods (e.g. Moritz, 1987; De Stasio, 1990).

Analysis of diapause in 'Cladocera' has been the subject of many studies. Generally two major trends emerge. The first, dealing with diapause as a phenomena, comprises studies on the mechanisms of its induction and termination (Stross & Hill, 1965, 1968; Stross, 1966, 1969a, b, 1971, 1987; Carvalho & Hughes, 1983; Schwartz & Hebert, 1987; Kleiven *et al.*, 1992) and its evolution in 'Cladocera' (see Hebert, 1978, 1987). The second emphasizes consequences of diapause for temporal dynamics of populations and communities. These are studies on the pool of resting eggs in sediments, dormancy and emergence patterns and their contribution to changes in populations density, species composition, genetic structure and clonal diversity (Kankaala, 1983; Herzig, 1985; Moritz, 1987; Carvalho & Wolf, 1989a, b; De Stasio, 1990). Effects of diapause from a long-time scale perspective tend to be analyzed by these studies.

As long ago as 1880, Weismann (cited from Alekseev, 1990) reported that exephippial females substantially differ from those produced by parthenogenesis. Their fecundity is higher and they do not produce sexual eggs. Presently well documented (Makrushin, 1985; Zaffagnini, 1987) is that energy reserves stored in the cytoplasm of resting eggs are markedly greater than those in the parthenogenetic eggs. Different from those of subitaneous eggs, the structure of the cytoplasm, and the concentration and chemical composition of reserves in sexual eggs are thought to be responsible for the endurance of resting eggs during harsh environmental conditions, and the dissimilarity of individual traits of hatchlings of exephippial and parthenogenetic generations. However, until recently there has been a lack of information regarding the life history traits of exephippial animals in comparison with those produced by parthenogenesis, and the contribution of these differences to population dynamics. Such information is of particular relevance in the study of the onset of seasonal development of populations with periodical re-establishment from resting eggs. This relatively short time duration may have critical significance for the seasonal success of a species.

The purpose of this research is as follows: (1) an investigation of the growth, size at maturity, time to maturity, and fecundity of exephippial and parthenogenetically produced generations of *Daphnia*; and (2) an assessment of the impact of potential differences of these traits on the development of a population of daphniids, namely the impact on the intrinsic rate of increase for a population after diapause. Life history characteristics of both generations were studied in the laboratory as well as in the field. Estimations of the intrinsic rate of increase have been accomplished for two field *Daphnia* populations. These populations originate each year by hatchlings from resting eggs. During study seasons in 1986 and 1987, even at their onset, the populations were experiencing different size-specific mortality due to the pressure of predators with dissimilar size-specific selectivity of feeding (Arbačiauskas, 1992).

Study sites

Field investigations were conducted in two small neighbouring forest lakes during the growing seasons of 1986 and 1987. These lakes, the Bolshoy Okunyonok and Maly Okunyonok (referred to below as BO and MO) are situated in the SW. part of the St.-Petersburg Region, Russia (58 °50′N, 29 °30′E). The water surface area of BO is 5.8 ha, with a maximum depth of 3.0 m. The area of MO is 2.7 ha, with a maximum depth of 2.1 m.

Deprived of aboriginal fish populations in 1963, Lakes BO and MO were used for rearing fingerlings of whitefishes and carp. Originally dystrophic lakes, they were limed and fertilized during the period of exploitation. The summer phytoplankton community was dominated by Cyanobacteria. Each year Lake BO was stocked with larval whitefishes (*Coregonus sp.*) and carp (*Cyprinus carpio*) fry. MO was devoid of fish. More information on the phytoplankton and further details of these lakes can be found in Alimov *et al.* (1992).

No overwintering daphniids were found in either lake. The dominant zooplankton species in Lake MO was *Daphnia longispina*. During the summer, on average, 97–99% of the biomass of planktonic crustaceans were comprised of *D. longispina*. Other constant species were the copepods *Eudiaptomus gracilis*, *Thermocyclops crassus*, and *Mesocyclops leuckarti*. *Chaoborus flavicans* larvae were the planktivorous predators in this lake. 'Cladocerans' in the zooplankton of Lake BO, among which the most abundant were *Daphnia galeata* and *Bosmina longirostris*, comprised 20–25% of the biomass. Copepods, namely *Thermocyclops crassus* and *Mesocyclops leuckarti*, were dominant in the plankton of this lake. The mean seasonal individual weight of planktonic crustaceans here was only one tenth of that in Lake MO (Arbačiauskas, 1992; Krylov *et al.*, 1993).

Resting eggs for the laboratory study of exephippial and parthenogenetic generations of *Daphnia pulex* in 1992, were collected from a small permanent pond in Vilnius, Lithuania. The area of the pond is some 40 m^2, with a maximum depth of 1.5 m. The pond is sustained by underground waters. Surface and near bottom water temperatures usually do not exceed 16–18 °C and 10–12 °C respectively. During 1992, the zooplankton of the pond in early spring was dominated by *Cyclops strenuus*. In May, when water temperatures reached 8–10 °C, *D. pulex* appeared. Later, *D. pulex* and *Chaoborus flavicans* larvae were the most abundant zooplankton species.

Materials and methods

Laboratory

Ephippia of *Daphnia pulex* were collected in the pond after the ice cover had melted on March 27, 1992. They were immediately taken to the laboratory and approximately 150 ephippia were placed into each of two 3-litre vessels filled with pond water. The temperature of one vessel was maintained at 11 °C, and the other, at 18 °C. Hatching at both levels of temperature proceeded the next day.

During March 29–30, three exephippial cohorts with 12–15 individuals each, one at 11 °C, and two at 18 °C, were initiated. Offspring of exephippial individuals formed the first two parthenogenetic cohorts, initiated on April 6 at the higher, and on April 14 at the lower temperature. Young of these started, accordingly, the second two parthenogenetic cohorts, on April 24 at 18 °C and on May 8 at 11 °C. Analogically, the third parthenogenetic cohort at 18 °C was initiated on May 3. The number of animals within cohorts was the same. Each cohort was maintained in a one litre vessel, under a field photoperiod. The *Daphnia* were nourished with natural food, in other words water for cultivation was taken from the pond and changed twice a week. Deviations of temperature did not exceed 11–12 °C and 17–19 °C for the duration of each treatment.

The animals usually were measured every 2–3 days. No less than 8 randomly selected individuals of each cohort were measured on each occasion to the nearest 0.025 mm. The appearance of the first eggs in the brood pouch was deemed to be the moment of maturing. Mean clutch sizes were assessed by the number of young, which were regularly removed from vessels, and by the number of eggs per female, during occasions when it was possible to count eggs.

Student's *t*-tests were used to assess differences between selected life history characteristics of exephippial and parthenogenetic generations. Growth curves fitted to the experimental data on the cohorts' growth are presented as

$$L_t = L_\infty - (L_\infty - L_0) \cdot e^{-k \cdot t}, \qquad (1)$$

where L_t is the length at time t, and L_0 is the length of newborns. Parameters L_∞ and k were estimated using the Marquardt method.

Field

Sampling. – Life history traits of exephippial and parthenogenetic generations of *Daphnia* in Lakes BO and MO were studied during late May and June of 1987. Zooplankton was sampled every three days from both lakes. Quantitative sampling was performed with a modified 8-litre Schindler-Patalas trap equipped with a 120 μm plankton net, at three stations, randomly distributed within the central part of the lake. At each station, samples taken from three depths (surface, 1.5 metre and near the bottom) were combined into one composite sample. Analysis of the mature part of the *Daphnia* populations was on the basis of zooplankton sampled with a plankton net. Live samples were immediately transported to the laboratory. Subsamples were placed into a Bogorov tray and fixed with 4% formalin. Depending on daphniid abundance, from 30 to 160 adults per sample were measured from the anterior edge of the head to the base of the tail spine, and the eggs in their brood pouches counted. The smallest gravid female on a sampling date was regarded as the size at maturity. Size-frequency distributions of adult females were established. Fecundity (eggs per adult female), clutch size (eggs per clutch) and linear regressions of clutch size on body length for both *Daphnia* generations were calculated.

Growth. – Size-specific growth rates of *Daphnia longispina* and *D. galeata* in the lakes during the early season were estimated over periodically conducted short-term experiments. 'Cladocerans' and water for the study were collected in the central part of the lake. For each experiment, 50–100 individuals representing all sizes of *Daphnia* in the lake on the date of experiment initiation were measured to the nearest 0.025 mm and subdivided into 5–9 size classes with 8–12 individuals with similar body size in each class. Individuals of each size class were placed in a 250 ml vessel filled with fresh lake water previously strained through 0.3 mm mesh net to remove other daphniids. For treatment to occur, 'cladocerans' were narcotized by a 0.5% urethane (ethyl carbamate $CO(NH_2)OC_2H_5$) solution. Experimental vessels were maintained in the lake for 2, with the youngest animals, or 3 days, and were turned over 5–6 times a day to prevent food particles from settling. After the termination of the experiment individuals were measured and the fecundity of the adults was counted. Such short-term experiments adequately portray the size-specific growth rates of *Daphnia* in natural conditions (Arbačiauskas, 1988).

Experimental size increments and initial sizes were used to calculate the growth curves (1) parameters L_∞ and k:

$$L_\infty = \frac{A}{B} \quad \text{and} \quad -k = \ln(1 - B),$$

where A and B are the intercept and slope of linear regression of L_{t+1}-L_t on L_t (Arbačiauskas, 1992).

Essential to specify is that obtained growth curves derived by the method described are not traditional growth curves depicting somatic growth of a cohort or

Table 1. Time to maturity (TTM, days), length at maturity (LAM, mm) and mean clutch size (CS, young per clutch) for the first three broods for exephippial (EX) and parthenogenetic (PA) cohorts of *Daphnia pulex* in growing experiments at different temperatures. Values are means ± 1 SD.

	11–12 °C			17–19 °C		
	TTM	LAM	CS	TTM	LAM	CS
EX	10.2 ± 0.7	2.62 ± 0.10	38.7 ± 0.6	5.1 ± 1.0	2.31 ± 0.14	37.8 ± 1.3
PA	14.0 ± 1.7	2.00 ± 0.13	10.0 ± 4.2	9.0 ± 1.5	1.98 ± 0.14	9.7 ± 4.6
t	6.4	12.7	11.3	8.9	7.3	14.4
df	22.8	20.5	5.4	37.0	33.4	9.9
P	< 0.001	< 0.001	< 0.001	< 0.001	< 0.001	< 0.001

individual from birth to definitive sizes. Instead, these growth curves are integrated estimates of size-specific growth rates of various sized animals during the study period.

Intrinsic rate of increase

Estimates of size-specific mortality for both *Daphnia* species during the early season were obtained from the instantaneous compartmental model of a population with continuous recruitment (Arbačiauskas, 1992; Krylov & Arbačiauskas, 1994). This model for the analysis of populations of planktonic crustaceans was first proposed by Argentesi *et al.* (1974). Size-specific fecundity (m_L = birth rate) for exephippial and parthenogenetic generations was assessed as: $m_L = C_L/D_E$, where C_L is the clutch size for a female of length L, calculated from obtained regressions of clutch size on body length, and D_E is the egg development time, estimated according to Botrell *et al.* (1976) as a function of water temperature.

Having estimates of size-specific growth rates and size at maturity, size-specific mortality and size-specific fecundity can be readily transformed into the survivorship function (l_x) and age-specific fecundity (m_x) (Arbačiauskas, 1992). It was assumed that all adult females produce eggs. The life-tables for hypothetical cohorts reflecting the life history traits of exephippial and parthenogenetic generations of *Daphnia* during the early season, were obtained in this manner. By following this method intrinsic rate of increase (r) can be readily calculated using the equation (Paloheimo & Taylor, 1987)

$$1 = \sum_{0}^{\infty} l_x \cdot m_x \cdot e^{-r \cdot x} \cdot \frac{1 - e^{-(r+d_x)}}{r + d_x}, \qquad (2)$$

Table 2. Fecundity (F, eggs per adult female) and clutch sizes (eggs per clutch), for exephippial (EX) and parthenogenetic (PA) generations of *Daphnia longispina* in Lake MO and *Daphnia galeata* in Lake BO during the early part of the 1987 season. Values are means ± 1 SD. Number of animals counted in parentheses.

Date	F	Clutch size EX	Clutch size PA
		Daphnia longispina	
20/05	32.1 ± 10.7(58)	32.7 ± 9.9(57)	–
23/05	30.9 ± 16.8(31)	35.5 ± 12.5(27)	–
26/05	13.5 ± 9.1(55)	25.6 ± 10.9(14)	9.3 ± 2.3(41)
30/05	12.6 ± 11.6(53)	32.4 ± 9.5(11)	8.4 ± 3.0(37)
02/06	10.1 ± 9.7(86)	20.7 ± 7.5(31)	5.7 ± 3.5(39)
05/06	7.6 ± 7.8(74)	16.9 ± 7.6(21)	6.2 ± 3.1(33)
08/06	1.5 ± 3.0(161)	7.9 ± 3.7(18)	3.3 ± 1.7(32)
		Daphnia galeata	
30/05	13.2 ± 9.2(9)	14.9 ± 8.3(8)	–
02/06	16.1 ± 15.6(39)	19.0 ± 15.2(33)	–
05/06	24.3 ± 17.2(35)	24.3 ± 17.2(35)	–
08/06	15.8 ± 13.6(96)	22.3 ± 12.5(62)	7.7 ± 1.7(17)
11/06	14.4 ± 12.8(74)	26.7 ± 8.3(33)	6.5 ± 4.1(29)
14/06	7.2 ± 9.9(85)	28.2 ± 7.6(12)	6.3 ± 4.2(44)

where $d_x = \ln(l_x/l_{x+1})$ is the instantaneous age-specific mortality rate.

Results

Laboratory experiments

The somatic growth rates and maximum body sizes for the exephippial generation were significantly higher in comparison with the parthenogenetic generation (Fig. 1). Neonates which had hatched from resting eggs were bigger than those from subitaneous eggs, with a mean length value of 0.76 mm *versus* 0.68 mm. The size at maturity of exephippial individuals was 1.2–1.3 times larger than of individuals produced by parthenogenesis (Table 1). In spite of this, they mature earlier. The first eggs in the brood pouches of exephippial animals were recorded on the 4th–5th day at 18 °C and on the 10th day at 11 °C from the cohort's initiation. The earliest eggs in cohorts of parthenogenetic origin appeared on the 7th–8th and 12th–13th day at corresponding temperatures. On average the time to maturity for the animals hatched from resting eggs was shorter by 4 days (Table 1). Size at maturity of exephippial daphniids at 11 °C was significantly ($t=6.0$, df$=21.3$, $P<0.001$) larger than at 18 °C. Parthenogenetically derived cohorts matured at the same sizes at 11 °C, but at 18 °C size at maturation for the second cohort (1.78 ± 0.10 mm) was significantly smaller than for the first (2.10 ± 0.08 mm, $t=5.62$, df$=13$, $P<0.001$). The third cohort matured at 1.98 ± 0.12 mm. This size did not differ significantly either from the first ($t=2.40$, df$=10$, $P=0.032$) or from the second cohort ($t=2.13$, df$=12$, $P=0.053$). The range of deviation of size at maturity for the first parthenogenetic cohort at 18 °C (2.0–2.2 mm) fully overlapped with those obtained for the exephippial animals at corresponding temperature (2.0–2.5 mm), but on average difference was highly significant ($t=4.04$, df$=22$, $P<0.001$).

The fecundity of studied generations differed greatly. The mean clutch size for the first three broods of the exephippial generation was nearly 4 times larger than that of individuals of parthenogenetic origin (Table 1). Amongst the subitaneous cohorts, this characteristic was markedly lower only for the second one at 18 °C. It was 6.3 ± 2.5 young per clutch in comparison with the first and the second cohorts, with 12.0 ± 6.1 and 11.0 ± 3.6 young respectively. The variance between separate broods for the sexual cohorts, as well as their somatic growth, was significantly lower than for the others (Fig. 1, Table 1). Food conditions during juvenile development and early maturity of exephippial cohorts which had been started simultaneously, were much more alike than for those of parthenogenetic origin which had been started on different dates. Data for cohorts from subitaneous eggs indicated this. Nevertheless, the opinion regarding this finding is that it is an indication of the greater independence of life history traits of exephippial animals from environmental conditions, than in those derived from parthenogenetic eggs.

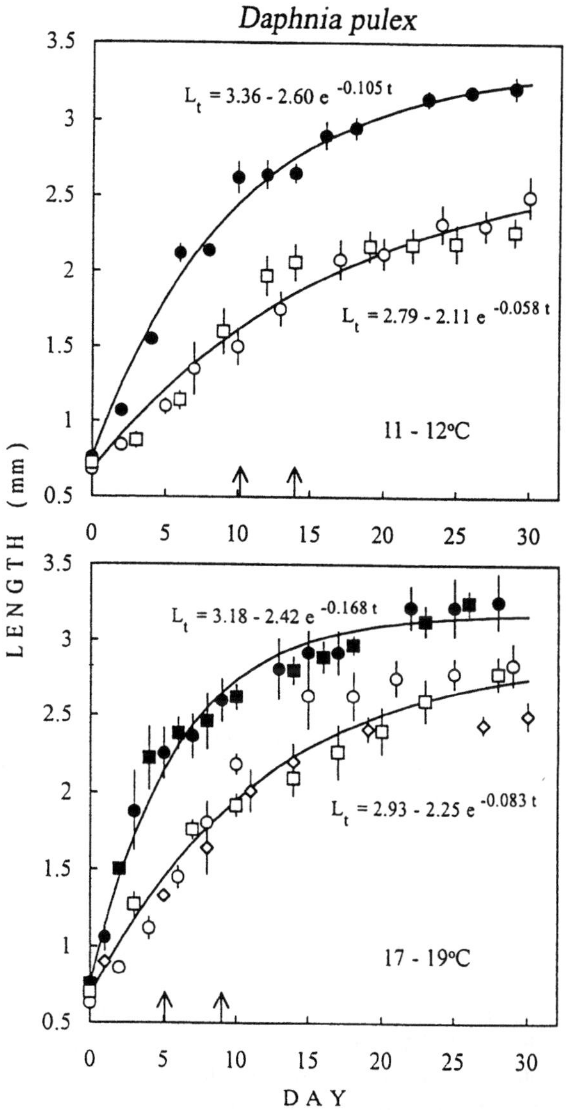

Fig. 1. Growth of exephippial (filled symbols) and parthenogenetic (open symbols) cohorts of *Daphnia pulex* in growth experiments at different temperatures. Arrows (left for exephippial cohorts) indicate the mean time to maturity. Exephippial cohorts, two at 18 °C and one at 11 °C, were started during March 29–30. The first parthenogenetic cohorts (circles) were started on April 6 at 18 °C and on April 14 at 11 °C, the second (squares) – on April 24 at 18 °C and on May 8 at 11 °C, and the third (rhombuses) – only at 18 °C on May 3. Values are means ± 1SD.

Field

Field data analysis presented here comprises only the period from the appearance of *Daphnia* species in the

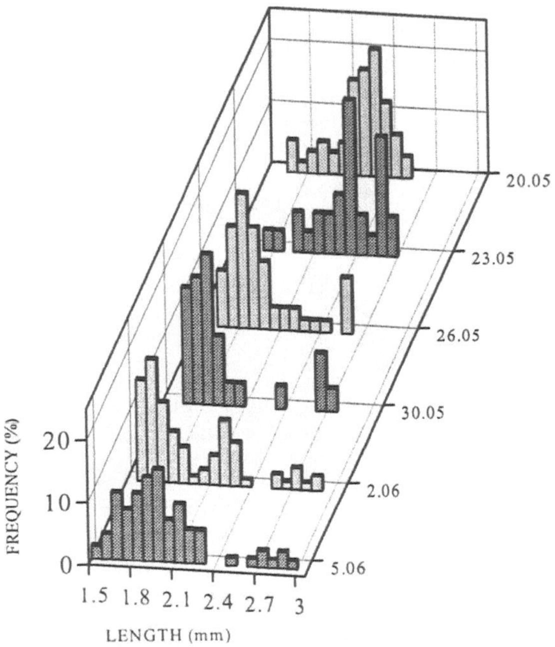

Fig. 2. Size-frequency distributions of the mature segment of the *Daphnia longispina* population in Lake MO during the early part of the 1987 season.

central part of the lakes up to the time of disappearance of exephippial individuals from the zooplankton samples.

During 1987, the first *Daphnia longispina* individuals in Lake MO were recorded in early May. Surface water temperature had warmed to 12–14 °C by this time. In early May, only juveniles with body sizes 0.50–1.12 mm were present. Their density was around 2 ind. l^{-1}. To May 20, density of daphniids increased to 15 ind. l^{-1}. Adult females constituted 5% of the entire population. By June 2, density and adult contribution were rising to 120 ind. l^{-1} and 23%. Latter, until nearly complete disappearance of exephippial animals on June 11, density varied between 80–110 ind. l^{-1} and the proportion of mature females decreased to 5%. Size-frequency distributions of *D. longispina* adult females during this period are presented in Fig. 2. The distributions illustrate that separate cohorts could be distinguished. The size at maturity was decreasing during this time from 1.70–1.75 mm, on May 20–23, to 1.68, on May 26–30, 1.60, on June 2, and 1.50 mm, on June 5–8. The maximum sizes of daphniids were continuously increasing (Fig. 2).

The number of eggs per adult female during the period of analysis fell dramatically from 32 to 1 egg (Table 2), and later was even below one egg per female (Arbačiauskas, 1992). On June 2, males and resting eggs were recorded in the *D. longispina* population. On June 5–8, 2–4% of the population was comprised of males, and 12–14% of adult females had ephippia.

Daphnia galeata abundance in the central part of Lake BO during late May was very low, no more than 0.1 ind. l^{-1}. Mainly neonates were present and only two gravid females were recorded. From May 30 to June 14, numbers increased from 0.5 to 16.0 ind. l^{-1}. Later on, because of severe *Coregonus* fry predation (Arbačiauskas, 1992; Krylov *et al.*, 1993), exephippial large-sized daphniids disappeared. The proportion of adult females in the population varied from 13% on May 30, to 20%, on June 2, and by mid-June had declined to 8%. Size-frequency distributions of mature daphniids are presented in Fig. 3. Decrease in maturation size of *D. galeata* in the early season was greater than for *D. longispina* in Lake BO. This life history characteristic changed from 1.55 mm on May 30, to 1.35 on June 2–5, to 1.20 on June 11, and to 1.10 mm on June 14. During July, values of size at maturity fell even to 0.78 mm. Maximum sizes were up to 2.25–2.50 mm (Fig. 3).

The number of eggs per adult female during the early part of June was increasing. From 13 on May 30, it grew to 24 on June 5, then decreased to 7 by mid-June (Table 2). Males appeared on June 8. On June 14, they constituted 12% of the total number of daphniids. No resting eggs were recorded.

The growth rates of various sized animals in Lake MO were studied in four experiments for *D. longispina*: on 20–23; 23–26; and 27–30 of May; and June 8–11. Three experiments in Lake BO were conducted for *D. galeata*: on 3–6; 5–8; and 9–12 of June. Temperatures in the surface water, where the experiments were performed, varied between 12–18 °C in Lake MO and 14–18 °C in Lake BO during this time. The mean water column temperature was 12–15 °C in the former lake and 13–14 °C in the latter.

The results of experiments showed that, in some vessels, especially in the first two experiments for both species, the length increments of individuals initially of the same size differed significantly. Concurrently, these increments divided into two groups with little deviation within them. For example, in the first experiment with *D. longispina* daphniids with an initial body length (± 1SD) 0.56 ± 0.02 mm divided

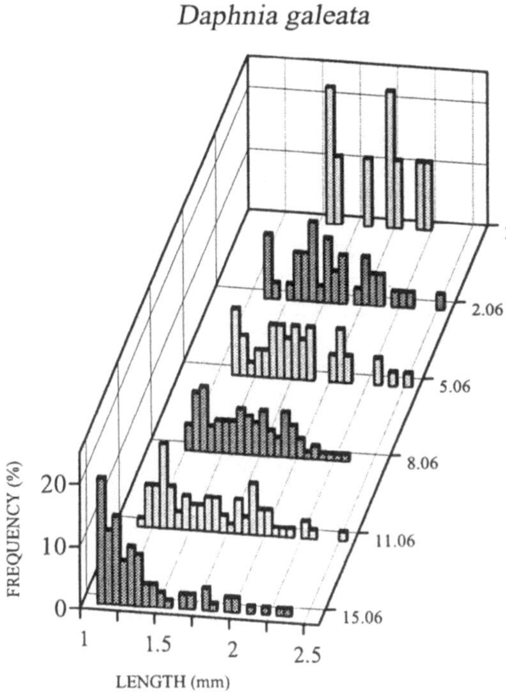

Fig. 3. As Fig. 2, but for *Daphnia galeata* in Lake BO.

into two length classes with mean values 0.75 ± 0.03 and 1.03 ± 0.03 mm ($t = 16.6$, df$= 12$, $P<0.001$) after two days. In the second experiment with *D. galeata*, initially at 0.82 ± 0.07 mm length, individuals were 1.03 ± 0.06 and 1.37 ± 0.08 mm length ($t = 6.6$, df$= 5$, $P<0.001$) after two days.

For the assessment of the average size-specific growth rates in the population, mean increments were utilized to calculate parameters of the growth curves (Arbačiauskas, 1992). The purpose of this was to investigate differences between daphniids, hatching from resting and subitaneous eggs. Experimental results presented in the previous section indicated substantially differing growth rates of exephippial and parthenogenetically derived animals. With this in mind, it was hypothesized that recorded differences in length increments were related to the type of egg from which the individual had been derived. Under this assumption, and the undoubted derivation of large bodied animals in the first and second experiments from the exephippial generation, recorded length increments were divided into two sets. One was assumed to portray the growth of the exephippial generation, and the other, the growth of the parthenogenetic generation during the early season.

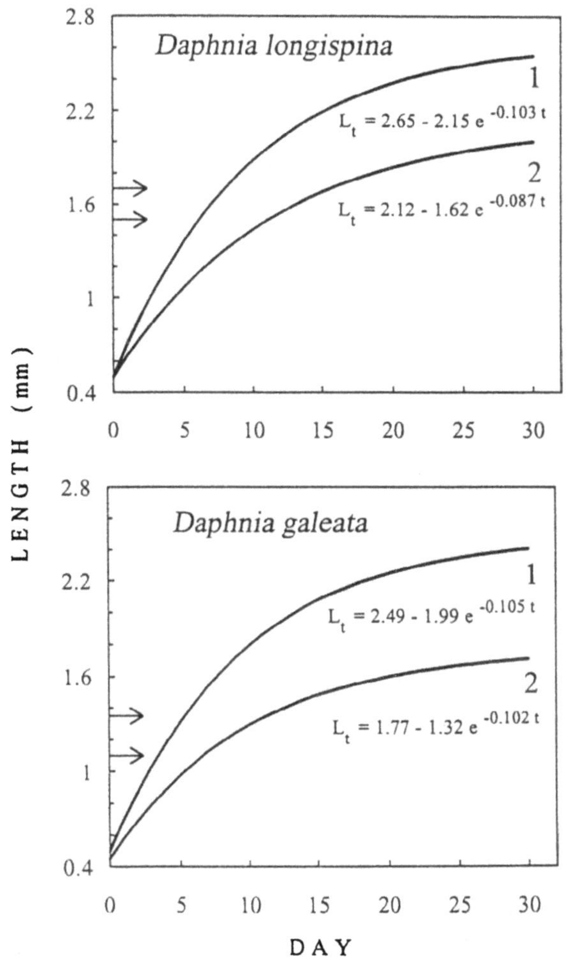

Fig. 4. Estimates of size-specific growth of exephippial (1) and parthenogenetic (2) generations of *Daphnia longispina* in Lake MO and *Daphnia galeata* in Lake BO during the early part of the 1987 season. Arrows (upper for exephippial generations) indicate the size at maturity.

The calculated growth curves (equation 1), as estimates of size-specific growth for *D. longispina* in Lake MO and *D. galeata* in Lake BO are presented in Fig. 4. The typical sizes at maturity for exephippial and parthenogenetic generations were determined to be the sizes at the beginning and end of the analyzed time periods, 1.70 mm and 1.50 mm for *D. longispina* and 1.35 mm and 1.10 mm for *D. galeata*. The time to maturity at these lengths was shorter for exephippial than for parthenogenetic daphniids, 7.9 and 11.0 days respectively for *D. longispina*, and 5.3 compared to 6.6 days for *D. galeata*.

To assess the fecundity of exephippial and parthenogenetic generations in the field, adult females were subdivided according to their derivation from one or other of these generations. Naturally, in the field these generations overlap and no clear boundary exist between them. Nevertheless, it is obvious that in lakes with no overwintering individuals, the first adult females are exephippial animals.

Fecundity was high in the *D. longispina* population on May 20, and 85% of the total population were individuals with lengths less than 0.85 mm. No doubt many of them were of parthenogenetic origin. Least plentiful were preadult individuals with lengths of 1.15–1.50 mm. They constituted only 0.7% of the population. On May 23, the abundance of preadult females had substantially increased, and on May 26, the first mature parthenogenetic females appeared. From this date, subdivision of all adult females was based on obtained growth curves and size-frequency distributions of mature daphniids.

The increase of preadult *D. galeata* females was recorded on June 5, and the first adult parthenogenetic individuals appeared on June 8, when the size at maturity had begun to fall. Adult females were subdivided, by analogy with *D. longispina*, into the same two categories.

The clutch size of exephippial animals was significantly larger than that of those of parthenogenetic origin (Table 2). To test whether such was merely the result of larger body sizes of individuals which had hatched from resting eggs, the relationships between clutch size and body length for both generations were established. Values, recorded over the separate dates, which had not clearly indicated this relationship difference, were pooled by the nearest sampling dates, and linear regressions calculated (Figs 5, 6). A plot of intercepts *versus* slopes of obtained regressions (Fig. 7) showed that relationships between clutch size and body length of the exephippial generation differed significantly from those of the parthenogenetic generation. The relationships for both generations of *D. longispina* were the same only on July 8.

The values of the slope and the intercept were decreasing for *D. longispina* during the study period. This, as well as the decline in clutch size for both generations (Table 2), indicated reduced availability of food for this species in Lake MO. During May 20–30, when, apparently, there was no limiting of fecundity due to any decrease in food supply, the mean clutch size for exephippial females was 3.6 times larger than that of females of parthenogenetic origin. Later, this difference between generations decreased to 3.2 times on June 2–5, and 2.4 times on June 8.

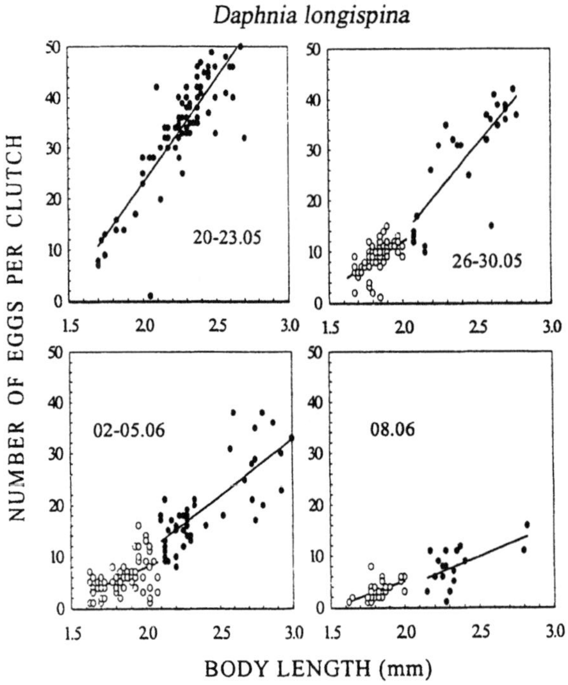

Fig. 5. Relationships between clutch size and body length for exephippial (filled circles) and parthenogenetic (open circles) generations of *Daphnia longispina* in Lake MO during the early part of the 1987 season.

For *D. galeata*, of which the abundance in Lake BO was only about a tenth of that of *D. longispina*, and where, therefore, food conditions were better, the mean clutch size of exephippial females was increasing (Table 2). This was due to the growth of mean body size of gravid daphniids derived from resting eggs. Mean clutch size for females of parthenogenetic origin at the time was approximately the same (Table 2). The decrease in values of the slope and intercept for the second distinguished time period was apparently caused by the presence of very large females with up to 76 eggs per clutch during the first time period (Figs 6, 7). Later, such females were not recorded. The mean clutch size for the exephippial generation of *D. galeata* was 3.5 times larger than for the parthenogenetic generation.

Intrinsic rate of increase

An attempt was made to assess the intrinsic rate of increase for exephippial and parthenogenetic generations at the beginning of population development after diapause, when animals of both generations can usually fully realize their abilities to grow and repro-

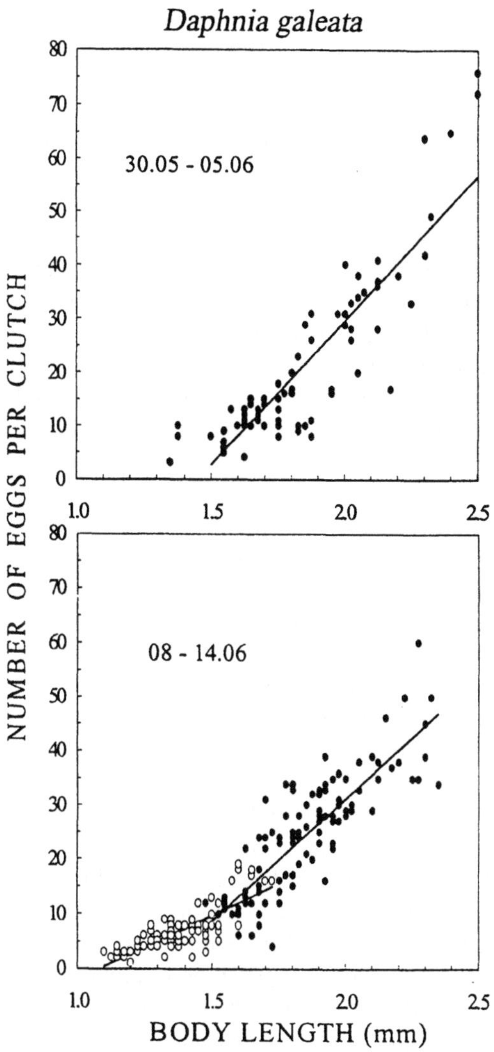

Fig. 6. As Fig. 5, but for *Daphnia galeata* in Lake BO.

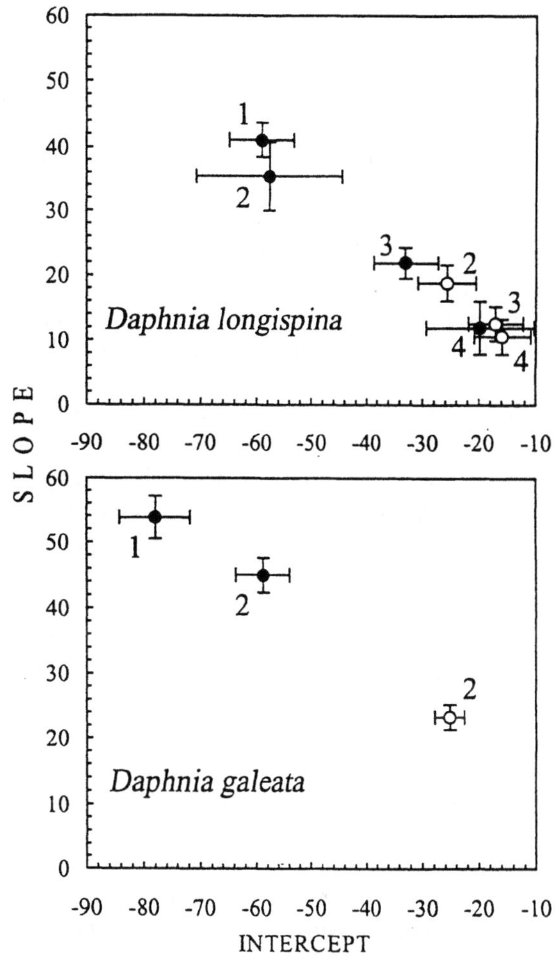

Fig. 7. Plots of slopes *versus* intercepts (± 1SE) of linear regressions of clutch size on body length for exephippial (filled circles) and parthenogenetic (open circles) generations of *Daphnia longispina* in Lake MO during May 20–23 (1), May 26–30 (2), June 2–5 (3) and June 8 (4); and *Daphnia galeata* in Lake BO during May 30–June 5 (1) and June 8–14 (2). Note, that these regressions are presented in Figs 5 and 6.

duce, and no suppression exists due to food limitations. Fecundity data for *Daphnia longispina* showed a decline in feeding conditions during the period of analysis. Therefore, the typical duration for the beginning of population development of this species was determined to be the period from 20 to 30 of May. General regressions of clutch size on body length during this time had intercepts equal to -43.6 ± 6.8 for the exephippial and -25.7 ± 5.1 for the parthenogenetic generation, and slopes equal to 33.1 ± 2.9 and 18.8 ± 2.8 respectively. These relationships for *D. galeata* were calculated for all data. Intercepts and slopes obtained were -68.6 ± 4.0 and 49.6 ± 2.2 for exephippial, and -25.2 ± 2.6 and 23.3 ± 1.9 for parthenogenetic animals. For the purpose of r calculations, the smallest clutch was determined to be at no less than 7 eggs for females from resting eggs, and 2 eggs for females from subitaneous eggs.

Water temperatures of 13 °C and 14 °C were assumed to be typical for early *Daphnia* population development in the central parts of Lakes MO and BO, respectively. Under these temperatures egg development times were equal to 5.5 days for *D. longispina* and 4.9 days for *D. galeata*. The size-specific growth rates and sizes at maturity for both generations used are presented in Fig. 4.

During 1986 and 1987, conditions for *Daphnia* populations in both lakes were widely different. *D. longispina* in Lake MO was preyed upon by *Chaoborus* larvae, and *D. galeata* was pressured by peled fingerlings. The influence of these planktonic predators, having dissimilar size-specific selectivity of feeding, resulted in different size-specific mortality of daphniids, as early as the beginning of the growing season (Arbačiauskas, 1992). For calculating *r*, estimates of size-specific mortality for the early part of the 1986 and 1987 seasons were utilized (Fig. 8). These estimates were expected to be typical of the beginning of the daphniids' population development in the lakes. Size-specific mortalities for adults of *D. longispina* in 1986 and 1987 differed significantly, primarily due to the fact that the estimate for 1986 was obtained for the period following the disappearance of the exephippial generation. A zero mortality of mature daphniids was assumed, therefore, for the purpose of comparability of *r* assessments for this species over both years.

The calculated values of the intrinsic rate of increase (equation 2) for the exephippial and parthenogenetic generation, and for the hypothetical generation with somatic growth and size at maturity of the first, and size-specific fecundity of the second generation, are presented in Fig. 8. Standard errors of the intercept and slope of regressions of clutch size on body length were utilized for estimations of maximum and minimum *r* values.

The highest intrinsic rates of increase were obtained for the exephippial generation of *D. longispina* with mean values of 0.269 d^{-1} for 1986 and 0.230 d^{-1} for 1987. The parthenogenetic generation at the beginning of population development increased with an *r* value equal to 0.126 and 0.076 d^{-1} during both years, respectively. Intermediate to these was the value for the exephippial *D. galeata* generation in Lake BO in 1986 of 0.106 d^{-1}. Daphniids of this species from subitaneous eggs under the size-specific mortality of 1986, indicated an *r* value equal to 0.026 d^{-1}. Only for *D. galeata* in 1987, intrinsic rates of increase for both generations had negative values. Moreover, *r* was a somewhat higher for the parthenogenetic than for the exephippial generation. This parameter on the average was equal to −0.142 *versus* −0.151 d^{-1} (Fig. 8).

Discussion

Laboratory and field investigation revealed that the life history traits of exephippial and parthenogenetic generations of *Daphnia* differ substantially. Daphniids hatching from resting eggs grow faster, and their definitive body sizes are bigger, than of hatchlings from subitaneous eggs. Size at maturity for exephippial animals is significantly larger. In spite of this, they mature a few days earlier than parthenogenetic females (Figs 1, 4; Table 1). In this study, the difference was 3–4 days for the laboratory experiments and 1–3 days for the field. The higher fecundity of animals hatching from resting eggs than of hatchlings from subitaneous eggs is well known for *Daphnia* species (see Alekseev, 1990). Here, the clutch size of exephippial females was up to 3.5–4.0 times as large as for those of parthenogenetic origin. Moreover, obtained results suggest that the relationship between clutch size and body length for the exephippial generation significantly differ from the relationship for the parthenogenetic generation (Figs 5, 6 and 7).

Different quality and higher energy resources of *Daphnia* resting eggs in comparison with subitaneous eggs (Zaffagnini, 1987) are at least in part responsible for dissimilar life history traits for both generations. However, a physiological explanation for such differentiation is not yet available.

The cohorts of parthenogenetic origin in this study were initiated from the first clutches. It has been shown (Green, 1956; Lynch, 1989; Tessier & Consolatti, 1989; Ebert, 1991; Glazier, 1992) that neonates released from this clutch of a daphniid normally are smaller than those from later broods, and this results in smaller size at first reproduction. Lampert (1993) has reported a mean difference of 10 and 17% in size at maturity of the first-brood animals as compared with the second and the fifth brood in a single clone of *Daphnia magna*. The first-clutch neonates, in the cited study, were 10 and 8% smaller than the neonates released from the former and the latter brood respectively. Investigation under controlled food and temperature conditions of two clones of *Daphnia pulex* isolated from two ponds in Vilnius, other than those from which ephippia for study were collected, (Arbačiauskas, *unpublished data*) revealed that the mean size of neonates from subsequent broods, according to the generally accepted notion, increased, but at the same time the individual lengths of animals from different broods strongly overlapped. This life history characteristic for the first four broods was on average 0.62, 0.65, 0.66 and 0.68 mm, with a range of intra-clutch deviation of 0.43–0.67 mm, basically 0.57–0.67 mm and a few very small neonates of 0.42–0.45 mm length, for the first, and 0.55–0.70, 0.60–0.70 and 0.60–0.71 mm for the later broods respectively. On the other hand, the mean difference of 11% in length of

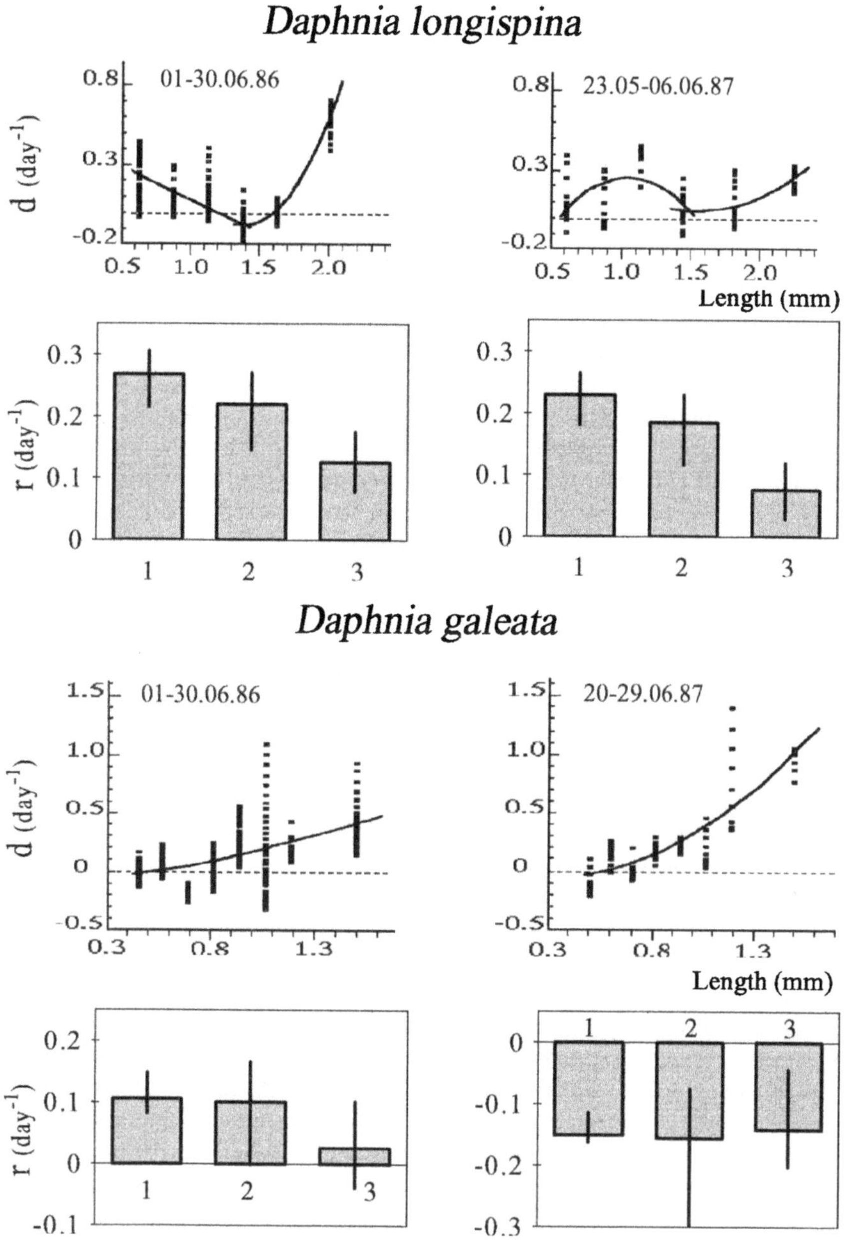

Fig. 8. Estimates of size-specific mortality (*d*) and intrinsic rate of increase (*r*) during the early part of the season for different generations of *Daphnia longispina* and *Daphnia galeata*: exephippial (1), parthenogenetic (3) and hypothetical (2), with growth and size at maturity of the first, and fecundity of the second generation. Size-specific mortalities used are estimates for 1986 and 1987 (data from Arbačiauskas, 1992). Size-specific growth and fecundity rates used are estimates for 1987. Vertical bars represents maximum and minimum estimates of *r* derived by using in calculations standard errors of the intercept and slope of regressions of clutch size on body length, with the values of other parameters being the same. See materials and methods, and results for calculating details.

neonates between the first and the fourth brood resulted in an 11% smaller mean size at first reproduction, 1.72 *vs* 1.93 mm respectively. Concurrently, the number of juvenile instars and time to maturity in both cases were the same.

Returning to the initiation of parthenogenetic cohorts from the first broods, it is likely that in spite of selection of the most viable neonates, which usually are the largest, the observed sizes at maturity in the subitaneous cohorts were slightly smaller than they should have been in cases where the later broods were used. However, it is hardly probable that employment of later broods had abolished the significant difference in size at first reproduction between the cohorts of exephippial and parthenogenetic origin, at least in the treatment at lower temperature where this difference was as great as 24%.

The purpose of this research was to assess differences of life history traits between exephippial and parthenogenetic daphniids in conditions reflecting natural circumstances. Therefore, natural pond water was used for laboratory experiments. Data on the phytoplankton dynamics in the pond during spring are not available. However, there are reasons to suppose that food reserves for *Daphnia* in the pond water were not very poor. For the duration of the experiment, daphniids, which are the most effective planktonic filtrators, capable of exhausting nutritional resources in a short time, were absent in the pond, or their population was increasing from 40 on May 8 to 630 ind. l^{-1} on June 2, when the experiment was already finished.

It is obvious that food supply for *Daphnia* during the experiment varied. These changes were most pronounced in the cohorts of parthenogenetic origin at 18 °C. Variable availability of food, as it is well established (Green, 1954; Lei & Clifford, 1974; Orcutt & Porter, 1984; Lynch, 1989; McCauley *et al.*, 1990), resulted in differences of growth rate, size at first reproduction and clutch size. On the whole, deviations of somatic growth rate and fecundity were smaller in exephippial than in parthenogenetic cohorts in this study (Table 1; Fig. 1). This finding, although food conditions in the experiment were not controlled, leads to a hypothesis that exephippial animals are less dependent upon environmental conditions than are parthenogenetically derived individuals. Nevertheless, the degree of such independence remains unknown and warrants further investigation.

During the early development of *Daphnia* populations in the field, size at maturity was gradually declining (Figs 2, 3). No abrupt decrease of this life history trait was observed, as might be expected from laboratory experiments. The most probable explanation is that biotic impacts had governed the gradual change of size at maturity, because the water temperature effecting the size of primiparae (Lei & Clifford, 1974; Brambilla, 1982) over the period of analysis was approximately the same. It is well documented (Green, 1954; Lei & Clifford, 1974; Orcutt & Porter, 1984; Lynch, 1989; McCauley *et al.*, 1990) that size at first reproduction increases and decreases in response to food limitation. Resource depression was obvious during the early development of the *D. longispina* population in Lake MO. Chemical substances released by fish or invertebrate predators could alter life history traits in 'Cladocera' as well (Vanni, 1987; Stibor, 1992; Weider & Pijanowska, 1993). It seems possible that such a chemical cue was responsible for the larger decrease in the size at maturity of *D. galeata* in Lake BO. It is unlikely that the shift in maturing size could be the result of selection for different genotypes in such a short period of time. Moreover, enclosure experiments in this lake demonstrated an increase in size at maturity of *D. galeata* from 0.92 to 1.15 mm in a week due to the removal of fish (Arbačiauskas, 1992). Even under such a decrease of maturation size, juvenile development time for exephippial daphniids in Lake BO was shorter by 1.3 days as compared with animals from subitaneous eggs (Fig. 4).

Estimates of the intrinsic rate of increase for field *Daphnia* populations revealed that, in comparison with parthenogenetic females, life history parameters of exephippial animals lead to a more than a two or threefold higher rate of increase under the conditions of invertebrate predation pressure (Fig. 8). Under such circumstances, around 60–70% of this increase is due to faster growth and maturation enabling body sizes to pass through those most vulnerable to predation more rapidly.

Under moderate fish pressure, as had been prevalent in Lake BO during the early season of 1986 (Arbačiauskas, 1992), life history traits of exephippial individuals also result in higher *r* value, 4 times larger by our assessments. Nearly the full impact on *r* was due to earlier maturation, but not to higher fecundity. High growth rate of daphniids hatching from resting eggs is disadvantageous only under the conditions of severe pressure by fish, as had been typical in 1987. Only in these conditions was the estimate of intrinsic rate of increase for the parthenogenetic generation slightly larger than for the exephippial (Fig. 8). Apparently, assessments of size-specific mortality used for *r* calculations for 1987 were higher than they had been during the beginning of *D. galeata* population development, because the population was growing during this time period.

In summary, the results suggest that diapause permits *Daphnia* not only to disperse and survive unfavorable conditions, but enables an acceleration of population increase by several times during the beginning of the development of a population periodically re-established from resting eggs. A difference in the quality of subitaneous and diapausing eggs is at least in part responsible for the faster growth and development as well as the higher fecundity of hatchlings, which results in the more rapid increase of populations derived from exephippial than from subitaneous eggs.

Acknowledgments

We thank A. A. Umnov for consulting on data analysis and P. I. Krylov for helpful discussions. We are grateful to an anonymous reviewer for his valuable critical comments and G. Fryer for considerable improvement of the language of the manuscript. This research was performed at the Zoological institute, St.-Petersburg, Russia, and Institute of Ecology, Vilnius, Lithuania.

References

Alekseev, V. R., 1990. Diapauza rakoobraznykh: Ekologo-fiziologicheskiye aspekty [Diapause in crustaceans: Ecologo-physiological aspects]. Nauka, Moscow, 144 pp (in Russian).

Alimov, A. F., L. E. Anokhina, E. V. Balushkina, S. M. Golubkov, T. V. Khlebovich, P. I. Krylov, E. V. Paveljeva, I. V. Telesh & L. P. Umnova, 1992. Fish impact on the structure, abundance, and biomass of plankton and benthos of two small shallow lakes: A three-year study. Russian Journal of Aquatic Ecology 1: 9–37.

Arbačiauskas, K., 1988. Dinamika rosta *Daphnia longispina* O. F. Müller v ozere Maly Okunyonok [The dynamics of growth of *Daphnia longispina* O. F. Müller in Lake Maly Okunyonok]. In L. A. Kutikova (ed.), Ghidrobiologicheskie issledovaniya morskikh i presnykh vod [Hydrobiological investigations of marine and fresh waters]. Zoological Institute, Leningrad: 73–88 (in Russian).

Arbačiauskas, K., 1992. Dinamika populyatsiy i functsional'niye svyazi planktonnykh rakoobraznykh v ekosisteme ozer [Population dynamics and functional relations of planktonic crustaceans in lake ecosystems]. Diss.... kand. biol. nauk [Ph. D. Thesis]. Zoological Institute, St.-Petersburg, 246 pp (in Russian).

Argentesi, F., R. de Bernardi & G. Di Cola, 1974. Mathematical models for the analysis of population dynamics in species with continuous recruitment. Mem. Ist. ital. Idrobiol. 31: 245–275.

Botrell, H. H., A. Duncan, Z. M. Gliwicz, E. Grygierek, A. Herzig, A. Hillbricht-Ilkowska, H. Kurasawa, P. Larsson & T. Weglenska, 1976. A review of some problems in zooplankton production studies. Norw. J. Zool. 24: 419–456.

Brambilla, D. J., 1982. Seasonal variation of egg size and number in a *Daphnia pulex* population. Hydrobiologia 97: 233–248.

Carvalho, G. R. & R. N. Hughes, 1983. The effect of food availability, female-density and photoperiod on ephippia production in *Daphnia magna* Straus (Crustacea, Cladocera). Freshwat. Biol. 13: 37–46.

Carvalho, G. R. & H. G. Wolf, 1989a. Resting eggs of lake-*Daphnia*. I. Distribution, abundance and hatching of eggs collected from various depths in lake sediments. Freshwat. Biol. 22: 459–470.

Carvalho, G. R. & H. G. Wolf, 1989b. Resting eggs of lake-*Daphnia*. II. In situ observations on the hatching of eggs and their contribution to population and community structure. Freshwat. Biol. 22: 471–478.

De Stasio, B. T., Jr., 1990. The role of dormancy and emergence patterns in the dynamics of a freshwater zooplankton community. Limnol. Oceanogr. 35: 1079–1090.

Ebert, D., 1991. The effect of size at birth, maturation threshold and genetic differences on the life-history of *Daphnia magna*. Oecologia (Berlin) 86: 243–250.

Glazier, D. S., 1992. Effect of food, genotype, and maternal size and age on offspring investment in *Daphnia magna*. Ecology 73: 910–926.

Green, J., 1954. Size and reproduction in *Daphnia magna* (Crustacea: Cladocera). Proc. zool. Soc. Lond. 124: 535–545.

Green, J., 1956. Growth, size and reproduction in *Daphnia* (Crustacea: Cladocera). Proc. zool. Soc. Lond. 126: 173–204.

Hebert, P. D. N., 1978. The population biology of *Daphnia*. Biol. Rev. 53: 387–426.

Hebert, P. D. N., 1987. Genetics of *Daphnia*. In R. H. Peters & R. De Bernardi (eds), *Daphnia*. Mem. Ist. ital. Idrobiol. 45: 439–460.

Herzig, A., 1985. Resting eggs – a significant stage in the life cycle of the crustaceans *Leptodora kindti* and *Bythotrephes longimanus*. Verh. int. Ver. Limnol. 22: 3088–3098.

Kankaala, P., 1983. Resting eggs, seasonal dynamics and productivity of *Bosmina longispina maritima* (P. E. Muller) (Cladocera) in the northern Baltic proper. J. Plankton Res. 5: 53–69.

Kleiven, O. T., P. Larsson & A. Hobaek, 1992. Sexual reproduction in *Daphnia magna* requires three stimuli. Oikos 65: 197–206.

Krylov, P. I. & K. Arbačiauskas, 1994. Are invertebrate predators capable to control populations of large herbivores? Contribution of *Chaoborus* to the mortality of *Daphnia* in a small fishless lake. Russian J. aquat. Ecol. 3: 1–13.

Krylov, P. I., K. Arbačiauskas & B. L. Gutelmacher, 1993. Vliyanie ryb na dinamiku biomassy, chislennosti i produktsii populyatsiy zhivotnykh: Populyatsii planktonnykh rakoobraznykh i lichinok *Chaoborus* [Effects of fish on the dynamics of biomass, abundance, and productivity of populations of animals: Populations of planktonic crustaceans and *Chaoborus* larvae]. In A. F. Alimov (ed.), Bioticheskiye vzaimosvyazi v ekosistemakh ozer-pitomnikov [Biotic interactions in the ecosystems of nursery lakes]. Ghidrometeoizdat, St.-Petersburg: 224–237 (in Russian).

Lampert, W., 1993. Phenotypic plasticity of the size at first reproduction in *Daphnia*: the importance of maternal size. Ecology 74: 1455–1466.

Lei, C.-H. & H. F. Clifford, 1974. Field and laboratory studies of *Daphnia schoedleri* Sars from a winterkill lake of Alberta. Natl. Mus. Canada, Publ. Zool. 9: 1–53.

Lynch, M., 1989. The life history consequences of resource depression in *Daphnia pulex*. Ecology 70: 246–256.

Makrushin, A. V., 1985. Anghidrobioz pervichnovodnykh zhivotnykh [Anhydrobiosis in primary aquatic animals]. Nauka, Leningrad, 104 pp (in Russian).

McCauley, E., W. W. Murdoch & R. M. Nisbet, 1990. Growth, reproduction and mortality of *Daphnia pulex* Leydig: life at low food. Functional Ecology 4: 505–514.

Moritz, C., 1987. A note on the hatching and viability of *Ceriodaphnia* ephippia collected from lake sediments. Hydrobiologia 145 (Dev. Hydrobiol. 35): 309–319.

Orcutt, J. D., Jr. & K. G. Porter, 1984. The synergistic effects of temperature and food concentration on life history parameters of *Daphnia*. Oecologia 63: 300–306.

Paloheimo, J. E. & W. D. Taylor, 1987. Comments on life-table parameters with reference to *Daphnia pulex*. Theoretical Population Biology 32: 289–302.

Schwartz, S. S. & P. D. N. Hebert, 1987. Methods for the activation of the resting eggs of *Daphnia*. Freshwat. Biol. 17: 373–379.

Stibor, H., 1992. Predator induced life-history shifts in a freshwater cladoceran. Oecologia (Berlin) 92: 162–165.

Stross, R. G., 1966. Light and temperature requirement for diapause development and release in *Daphnia*. Ecology 47: 368–374.

Stross, R. G., 1969a. Photoperiod control of diapause in *Daphnia*. II. Induction of winter diapause in the arctic. Biol. Bull. 136: 264–273.

Stross, R. G., 1969b. Photoperiod control of diapause in *Daphnia*. III. Two-stimulus control of long-day, short-day induction. Biol. Bull. 137: 359–374.

Stross, R. G., 1971. Photoperiod control of diapause in *Daphnia*. IV. Light and CO_2-sensitive phaseu within the cycle of activation. Biol. Bull. 140: 137–155.

Stross, R. G., 1987. Photoperiodism and phased growth in *Daphnia* populations: coactions and perspective. In R. H. Peters & R. De Bernardi (eds), *Daphnia*. Mem. Ist. ital. Idrobiol. 45: 413–437.

Stross, R. G. & J. C. Hill, 1965. Diapause induction in *Daphnia* requires two stimuli. Science 150: 1462–1464.

Stross, R. G. & J. C. Hill, 1968. Photoperiod control of winter diapause in freshwater crustacean *Daphnia*. Biol. Bull. 134: 176–198.

Tessier, A. J. & N. Consolatti, 1989. Variation in offspring size in *Daphnia* and consequences for individual fitness. Oikos 56: 269–276.

Vanni, J. M., 1987. Effects of food availability and fish predation on a zooplankton community. Ecol. Monogr. 57: 61–88.

Weider, L. J. & J. Pijanowska, 1993. Plasticity of *Daphnia* life histories in response to chemical cues from predators. Oikos 67: 385–392.

Zaffagnini, F., 1987. Reproduction in *Daphnia*. In R. H. Peters & R. De Bernardi (eds), *Daphnia*. Mem. Ist. ital. Idrobiol. 45: 245–284.

Hydrobiologia **320**: 223–227, 1996.
V. R. Alekseev & G. Fryer (eds), Diapause in the Crustacea.

Variable life history of a cyclopoid copepod: the role of food availability

Anne-Mette Hansen
National Environmental Research Institute, Department of Freshwater Ecology, Denmark
Present address: Institute of Biology, Odense University, DK-5230 Odense M, Denmark

Key words: cyclopoid copepods, life cycles, food availability, diapause, lakes, *Cyclops vicinus*

Abstract

Populations of the copepod *Cyclops vicinus* in two Danish lakes differed in their life cycles. In Lake Væng *C. vicinus* was absent during summer, whereas in Lake Søbygård it continued producing distinct cohorts throughout summer. A comparison between the lakes showed that in the lake where *C. vicinus* was absent during summer, food limitation generally existed. In contrast, abundance of *C. vicinus* during summer in Lake Søbygård always coincided with available resources sufficient for the whole life cycle. Thus, summer diapause of *C. vicinus* is suggested to be a strategy to avoid food limitation. Avoidance of fish predation does not seem substantially to influence life cycles.

Introduction

According to Nilssen (1978, 1980) and Elgmork (1980) the diapause stage of cyclopoid copepods has evolved to optimize reproduction and growth and to avoid food scarcity, increased competition and predation. However, among cyclopoids known to have a summer diapause, deviations from the main development pattern have been recorded. In most lakes *Cyclops vicinus* is recorded as remaining in diapause during the summer (e.g. Einsle, 1967; Mittelholzer, 1970; Vijverberg, 1977; Hansen & Jeppesen, 1992a); whereas in some lakes it produces generations in midsummer (Hrbáček, 1962; George, 1973, 1976; Maier, 1989; Hansen & Jeppesen, 1992b).

Presumably, the variable life history of *C. vicinus* is the response to different environmental conditions. The fact that induction of diapause of cyclopoid copepods in late spring often coincides with a clear water phase and fish recruitment, stresses that the diapause phase is an advantageous life history strategy. Otherwise, when fish predation is low and food concentration high, diapause is obviously less important. Maier (1989) concluded from studies in two German lakes that differences in predation pressure caused the variable life cycles of *C. vicinus*. In contrast, Hansen & Jeppesen (1992b) suggested that high densities of *C. vicinus* during summer in the Danish Lake Søbygård were due to the concurrence of low predation and the abundance of food.

Recently, laboratory studies have elucidated the food requirements of *C. vicinus*. Results from these experiments are summarized below. Flagellate algae are sufficient resources for a successful naupliar development whereas diatoms and nonflagellated green algae result in death before metamorphosis (Santer & Van den Bosch, 1994; Hansen & Santer, 1995). Furthermore, high food requirements are necessary for a successful juvenile development and for the reproduction of *C. vicinus*. Threshold food concentration for a successful naupliar development is about 0.3–0.5 mg C l^{-1} of phytoflagellates, and the juvenile development time is dependent on food concentration, thus food limitation takes place when phytoflagellate concentration is below 1 mg C l^{-1} (Santer & Van den Bosch, 1994; Hansen & Santer, 1995).

Though *C. vicinus* can develop to the adult stage and reproduce on a pure algal diet (Santer, 1993) it benefits from a food mixture of rotifers and phytoflagellates, i.e. increasing copepodite survivorship (Hansen & Santer, 1995) and egg production (Santer & Van den Bosch, 1994). That rotifers are among preferred prey of *C. vicinus* is substantiated from selection experiments (Brandl & Fernando, 1978, 1981; Zankai, 1984; Adrian, 1991).

In this paper variation in the life cycle of *C. vicinus* in two Danish lakes is discussed in relation to the available food resources and to food requirements as evaluated in laboratory experiments. In order to do this, phytoflagellate biovolumes were converted to mg C l^{-1} by assuming algal dry weight to be 20% of wet weight and carbon content to be 50% of dry weight. Thus C l^{-1} is equal to approximately 10% of biovolume. Therefore, phytoflagellate biovolumes of less than 4 mm^3 l^{-1} correspond to food concentrations below threshold. Food limitation of *C. vicinus* takes place with phytoflagellate biovolumes in the range of 4–10 mm^3 l^{-1}. When phytoflagellate biovolume is above 10 mm^3 l^{-1} optimum food conditions for the herbivorous stages of *C. vicinus* exist. Altogether, a mixed food supply of rotifers (>500 ind l^{-1}) and phytoflagellates (>10 mm^3 l^{-1}) seem to be adequate for the life cycle development of *C. vicinus.*

Study sites

The two lakes are situated less than 50 km apart in the central part of Jutland, Denmark. The morphometry of the two lakes is rather similar. Lake Søbygård, the lake with *C. vicinus* summer-cohorts (Hansen & Jeppesen, 1992b), has a surface area of 0.4 km^2 and a maximum and mean depth of 1.9 and 1.0 m, respectively. Lake Væng, without *C. vicinus* in summer (Hansen & Jeppesen, 1992a), has a surface area of 0.15 km^2 and a maximum and mean depth of 1.9 and 1.2 m, respectively. The lakes are completely mixed unless ice-covered and show similar water temperatures throughout the year. Lake Søbygård is hypertrophic whereas Lake Væng is eutrophic. In both lakes planktivorous fish predation was highly variable within the investigation period. In Lake Søbygård no fish recruitment was observed from 1984 to 1987 whereas fry of roach (*Rutilus rutilus*), rudd (*Scardinius erythrophthalmus*) and three-spined stickleback (*Gasterosteus aculeatus*) were observed in 1988 and 1989. In Lake Væng, fish biomanipulation was carried out from autumn 1986 to spring 1988 reducing the planktivorous fish biomass by approximately 50% by removal of roach and bream (*Abramis brama*). The only copepod present in Lake Søbygård was *C. vicinus*, whereas *Mesocyclops leuckarti* replaced *C. vicinus* in summer in Lake Væng. Occasionally other cyclopoids occurred during summer in Lake Væng.

Methods

Phytoplankton and zooplankton were sampled at a mid-lake station from two depths (subsurface and 1 m), using a 3.3-L Patalas sampler, and the samples were pooled. Sampling frequency varied between lakes. In Lake Søbygård sampling was carried out two to three times per week: in Lake Væng weekly.

Phytoplankton samples were preserved in Lugol's solution. In the laboratory samples were quantified and the biovolume of each species was determined. Biovolume of each algal species was determined by multiplying mean cell volume by cell population density. Biovolume of phytoflagellates was determined by total biovolume of cryptophytes and green flagellates.

Zooplankton samples were filtered through 20 μm mesh net and fixed in Lugol's solution. Copepodids and nauplii were enumerated in complete samples when the density was less than 100 ind l^{-1}. Otherwise at least 100 individuals of either nauplii or copepodids were counted. Whereas copepodids were distinguished by species, nauplii were not. Abundance of rotifers was estimated on subsamples. As *C. vicinus* was the only cyclopoid copepod in Lake Søbygård naupliar abundance was used to interpret the seasonal pattern of *C. vicinus*. Whereas in Lake Væng, due to the coexistence of several cyclopoid species, copepodids of *C. vicinus* were used for the description of seasonal abundance of *C. vicinus.*

Results

In Lake Søbygård, *C. vicinus* was very abundant during summer in most years (Fig. 1). Summer cohorts (June–July) were observed from 1986 to 1989 whereas *C. vicinus* was absent in June–July, 1985. In 1986 to 1989, through June and July, phytoflagellate biovolume of more than 10 mm^3 l^{-1} was observed for a more or less extended period, from few days up to 20 days. Often these high biovolumes were replaced by only short periods of biovolumes in the range of 4–10 mm^3 l^{-1}. In June–July 1985, however, phytoflagellate biovolume was entirely below 4 mm^3 l^{-1}. In Lake Søbygård, rotifers were periodically abundant (>500 ind l^{-1}) through the summer in 1985–1989. The summer cohorts of *C. vicinus* in Lake Søbygård always coincided with high biovolumes of phytoflagellates and an abundance of rotifers.

In Lake Væng, the number of *C. vicinus* copepodids decreased in late May and by the beginning of June

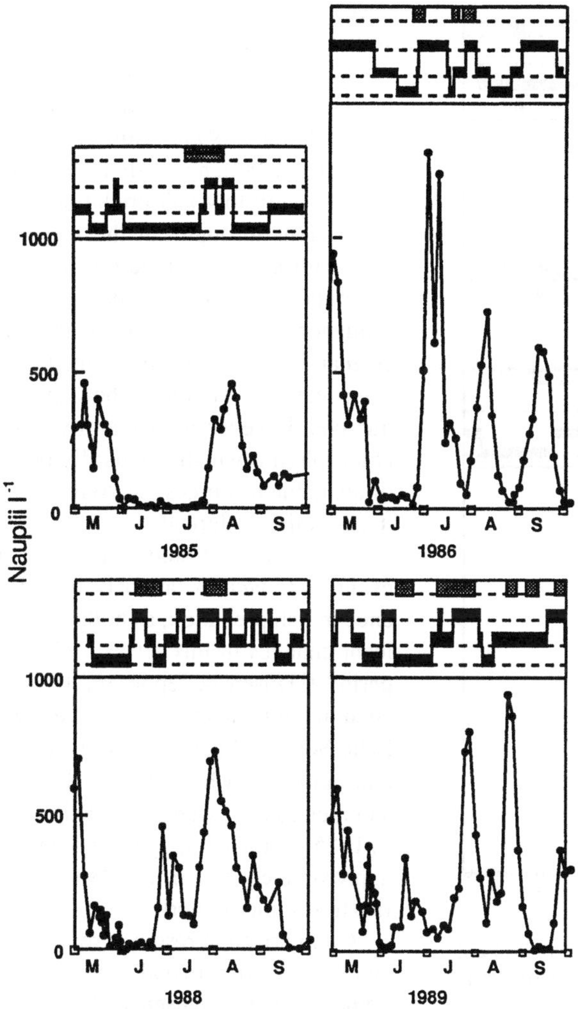

Fig. 1. Abundance of *Cyclops vicinus* nauplii (no. l^{1}) and food resources during May to September, 1985–1989, in Lake Søbygård. Light shaded horizontal bar, rotifer density >500 ind l^{-1}. Phytoflagellate biovolume divided up into 3 groups (dark shaded horizontal bars): lower bar, <4 $mm^3\ l^{-1}$; central bar, 4–10 $mm^3\ l^{-1}$; upper bar, > 10 $mm^3\ l^{-1}$.

were almost absent from the pelagic zone and continued so through June and July in all years (Fig. 2). In late summer *C. vicinus* reappeared in the water samples. In general, biovolumes of phytoflagellates were well below 10 $mm^3\ l^{-1}$; biovolumes below 4 $mm^3\ l^{-1}$ were just as common as values between 4–10 $mm^3\ l^{-1}$. Only in the summer of 1987 were values above 10 $mm^3\ l^{-1}$ observed for a short period. Rotifers were often abundant, but only in the summer of 1987 was a contemporary abundance of rotifers and phytoflagellates observed. Meanwhile, no increase in *C. vicinus* abundance was observed.

Discussion

In this study summer cohorts of *C. vicinus* always coincided with high concentrations of both phytoflagellates and rotifers. Thus, differences in food availability seem to explain the variable life cycles of *C. vicinus*. Resources were generally limiting or even below threshold in Lake Væng, whereas summer-cohorts in Lake Søbygård always arose when resources (i.e. phytoflagellates and rotifers) were sufficient for the whole life cycle. Hansen & Jeppesen (1992b) showed that resources were decisive for the abundance of *C. vicinus* in Lake Søbygård, as maximum density

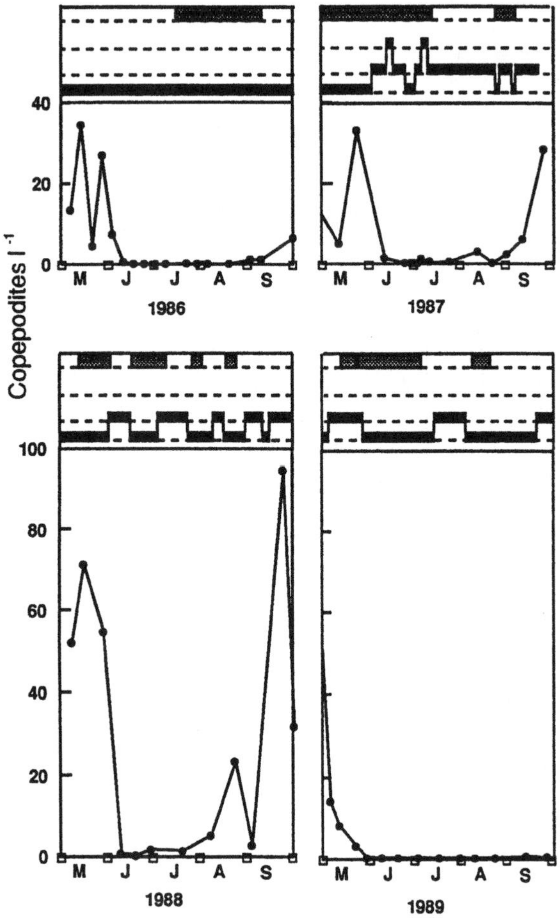

Fig. 2. Abundance of *Cyclops vicinus* copepodids (no. l^{-1}) and food resources during May to September, 1986–1989, in Lake Væng. Legends as in Fig. 1.

of copepodids and mean biovolume of cryptomonads were significantly positively correlated.

Furthermore, the study shows that the two populations have develop different life history strategies. In lakes like Væng where food limitation generally exists through the whole summer, summer diapause seems to be a good strategy. Santer (1991) showed that 50 μm filtered lake water from Schierensee was not sufficient for the development of *C. vicinus* in summer and she interpreted summer diapause as a strategy to avoid food limitation. Data on *Diacyclops thomasi* from Lake Oneida also support this hypothesis as the copepod entered diapause when ambient food became limiting, i.e. prolonged development time of the juvenile stages was observed in the laboratory (Hansen & Hairston, in prep.). Otherwise, when resources are abundant in summer, it may be advantageous to reproduce in summer, presumably depending on the intensity of fish predation.

Avoidance of fish predation has been suggested to be the main reason for diapause in cyclopoid copepods (Nilssen, 1977). This hypothesis was supported by Maier (1989) in a study from two German lakes. In the lake with low fish predation *C. vicinus* produced generations through out the summer whereas in the lake with a large population of planktivorous fish it responded by summer diapause. He also discussed the influence of food resources on the life history but stated that food availability seemed not to influence the life cycle of *C. vicinus* as both lakes were highly productive. However, food quantity was only measured as chlorophyll-*a* which can hardly be used as a measure of available resources due to the rather restricted food items of the cyclopoids.

In the present study, however, the variable life cycles of *C. vicinus* cannot be explained only by difference in fish predation, as both lakes underwent significant changes in this respect within the investigation period (Hansen & Jeppesen, 1992a, 1992b) without changing the life cycles of *C. vicinus*. For instance, in Lake Søbygård fish predation was rather low in 1985 and significantly higher in 1988–1989 but nevertheless summer-cohorts were observed in 1988–1989 and not in 1985. In Lake Væng fish predation was reduced by approximately half from 1986 to 1988 without giving rise to summer-cohorts.

Comparison of resource requirements of *C. vicinus* and *M. leuckarti* reveal dietary overlap in all development stages (Hansen & Santer, 1995). Therefore it cannot be excluded that interspecific competition of cyclopoid copepods also influenced the life cycle of *C. vicinus*. By entering summer diapause in Lake Væng, *C. vicinus* minimized temporal niche overlap, so exploitative competition with other cyclopoids only took place during a shorter period. However, elimination of interspecific competition cannot alone explain the presence of *C. vicinus* during summer in Lake Søbygård as it was not present during the summer of 1985 though no other cyclopoids were present. Thus high food availability must be the main explanation for the presence of *C. vicinus* during the summer in Lake Søbygård.

Genetic differentiation between populations with respect to ability to undergo diapause has been shown by Wyngaard (1988). Whether or not this is the case with the two *C. vicinus* populations studied here is not possible to determine. However, the fact that *C. vicinus* did not increase in abundance in Lake Væng in the summer 1987, when resources were available and fish predation rather low, suggests that this might be the case.

Long photoperiod of 15–16 h, which is known to induce diapause (e.g. Einsle, 1964; Spindler, 1970) was reached in mid-May contemporary with a decline in copepodid abundance in both lakes. Hansen & Jeppesen (1992b) proposed that only part of the *C. vicinus* population in Lake Søbygård did not enter diapause and that the summer cohorts of *C. vicinus* in Lake Søbygård arose from these non-diapausing individuals. If this is the case, genetic differentiation within the *C. vicinus* population of Lake Søbygård may have taken place.

Acknowledgments

I gratefully acknowledge comments on an earlier draft of the manuscript by H. S. Jensen, F. Ø. Andersen, K. Elgmork and A. Monakov. The staff of the National Environmental Research Institute are acknowledged for assistance.

References

Adrian, R., 1991. The feeding behavior of *Cyclops kolensis* and *C. vicinus* (Crustacea, Copepoda). Verh. int. Ver. Limnol. 24: 2852–2863.

Brandl, Z. & C. H. Fernando, 1978. Prey selection by the cyclopoid copepods *Mesocyclops edax* and *Cyclops vicinus*. Verh. int. Ver. Limnol. 20: 2505–2510.

Brandl, Z. & C. H. Fernando, 1981. The impact of predation by cyclopoid copepods on zooplankton. Verh. int. Limnol. 21: 1573–1577.

Einsle, U., 1967. Über einige Auswirkungen der Eutrophierung des Bodensee-Obersees auf seine planktisch lebenden Copepoden-populationen. Schweiz. Z. Hydrol. 29: 305– 310.

Elgmork, K., 1980. Evolutionary aspects of diapause in freshwater copepods. W. C. Keerfoot (ed.), Evolution and Ecology of zooplankton Communities. The University Press of New England, Hanover (N.H.): 411–417.

George, D. G., 1973. Diapause in *Cyclops vicinus*. Oikos 24: 136–142.

George, D. G., 1976. Life cycle and production of *Cyclops vicinus* in a shallow eutrophic reservoir. Oikos 27: 101–110.

Hansen, A.-M. & E. Jeppesen, 1992a. Changes in the abundance and composition of cyclopoid copepods following fish manipulation in eutrophic Lake Væng, Denmark. Freshwat. Biol. 28: 183–193.

Hansen, A.-M. & E. Jeppesen, 1992b. A five year study of the life cycle of *Cyclops vicinus* in relation to food availability, predation diapause and temperature. J. Plankton Res. 14: 591–605.

Hansen, A.-M. & B. Santer, 1995. The influence of food resources on the development, survival and reproduction of the two cyclopoid copepods: *Cyclops vicinus* and *Mesocyclops leuckarti*. J. Plankton Res. 17: 631–646.

Hrbáček, J., 1962. Species composition and the amount of zooplankton in relation to the fish stock. Rozpr. Cesk. Akad. Ved. Rada Mat. Prir. 72, 116 pp.

Maier, G. M., 1989. Variable life cycles in the freshwater copepod *Cyclops vicinus* (Uljanin 1875): Support for the predator avoidance hypothesis? Arch. Hydrobiol. 115: 203–219.

Mittelholzer, E., 1970. Populationsdynamik und Produktion des Zooplanktons im Greifensee und im Vierwaldstättersee. Schweiz. Z. Hydrol. 32: 90–149.

Nilssen, J. P., 1977. Cryptic predation and the demographic strategy of two limnetic cyclopoid copepods. Mem. Ist. ital. Idrobiol. 34: 187–196.

Nilssen, J. P., 1978. On the evolution of life histories of limnetic cyclopoid copepods. Mem. Ist. ital. Idrobiol. 36: 193–214.

Nilssen, J. P., 1980. When and How to reproduce: a dilemma for limnetic cyclopoid copepods. In W. C. Keerfoot (ed.), Evolution and Ecology of zooplankton Communities. The University Press of New England, Hanover (N.H.): 418–426.

Santer, B., 1991. The role of food limitation in summer diapause of cyclopoid copepods. Verh. int. Ver. Limnol. 24: 2851.

Santer, B., 1993. Potential importance of algae in the diet of adult *Cyclops vicinus*. Freshwat. Biol. 30: 269– 278.

Santer, B. & F. van den Bosch, 1994. Herbivorous nutrition of *Cyclops vicinus*: the effect of a pure algal diet on feeding, development, reproduction and life cycle. J. Plankton Res. 16: 171–195.

Vijverberg, J., 1977. Population structure, life histories and abundance of copepods of copepods in Tjeukemeer, The Netherlands. Freshwat. Biol. 7: 579–597.

Wyngaard, G. A., 1988. Geographical variation in dormancy in a copepod: evidence from population crosses. Hydrobiologia 167/168 (Dev. Hydrobiol. 47): 367–374.

Zankai, N. P., 1984. Predation of *Cyclops vicinus* (Copepoda: Cyclopoida) on small zooplankton animals in lake Balaton (Hungary). Arch. Hydrobiol. 99: 360–378.

Hydrobiologia **320**: 229–233, 1996.
V. R. Alekseev & G. Fryer (eds), Diapause in the Crustacea.

Diapause of copepods as an element for stabilizing the parasite system of some fish helminths

Nadiya V. Evseeva
Institute of Biology, Karelian Research Centre, Rus. Ac. Sci., Pushkinskaya 11, Petrozavodsk, 185610, Russia

Key words: copepods, cestodes, intermediate hosts, infection, freshwater fishes, diapause

Abstract

The analysis of infection dynamics of copepods, which are the intermediate hosts of three helminth species of freshwater fishes (*Triaenophorus crassus*, *Proteocephalus exiguus* and *P. percae*), has shown that diapause in the life cycle of the copepods is favourable for preserving the infection in the waterbody until physiological prerequisites for successful infection of the final host are acquired. The peculiarity of a copepod's life cycle may determine the strategy of a parasite in its preimaginal phase as a waiting stage, and the duration of the residence of helminth larvae in copepods that have an obligatory diapause, is one of the elements the provide stability in parasite systems.

Introduction

The existence of any parasite system is determined by a large number of abiotic and biotic factors. The latter include the spatial-temporal structure of intermediate hosts for parasites with complicated cycles of development. Tapeworms of fishes (Cestoda) usually have an annual cycle of development. In the spring–summer period the ova are excreted from the intestine of the definitive host and are disseminated within the water body. Copepods of the orders Calanoida and Cyclopoida swallow the ova (*Proteocephalus*, Weinland, 1858) or the free-swimming larvae (coracidia) developed from them (*Triaenophorus* Rudolphi, 1793). Inside the copepod intestine the parasite embryos become free from the membranes and penetrate the body cavity where one or two stages of development occur. Later, infected copepods are eaten by, and enter the intestine of, a definitive host where they become sexually mature. A second intermediate host, a coregonid fish, participates in the life cycle of the genus *Triaenophorus*.

Much information has been accumulated on the role of copepods in the life cycle of freshwater cestodes. It includes the identity of the intermediate host, infestation rates, population dynamics and age structure, and the impact of different factors on infection levels (Michajlow, 1962; Sysoev, 1982; Kuperman, 1973; Evseeva, 1991; Evseeva & Anikieva, 1994). Researchers who studied the range of intermediate hosts in experiments, noted that the same crustacean species are infected with cestode larvae in different ways, depending on the stage of the copepod's development and on the season of the year. Many inconsistencies were detected while assessing the role of intermediate hosts in experiments and in nature: some species which were not infested or were poorly infested in experiments, were strongly infested in nature, and vice versa. Analysis of the trophic relationship of fish and copepods (Watson & Lawler, 1965; Sysoev & Yakushev, 1985; Evseeva, 1988, 1991) revealed a discrepancy in the rate of infection of fish and plankton Crustacea. The presence of a large number of invasive larvae in the plankton was not attended by an increase in the rate of fish infection by juvenile helminths. Such observations indicated the need for a more detailed analysis of copepod ecology as an integral component of a parasite system. The life cycle of certain copepods is an alternation of active growth and development with diapause in a copepodid stage. We investigated the role of diapause in the intermediate hosts on the functioning of the parasite system of freshwater cestodes of fish.

The research was conducted in 1984–1987 in Otradnoye and Rindozero Lakes, in north-western European Russia. Lake Otradnoye (6910 ha) of the Leningrad region, belongs to the Ladoga Lake basin. In 1984–1986 the life cycle of *Triaenophorus crassus* Forel, 1868, a parasite of vendace (*Coregonus albula* L) and cisco (*C. lavaretus pallasi* (Val.) was studied. The intermediate hosts in Otradnoye Lake are: *Cyclops kolensis* Lilljeborg, 1901; *C. strenuus* Fischer, 1851; *Thermocyclops oithonoides* Sars, 1863; *Acanthocyclops viridis* (Jurine, 1820) and *Eudiaptomus gracilis* (Sars, 1863). Rindozero Lake is a small waterbody (186 ha) in southern Karelia, in the Onega Lake drainage area near Petrozavodsk. Studies were performed in 1987. The cestodes were *Proteocephalus exiguus* La Rue, 1911, a widely spread parasite of vendace, and *P. percae* (Müller, 1780), a parasite of perch (*Perca fluviatilis* L.). Six copepod species: *C. scutifer* Sars, 1863; *C. vicinus* (Uljanin, 1875; *C. kolensis*; *T. oithonoides*; *E. graciloides* (Lilljeborg, 1883) and *E. gracilis* are intermediate hosts to both species.

Materials and methods

Parasites were collected and treated according to generally accepted methods (Bykhovskaya–Pavlovskaya, 1985). Infection of fishes by *Proteocephalus* and adults of *T. crassus* was studied by examination of the digestive tract. The plerocercoid phase of *T. crassus* was studied by examining sections of skeletal muscles.

The infection in copepods was investigated on live material by crushing between glasses at a 12.5 × 4 magnification. From 250 thousand copepods of the orders Calanoida and Cyclopoida examined from each water body, about 350 cestode larvae were found and identified. The percentage of infection was calculated individually for each copepod species proceeding from its number at the time of sampling.

Samples for the analysis of zooplankton populations were collected every ten days from May to October, in other periods monthly, and were treated according to conventional methods (Kiselev, 1969).

To reveal the means of parasite transfer the contents of fish digestive tracts were studied according to 'The Manual on Study of Fish Feeding under Natural Conditions' (1961).

In order to clarify the possibility of infection of copepods during diapause, two series of experiments were performed with *C. kolensis*. Stage IV copepodites were caught in June 1994 in a small forest Lake, Lindalamba (8.6 ha), near Petrozavodsk and kept in 500 ml vessels of lake water filtered through gauze 77. These copepods were infected by introducing into the vessels coracidia of *T. nodulosus* (Pallas, 1781) and mature ova of *P. percae* with a pipette. The experiment was conducted at 16–17 °C and natural lighting. The copepods were regularly observed for 15 days from the moment of infection at a 10 × 10 magnification. The level of infestation, the number and stage in the development of their larvae, and their size and location were recorded.

Results

In Otradnoe Lake, of the five intermediate hosts of *T. crassus*, *C. kolensis* proved to be the most frequent. The peculiarities of copepod biology determined the terms and level of fish infection by the parasite. It was shown that vendace has two periods of infection – summer and autumn. At the time of the summer infection, *C. kolensis* are infected by larvae that have developed as a result of invasion by coracidia of *T. crassus* in May and June. At this time the population of *C. kolensis* is in the state of active growth and development. After that, copepodid stage IV enters diapause, which coincides with the time of intensive feeding of the vendace. But, the departure of *C. kolensis* from the plankton and its migration to the sediment interrupts the process of infection of the vendace, a typical plankton-eater. In autumn, when *C. kolensis* reappears in the plankton, a new period of vendace infection begins (Fig. 1).

The infection of cisco begins in July, and the process lasts all autumn with single finds of juvenile plerocercoids in winter and spring. Having a mixed type of nutrition, this host consumes diapausing as well as active copepods.

More complicated is the relationship between vendace and *P. exiguus* in Rindozero Lake. Of the six intermediate hosts, the heaviest functional loads in the total infection are shared by *E. graciloides* and *C. scutifer*. *E. graciloides* is monocyclic with a long period of sexual maturation from copepodid stage V to the adult. *C. scutifer* is not uniform in respect of realizing its life cycle. After reproduction (July and August) a part of the population develops to copepodid stage IV which hibernate within the water and sink to the bottom by March. For the majority of the population of *C. scutifer* a delay in the development was recorded at early metamorphosisis stages -nauplii and copepodid stage I, which reached copepodid stages II and III and

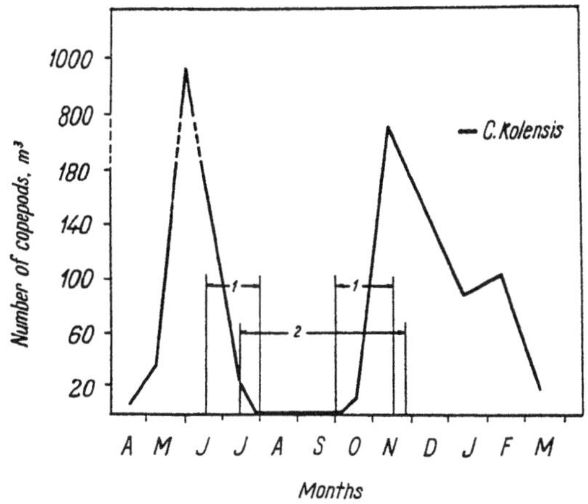

Fig. 1. Dynamics of the number of *Cyclops kolensis* (Cop. IV–V, ad) and period of infestation of vendace *Coregonus albula* (1) and cisco *Coregonus lavaretus pallasi* (2) by *Triaenophorus crassus* larvae in the Otradnoye Lake.

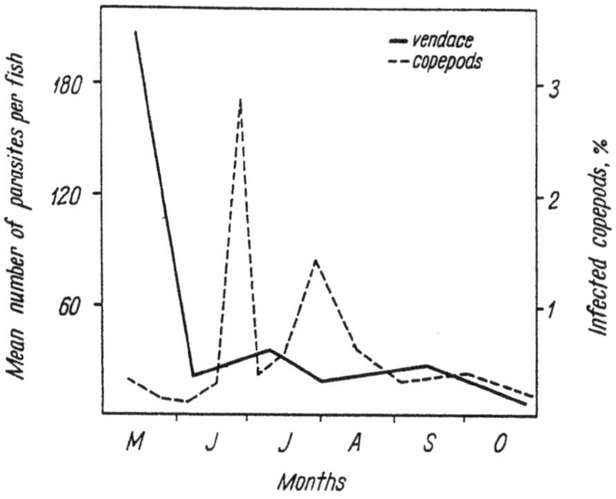

Fig. 2. Seasonal dynamics of infection rate of vendace *Coregonus albula* and the copepods *Cyclops scutifer* and *Eudiaptomus graciloides* (Cop. IV–V, ad) by the cestode *Proteocephalus exiguus* in Rindozero Lake.

inhabited the water. Such heterogeneity of the *C. scutifer* population in favourable for a constant infection with ova of the parasite. Eggs of the cestode are dispersed in the water body throughout the period of open water. Larvae of *P. exiguus* were recorded in spring and summer in *Diaptomus* and *Cyclops*. Two increases in the plankton infection rate were noted (Fig. 2). The first was provided chiefly by old hibernated helminth larvae that infected copepods in autumn, the second was the result of a summer infestation. In the summer-autumn period the definitive host, vendace, was invaded only slightly. Intensive infection and an accummulation of larvae took place in the winter months. The maximum invasion rate was noted in spring.

P. percae, a parasite of perch, differs from *P. exiguus* in that it has a short period of ova production (2 to 3 weeks in June). At this time the structure of the intermediate host population is represented by all ages, but nauplii and copepodid stages I–III of *E. graciloides* and *C. scutifer* make up the majority. The parasite larvae were chiefly concentrated in cyclops, which served as the main food source for perch. However, the rate of infestation was not high in summer. Only in September does a marked increase in the number of juvenile helminths in perch begin (Fig. 3).

As a result of experimental infection of *C. kolensis* with the ova of *T. nodulosus* and *P. percae*, it was shown that the cyclops, being in 'active phase of diapause' (Alekseev, 1990) at copepodid stage IV can be infected, but the rate of invasion is much less than that of actively developing individuals. In the first series of experiments the incidence of infestation was 9.6%,

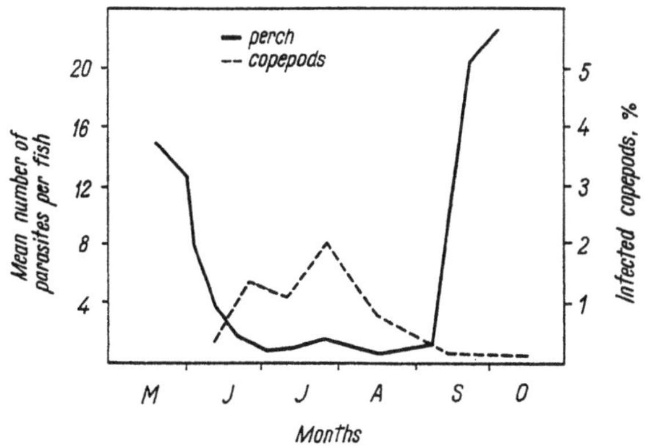

Fig. 3. Seasonal dynamics of infection rate of perch *Perca fluviatilis* and the copepods *Cyclops scutifer* and *Eudiaptomus graciloides* (Cop. IV–V, ad) by the cestode *Proteocephalus percae* in Rindozero Lake.

one week later it was 1.3%. In experiments with active copepods, this figure usually approaches 100%. However, in 10–12 days, normal larvae developed in the body cavity of the copepods. In addition, it was noted that some of infected *C. kolensis* individuals moulted to the next stage of development. Such moulting was not observed in uninfected specimens.

Discussion

Watson & Lawler (1965), were the first to call attention to the role of diapause in intermediate hosts of cestode parasites of coregonid fishes. Based on observed rates of infestation of *T. crassus* in copepods and later in the coregonids, they speculated that stage IV copepodids of *Diacyclops bicuspidatus thomasi* Forbes are a reservoir for the parasite during the summer. Our research has shown that cisco are infected by larvae of *T. crassus* in August–September, when *C. kolensis* has been in the state of diapause for 2 to 2.5 months. The introduced larvae grew quickly and developed to the next stage of plerocercoid. From late September the viability of larvae decreased somewhat, development became slower, and some of them were dead. We connect the fall in invasion rate with expended reserve nutrients, which provided for the existence of the parasite in copepods. A long stay of the parasite in an intermediate host (several months) can change the behaviour of a part of the copepod population. We made an experiment with the diapausing stages of *C. kolensis* and registered the persistence of some infected copepods from copepodid stage IV to stage V. I believe that the parasite, by speeding up metamorphosis, compels the host to complete diapause and enter the food chain to infect plankton-eating fish.

In the parasite system '*P. exiguus* – vendace' long development of the earliest age groups of *C. scutifer* is conductive to a high infestation of this fish in the following spring. Having been infected in late summer, the nauplii and copepodid stage I preserve the invasion for 3 to 4 months, remaining an unavailable food for vendace because of small size. This period (autumn) coincides with the development of an immune response in fish of the family Salmonidae within the temperature range of 4 to 10 °C, that causes elimination of a considerable part of the sexually immature helminths. In winter the intensity of immunity decreases and the vendace is strongly infested by *P. exiguus*, localized in copepods, which have reached stage IV and form dense aggregations near the bottom.

Perch are also lightly infected by *P. percae* in the summer months because larvae lose the ability to infect at temperatures higher than 10 °C. Adults and copepodids IV and V of *C. scutifer*, the main intermediate host of the parasite, were intensively consumed by perch during the first half of summer, but practically no infection took place. In autumn copepods, that were infected at early stages of development, enter the food chain of the perch. Coincidence of the period of ova dispersion by the helminth with the time of the maximum number of the earliest age groups of the copepod (nauplii and copepodids of stages I–III) can be considered as a profound adaptation of the parasite to the ecology and physiology of its intermediate hosts.

The results lead me to conclude that copepod populations in the state of active growth and development are the most susceptible to infection by parasite larvae. At the same time our initial idea about the impossibility of infection of diapausing copepods proved to be not entirely true. Copepods may be infected at the initial period of diapause – 'preparatory and induction phase' (Alekseev, 1990). As diapause becomes deeper, the ability to become infected decreases. The longer the diapause, the quicker the parasite loses the ability to infect copepods after they return to active development.

The analysis of parasite systems of three cestode species (*T. crassus*, *P. exiguus* and *P. percae*) has shown that among various species of intermediate hosts the heaviest functional load in total infection is borne by copepods with a long life cycle. The existence of obligatory diapause in *Cyclops* favours preserving the infection in a water body until ecological or physiological preconditions for successful infestation of the final host originate. The 'reserved supply' of parasite larvae in dormant stages of copepods is the basis for infecting fishes by helminths of a new generation. It enters the food chain at the moment when the definitive host is maximally ready for infestation, i.e. in the period of its lowest resistance to the parasite. The monocycly of intermediate hosts determines the parasite's strategy as a waiting stage in a larval phase of development, and helminth larvae remaining in copepods throughout diapause provide an element of stability in the parasite system.

Because of diapause in the life cycle of the intermediate host, the parasite disperses not only in time, but in space. Usually this function is performed by free-living stages in development: ovum, coracidium, etc. Dormant stages of cyclops can be borne by currents from the littoral zone, where copepod infection by helminth ova takes place, to deep sections of the lake, forming dense aggregations. This facilitates contact between the parasite and the final host and allows it to complete its life cycle successfully.

The existence of diapause in intermediate hosts can also be considered as a factor in the regulation of helminth numbers. Parasitic organisms have a very high fertility and, in spite of immense losses in the free-living stages, the total number of larvae in intermediate hosts is still adequate. Distribution of the parasite in copepods with different life cycles (with and without diapause, poly- and monocyclic) provides a gradual introduction of the invasive agent into fishes. During the long dormancy period, the invasion fails in copepods, and, as a consequence, the total number of parasites decreases. Immigration of the total 'burden of parasites', concentrated in the intermediate host, to the final host within a very short period of time can break the balance in a parasite system, and lead to a sharp increase in numbers of the parasite.

References

Alekseev, V. R., 1990. Diapausa rakoobrasnykh. Ekologo – phisiologicheskie aspecty. Nauka, Moskva, 144 pp. [Diapause of crustaceans. Ecologo-physiological aspects].

Bykhovskaya-Pavlovskaya, I. E., 1985. Parasity ryb. Rukovodstvo po izucheniyu. Nauka, Leningrad, 123 pp. [The parasites of fish. Manual on study].

Evseeva, N. V., 1988. Rol' planktonnykh rakoobrasnykh v tsirculyatsii tsestody *Triaenophorus crassus* v ozernom biotsenose. Ekologo-populyatsyonnyi analiz parasitov i krovososushikh chlenistonogikh. Karel'skii Filial Academii Nauk SSSR, Petrozavodsk: 56–68. [Role of plankton crustacea in circulation of cestode *Triaenophorus crassus* in lake biocenosis].

Evseeva, N. V., 1991. Rol' promezhutochnykh khozyaev v dinamike zarazhennosti okunya tsestodoi *Proteocephalus percae*. Ekologo-populyatsyonnyi analiz parasitov i krovososushikh chlenistonogikh. Karel'skii Nauchnyi Tsentr AN SSSR, Petrozavodsk: 93–103. [Role of intermediate hosts in dynamics of perch infestation with cestode *Proteocepalus percae*].

Evseeva, N. V. & L. V. Anikieva, 1994. Promezhutochnye khozyaeva tsestody *Proteocephalus exiguus*: structura, dinamika chislennosti, zarazhennost'. Ekologicheskaya parasitologiya. Karel'skii Nauchnyi Tsentr RAN, Petrozavodsk: 112–125. [Intermediate hosts of cestode *Proteocephalus exiguus*: structure, quantitative dynamics, infection rate].

Kiselev, I. A., 1969. Plankton morei i kontinental'nykh vodoemov. T. 1. Nauka, Leningrad, 319 pp. [Plankton of seas and inland waterbodies].

Kuperman, B. I., 1973. Lentochnye chervi roda *Triaenophorus* – parasity ryb. Nauka, Leningrad, 207 pp. [Tapeworms of genus *Triaenophorus* – parasites of fish].

Michajlow, W., 1962. Species of genus *Triaenophorus* (*Cestoda*) and their hosts in various geographical regions. Acta parasitol. Polon., 10: 1–36.

Rukovodstvo po izucheniyu pitaniya ryb v estestvennykh usloviyakh, 1961. AN SSSR, Moskva, 263 pp. [The Manual on Study of Fish Feeding under Natural Conditions].

Sysoev, A. V., 1982. Composition and dynamics of invasion of the first intermediate hosts of *Triaenophorus nodulosus* (Pallas) (*Cestoda:Trienophoridae*) under conditions in Karelia. Helminthologia, 19: 249–255.

Sysoev, A. V. & V. Yu. Yakushev, 1985. Sootnoshenie mezhdu pokazatelyami invasii tsestodami veslonogikh racoobrasnykh i ryb. Materialy konferentsyi VOG, 35. Nauka, Moskva: 152–159. [Correlation between indices of infection of copepods and fish by cestodes].

Watson, N. F. & G. H. Lawler, 1965. Natural infection of cyclopid copepods with procercoids of *Triaenophorus* spp. J. Fish. Res. Bd. Can, 22: 1335–1343.

Hydrobiologia **320**: 235–241, 1996.
V. R. Alekseev & G. Fryer (eds), Diapause in the Crustacea.

Ecology of diapausing copepodids of *Cyclops kolensis* Lill. in reservoirs of the Upper Volga

Irina C. Rivier
Laboratory of Aquatic Invertebrates Ecology, Institute for Biology of Inland Waters, Borok Nekouz, Yaroslavl, 152 742, Russia

Key words: Copepoda, *Cyclops kolensis*, seasonal cycle, diapause

Abstract

Cyclops kolensis Lill. forms dense populations in different water-bodies of the Upper Volga basin. In spring, when the other planktonic crustaceans have not yet appeared, *C. kolensis* is an important food object for fishes. Its active reproduction occurs from April to the beginning of June. A rapid accumulation of stage IV copepodids in the plankton occurs in June and at 12–14 °C they sink into the pelogene in a diapause stage. During the summer stagnation period the diapausing copepodids are distributed evenly over the bottom; their abundance here is 0.7–0.8 million ind. m^{-2} (Rybinsk reservoir). During storms and autumn active water mixing the copepodids together with detritus are disturbed and brought to the deepest, silt rich, part of the water-bodies. After the ice formation and at the beginning of bottom heating the diapausing copepodids are transported by near bottom currents and are concentrated in depressions; their biomass here reaches 60 g m^{-3}. After the thermo-oxy-cline formation they revive and begin to live actively. Copepodids feed, accumulate adipose matter and in February–March they begin to moult.

Introduction

Cyclops kolensis Lill. belongs to an Arctic complex of Palaeoarctic species. It is a typical inhabitant of lakes in Poland and the Kola Peninsula, Siberia from the northern Urals to the Novosibirsk Islands and Kamchatka (Kozminski, 1936; Wierzbicka, 1956; Elgmork, 1959; Kozhov, 1962; Mazepova, 1978). The species is not very variable and even in Lake Baikal it does not differ much from a typical form from lakes of southern Sweden (Kozhov, 1962; Lindberg, 1955). In the European part of Russia before the 1950's *C. kolensis* had not been found south of the taiga zone (Rylov, 1948). At present it is widely spread in reservoirs of the Volga River up to its delta (Kosova, 1970; Rivier, Dzyuban, 1978). It is one of the most active migrants from the northern water-bodies to the Volga reservoirs.

In the Volga reservoirs *C. kolensis* is a monocyclic species. Its active life — copulation, reproduction, growth of nauplus and copepodids — takes place in early spring and at the beginning of summer. In summer and autumn it is in a state of diapause on the bottom. Diapause occurs in *Cyclops scutifer* inhabiting deep lakes and is apparently associated with an effect of fish (Elgmork *et al.*, 1978; Nilssen, 1977). Probably, diapause in *C. kolensis* in the Volga reservoirs is also dependent on the effect of fishes. The copepodids are transported with currents to the deepest parts where they are less available as prey. Diapause in shallow water-bodies warming up to the bottom is associated with an unfavorable temperature regime. Temperatures of 20– 25 ° prevail from the Upper Volga reservoirs to its delta.

In Lake Baikal this species has no diapause at all. In summer its distribution is limited to the cold pelagic zone (Mazepova, 1978).

A similar situation prevails in Rybinsk reservoir in May during reproduction when the population is preserved only in the main part. Then the temperature rises above 12– 15 °, reproduction ceases, and the IV stage copepodids sink to the bottom.

During diapause copepodids are subject to the effects of abiotic and biotic factors. The aim of the investigations reported here was to study the ecology of copepodids of *C. kolensis* for 8–10 months (June–March) in different water-bodies.

Materials and methods

The ecology and life cycle of *Cyclops kolensis* were observed in Upper Volga water-bodies (Lakes: Beloe, Kubenskoe, Siverskoe, Borodaevskoe, Pleshcheevo, Selinger, Nero; Ivankovo and the Rybinskoe reservoirs) from 1978–1990. Weekly samples were taken from November–December to April in the Mologa Channel in Rybinsk reservoir in 1970–1983.

Quantitative samples were collected with a bathometer (V = 10 l) at one metre intervals from surface to bottom in the winter–spring period and with a microbenthometer (d = 6 cm) in the summer–autumn period when copepodids were concentrated in the pelogene. For study of the structure of populations large qualitative samples were collected with nets.

Simultaneously with plankton and micro-benthos sampling the necessary environmental factors: temperature, oxygen content, colour, transparency, numbers of Bacteria, Protozoa, and 'cladocerans' were determined. In the laboratory the zooplankton samples were analyzed microscopically. A total of 1107 samples of zooplankton have been examined and included in this study.

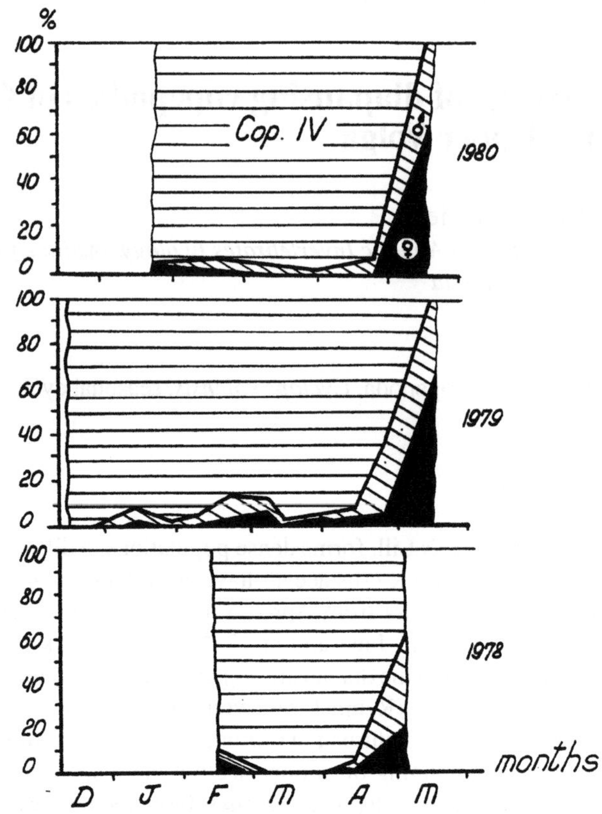

Fig. 1. Structure of population (%), of *Cyclops kolensis* in the Mologa channel in the Rybinsk reservoir.

Results

Cyclops kolensis belongs to a winter zooplankton complex (Rivier, 1986). It reproduces intensively in Spring (March, April — ice period, and May, June — in ice free period). Then an old generation is lost, juveniles grow and at copepodid stage IV become less active, hibernate and sink into the pelogene. For most of its annual cycle of development it is in a nonactive state, in diapause.

At the end of February to early March in the bottom biocoenosis copepodids of the IV stage begin to feed actively, moult, mature and copulate. Females with egg sacs and juvenile appear. After breakup of the ice in a water-body an active development and growth of juveniles continues at 5- 12 °C. As a water-body gets warmed from shore to centre the zone of habitation decreases. In large shallow water-bodies the most favourable conditions are preserved in central parts far from the shores; in large, deep stratified ones, in a cold hypolimnion.

However by the middle of June when the temperature rises to 12–15 °, reproduction stops. Copepodids at stage IV stop growing. Their abundance in the plankton increases, but the abundance of stages I–III decreases. During the second half of June stage IV copepodids disappear from the plankton.

During the period of the open water, copepodids are in an immovable state in pelogene. They are effected by runoff, wind, compensatory and density currents, become mixed with bottom particles, are repeatedly brought into suspension and are transported to deeper sites where they are deposited and concentrated.

During the long period of calm weather in June–August the copepodids are evenly distributed along the bottom of the open part of a reservoir; they are not found in shallow parts and in the zone of the offshore aquatic vegetation. Thus in the main part of the Rybinsk reservoir at the end of June–August at depths of 4–7 m the copepodids stay motionless on the silt surface covered with large light detritus particles of plant origin. Their average abundance is 0.2–0.4 million individuals m^{-2} maximum density up to 0.8 million ind. m^{-2}.

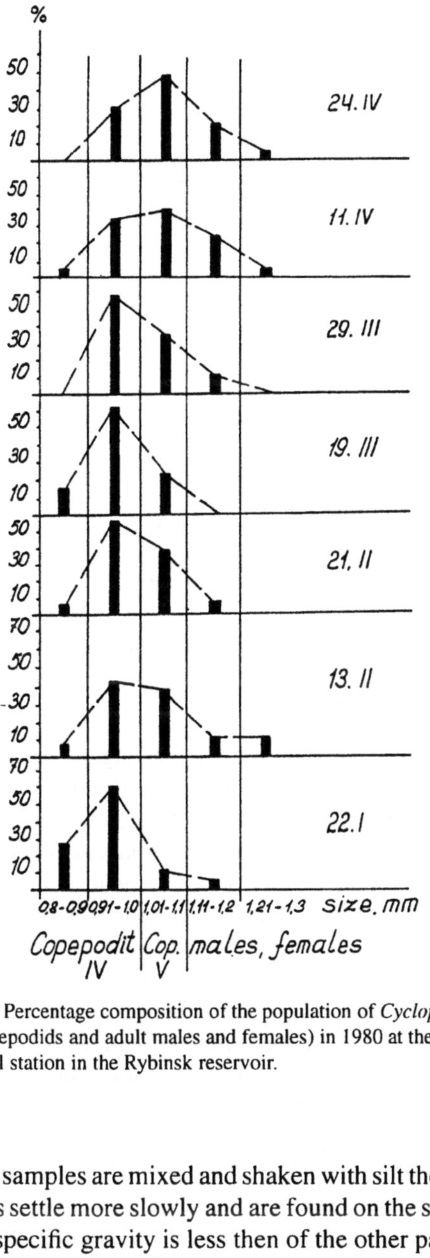

Fig. 2. Percentage composition of the population of *Cyclops kolensis* (copepodids and adult males and females) in 1980 at the Mologa Channel station in the Rybinsk reservoir.

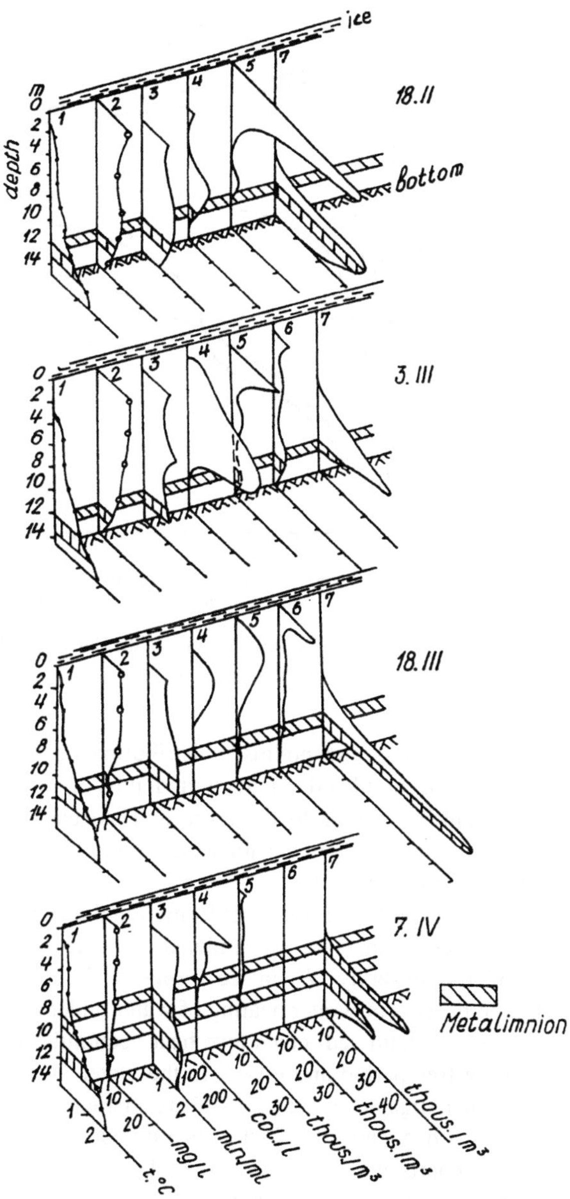

Fig. 3. Variations of vertical distribution of temperature (°C, 1), dissolved oxygen (mg l^{-1}, 2), total number of Bacteria (million cells ml^{-1}, 3), Protozoa (*Sphaeroeca volvax*, colonies, l^{-1}, 4) rotifers — Synchaeta (thousands m^{-3}, 5) and *Polyarthra* (6) and copepodids of *Cyclops kolensis* (7) in the ice covered Mologa Channel February–April 1980.

When samples are mixed and shaken with silt the copepodids settle more slowly and are found on the surface. Thus specific gravity is less then of the other particles and is close to that of water.

In June–August the copepodids are intensively eaten by juveniles of bottom fishes, especially ruff (Acerina cernua).

In autumn, wind mixing and bottom sediment mixing intensify. Waves reach a depth of 6–8 m, the rapid redistribution of light detritus particles, and motionless copepodids as the lightest among them, occurs. Copepodids with detritus enter the water column, are transported with wind and compensatory currents, and are deposited in deep parts where there are no waves. In August–September the copepodids are accumulated in bottom pits at depths of more than 7–8 m, where they form accumulations of up 2–3 million ind. m^{-2}. Here without light they are less available as prey for fishes.

During the time of heavy autumn storms, when wave turbulence occurs at a depth of 10 m, one more

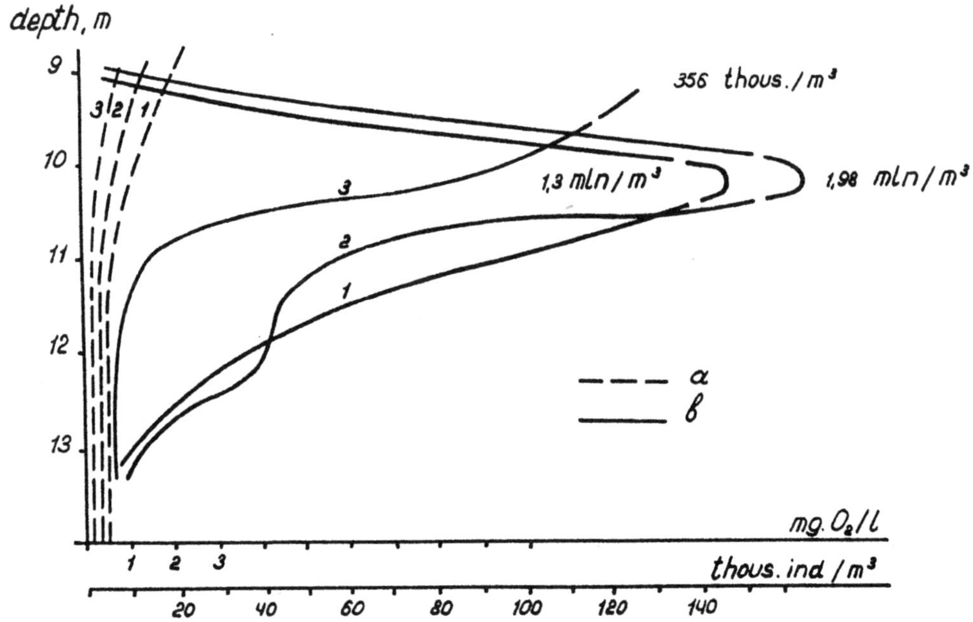

Fig. 4. The oxygen concentration (mg l^{-1}, (a) and the abundance of *Cyclops kolensis* (thousands m^{-3}, (b) in the bottom layers in the Rybinsk reservoir in spring 1980 (1–19 III, 2–29 III, 3–11 IV).

step of motionless copepodids redistribution occurs and they are accumulated in parts with depths of more than 10 m, in channels of flooded rivers and lake basins. There accumulations with the abundance up to 10 million ind. m^{-2} are formed.

During the period of ice cover formation the copepodids of the IV stage are accumulated in the deepest parts of a water-body. When ice formation is complete, the process of winter heating of water under the ice begins. Bottom sediments lose heat accumulated during ice free period and give it to the water. Sediments rich in organic matter (8–12%) are characterized by the most heat capacity. Such silts are deposited in the deepest parts of a water-body, where large accumulations of overwintering copepodids are located. Heat goes into a near bottom water layer whose temperature rises to 4–4.5 °, and can increase to 5–5.4 °, being held near the bottom by virtue of an increase in mineralization in winter by a factor of more than 2(from 90 to 230 mg l^{-1}) (Volga and its life, 1978) which increases its specific gravity. As water warms up near the bottom and its specific gravity increases the copepodids rise to the mud surface and form large aggregations. In winter copepodid density in these near bottom accumulations are maximum (biomass up to 60 g m^{-3}). Such a phenomenon was observed in the Mologa River channel in December–January. During an intensive water heating in the period of ice cover, water layers warmed up to 3–5 ° flow down forming strong currents towards the deepest parts. Copepodids are transported into channel accumulations by these currents. In the second half of winter in the bottom layers of a channel and in lake hollows a winter biocoenosis consisting of bacterio-phyto- and zooplankton is formed. In such biocoenosis the copepodids of *Cyclops kolensis* form a natural link of secondary producers.

In December — the end of March (before breaking up of the ice) the population structure of *C. kolensis* is constant in the bottom accumulation. Copepodids of stage IV form the basis (100–87%) (Fig. 1). Part of the population (up to 10– 13%) is of adult individuals of the older generation; they reproduce slowly during the winter. The quota of copepodids of stage IV begins to decrease from about the beginning of April. The transition from diapause to an active state occurs slowly. From mid February only part of the copepodid stage V population increases (Fig. 2). An increase in abundance of adults is observed before breaking up of the ice when snow melt water flows under the ice (Rivier, 1986).

In winter the vertical distribution of copepodids in deep parts of the Rybinsk reservoir and deep lakes is stable. They form near bottom accumulations. In December to early March protozoans, rotifers, and juveniles of diaptomids are found in the bottom accumulations before the copepodids are activated. By the

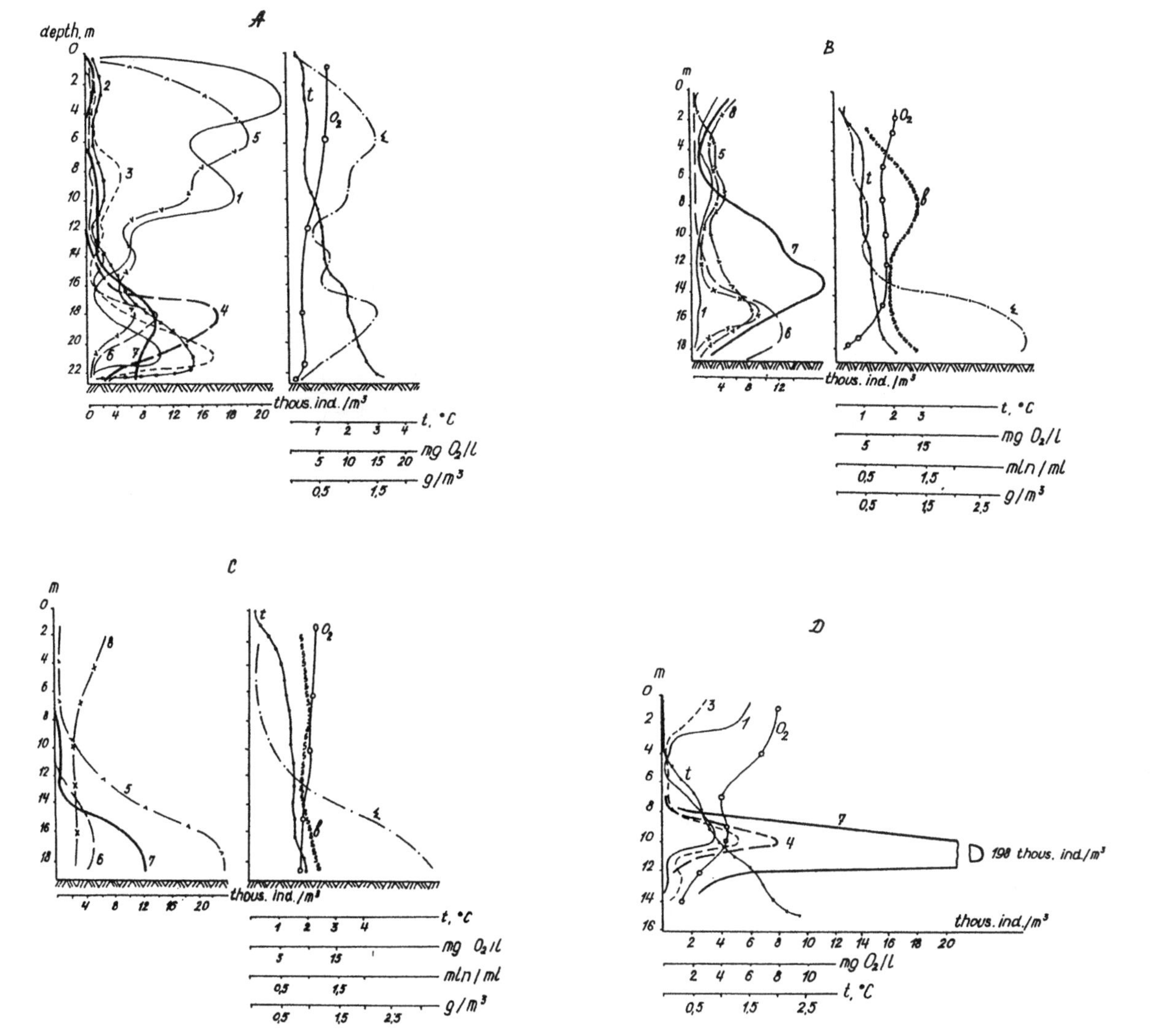

Fig. 5. Vertical distribution of the ecological conditions and abundance (thousands ind. m) copepodids and other zooplankters in Lakes Pleshcheevo (A), Siverskoe (B) in March 1980, Borodaevskoe (C) in March 1983 and Rybinsk reservoir in March 1985 (D). 1 — *Keratella cochlearis macracantha*, 2 — *Asplanchna priodonta*, 3 — *Filinia maior*, 4 — *Conchioloides natans*, 5 — *Eudiaptomus graciloides*, 6 — *Daphnia* (*D. cristata* + *D. longispina*), 7 — *Cyclops kolensis* copepodids stage IV, t — temperature (°C), O_2 — oxygen (mg l^{-1}), ε — total biomass of zooplankton (gm^{-3}).

end of the ice period a non-predatory part of a community is consumed by the cyclopoids and the numbers of the non-predatory zooplankters decrease (Fig. 3).

Overwintering copepodids endure well a deficiency of oxygen up to 2 mg l^{-1} and only at the beginning of winter kill do they scatter and move to the upper water layers (Fig. 4).

In winter in deep stratified lakes (Siverskoe, Pleshcheevo, Borodaevskoe) the populations of *C. kolensis* are accumulated in bottom layers where oxygen levels can fall to 2–3 mg l^{-1} (Fig. 5). In shallow water-bodies (Lake Nero, the Vesetski part of Lake Seliger) by the end of winter, mortality caused by a lack of oxygen increases; CH_4 and H_2S are found in the water. In such conditions the copepodids migrate to the upper layers near the ice (Rivier *et al.*, 1992).

After breaking up of the ice the population becomes more active and by the beginning of May the stage IV copepodids disappear. For 20–30 days intensive reproduction with accumulation of juveniles of stages I–IV in the population occurs and then there is a sharp rise in numbers of copepodids of stage IV and their transition into the motionless state are observed (Fig. 6a, b). The active period in the life cycle of the population is over and a period of hibernation (diapause) starts.

Discussion

Data on the ecology of diapausing copepodids of *C. kolensis* are scarce. Apparently, the first report was from Elgmork (1959): 'In Lake Kortowo in Poland copepodids IV of *C. kolensis* have been found in mud samples from the anaerobic profundal zone during summer'. Accumulations of copepodids on the bottom of deep parts can be detected during meiobenthos sampling.

In the studied reservoirs the durations of the active period and diapause were different. The copepodid behaviour in winter time was specific. In the shallow eutrophic lakes Nero and Seliger (2–3 m depth) *C. kolensis* is not numerous. Due to a lack of oxygen in December the copepodids are already depressed and accumulate near the water surface under the ice. In spring an active period lasts only 20–30 days and the transition of copepodids to diapause is rapid. In the stratified lakes Pleshcheevo, Siverskoye, and Borodayevskoe reproduction lasts 2 months until July and the transition of copepodids to diapause takes more time. The relationship between active and passive stages in the seasonal cycle of *C. kolensis* depends on the type of reservoir and on temperature. In Lake Baikal in the open pelagic zone with a depth of more than 100 m reproduction and development of copepodids lasts throughout the year. The duration of copepodid stages is the same and diapause is absent. The development of a stage lasts 6–8 days when the water temperature is 5–9 °C and 3–4 days at maximum summer temperatures of 14–17 °C. Adults and copepodids prefer a predatory type of feeding: in experiments their development stops without animal food.

In the shallow coastal zone of Lake Baikal during the period of maximum summer temperatures *C. kolensis* disappears from the zooplankton (Mazepova, 1978).

The fourth copepodids of *C. kolensis* are cold-loving, enter an active stage at 2–3 °C and diapause at 12– 14 °C. Temperature is one of the most important factors responsible for changes in cyclopoid activity (Monchenko, 1989).

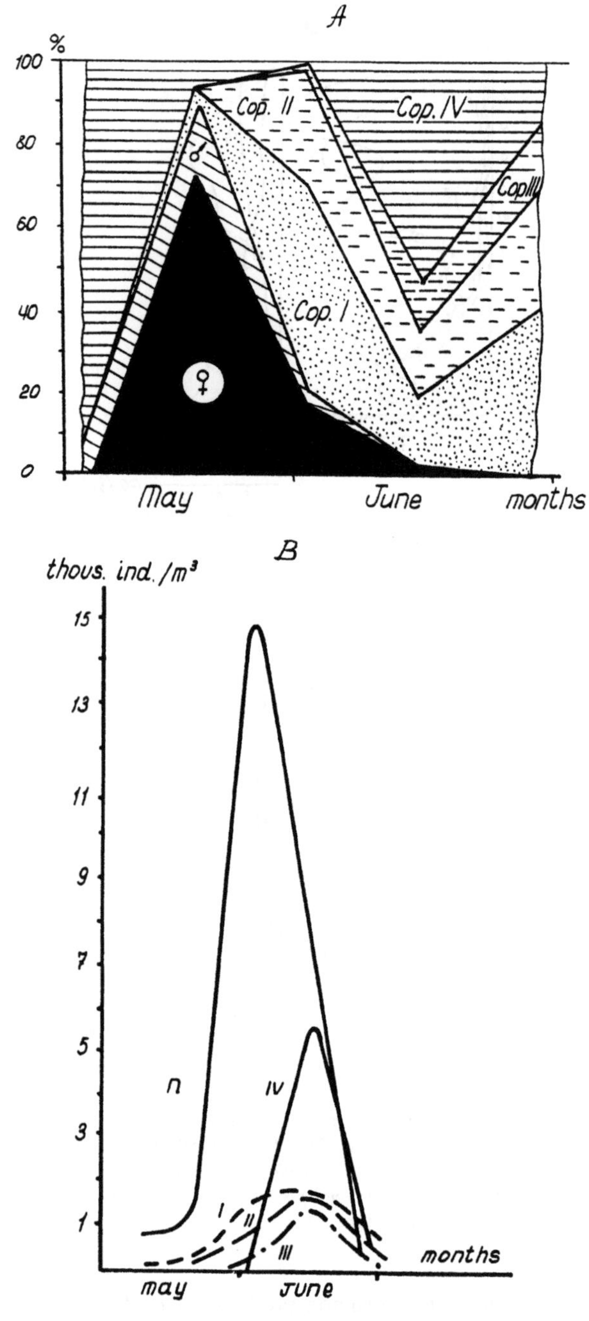

Fig. 6. Population structure (A) and variations of abundance (thousands m^{-3}), of nauplii (n) and copepodids stages I– IV (B) in the ice-free period — May–June 1982 in the main part of Rybinsk Reservoir.

Data collected from different lakes of the Upper Volga show that the duration of juvenile development, and the beginning of diapause depend on temperature, the abundance distribution and activity of copepodids of stage IV in winter, and on water-body morphometry, gas regime and feeding conditions.

Conclusion

In the Upper Volga water-bodies *C. kolensis* is characterized by a short period of intensive reproduction and growth (April–June) and a prolonged period of hibernation in a state of diapause at the fourth copepodid stage on the bottom (June–November) and in the near bottom layers (December–March).

At the beginning of summer, copepodids are evenly distributed over the bottom. As a result of wave mixing they are transported with detritus particles to the deepest part of the water-body where they form dense aggregations and are not readily available to fishes. Redistribution of copepodids during the period of open water is a result of water mass dynamics. After ice formation the temperature of the near- bottom layer increases and copepodids go up to the surface of the silt. In winter in these layers a biocoenosis consisting of Bacteria, Algae, Protozoans, rotifers, 'cladocerans' and cyclopoidids is formed. The vertical distribution of copepodids depends on the temperature and oxygen regime of the water-body.

In December–February the population structure of *C. kolensis* changes insignificantly. In March–April there is active growth and maturing, and the copepodids begin to influence the metalimnial biocoenosis.

Reproduction begins during the ice period in April and is especially intensive in May and early June. Growth stops at the copepodid IV stage and copepodids accumulate in the plankton. When the temperature rises to 14–15 °C they sink to the bottom and enter diapause.

References

Elgmork, K., 1959. Seasonal occurrence of *C. strenuus*. Folia Limnol. Scand. 11: 1–196.

Elgmork, K., J. Nilssen, T. Broch, R. Ovrevik, 1978. Life cycle strategies in neighboring populations of the copepod *Cyclops scutifer* Sars. — Verh. int. Ver. theor. angew. Limnol. 20, p 2518–2523.

Kosova, A., 1970. Seasonal changes in zooplankton 'protoca Bustroi' after hydrotechnical reconstructions of the Volga River flow. Trudy Astrakhan. zapovedn., 13, Astrakhan: 195–218. (In Russian).

Kozhov, M., 1962. Biology of Lake Baikal, Moscow, 312 pp. (In Russian).

Kozminski, Z., 1936. Morphometrische and okologische Untersuchungen and Cyclopoiden der strenuus-Gruppe. Int. Revue ges. Hydrobiol. Hydrogr. 33: 161–240.

Lindberg, K., 1955. Les Cyclopoids du lac Baikal.-Kungl. Fysiogr. sallskap — Lund Forhandl. 25, 5: 1–5.

Mazepova, G., 1978. Cyclopoida of Lake Baikal, Novosibirsk, 143 pp. (In Russian).

Nilssen, J., 1977. Cryptic predation and the demographic strategy of two limnetic cyclopoids copepodids. — Mem. Ist. ital. Idrobiol. 'Dott. M. Marcki', 34: 187–196.

Rivier, I., 1986. Composition and ecology of winter zooplankton communities, Leningrad, 160 pp. (In Russian).

Rivier, I., N. Zhgarev, A. Krylov, 1992. Mass development of Rotatoria during the first half of winter in the Nero Lake. — Biol. of Inl. water, Inform. Bull. 92. St Petersburg: 39–44. (In Russian).

Rylov, V., 1948. Cyclopoida of fresh water (Fauna of the USSR). 3, 3: 319.

The Rybinsk reservoir and its life, 1972. Leningrad, 357 pp. (In Russian).

The Volga and its life, 1978. Leningrad, 346 pp. (In Russian).

Wierzbicka, M., 1956. Wyniki sztucznego zarazenia *Cyclops vicinus* UCJ larwami Protocephalus percae (Muller) — Acta parasitol. pol. v. IV, fasc. 3, p. 88–100.

MIX
Papier aus verantwortungsvollen Quellen
Paper from responsible sources
FSC® C105338

If you have any concerns about our products,
you can contact us on
ProductSafety@springernature.com

In case Publisher is established outside the EU,
the EU authorized representative is:
Springer Nature Customer Service Center GmbH
Europaplatz 3, 69115 Heidelberg, Germany

Printed by Libri Plureos GmbH
in Hamburg, Germany